Manual of
Soil Laboratory Testing

Manual of Soil Laboratory Testing

Volume 2: Permeability, Shear Strength and Compressibility tests

Third Edition

K. H. Head, *MA (Cantab), C. Eng, FICE, FGS*
R. Epps, *B.Sc. A.R.S.M. C.Geol. FGS*

Whittles Publishing

CRC Press
Taylor & Francis Group

Published by
Whittles Publishing,
Dunbeath Mill,
Dunbeath,
Caithness KW6 6EG,
Scotland, UK
www.whittlespublishing.com

Distributed in North America by
CRC Press LLC,
Taylor and Francis Group,
6000 Broken Sound Parkway NW, suite 300,
Boca Raton, FL 33487, USA

ISBN 978-1904445-69-2
USA ISBN 978-14398-6988-8

Permission to reproduce extracts from BS 1377 is granted by BSI. British Standards can be obtained
in PDF or hardcopy formats from the BSI online shop: www.bsigroup.com/Shop or by contacting BSI
Customer Services for hardcopies only: Tel: +44 (0)20 8996 9001, Email: cservices@bsigroup.com.

The publisher and authors have used their best efforts in preparing this book, but assume no
responsibility for any injury and/or damage to persons or property from the use or implementation
of any methods, instructions, ideas or materials contained within this book. All operatiom
should be undertaken in accordance with existing legislation and recognized trade practice.
Whilst the information and advice in this book is believed to be true and accurate at the
time of going to press, the authors and publisher accept no legal responsibility or
liability for errors or omissions that may have been made.

Typeset by Datamatics

Printed in the UK by CPI Antony Rowe, Chippenham and Eastbourne

Preface to the third edition

This book is the second of a series of three, intended primarily to provide a working manual for laboratory technicians and others engaged in the testing of soils for building and engineering purposes. It is not meant in any way to be used as a substitute for the Standards referred to therein, but to augment their requirements by the provision of step-by-step procedures. This third edition has been revised to take account of the current requirements of BS 1377:1990, including recent amendments, and the impact of Eurocode 7 on sampling and testing practice.

This volume relates to Parts 1, 5 and 7 of BS 1377:1990, plus the California bearing ratio (CBR) test in Part 4. We have also included a description of the fall cone test in Chapter 12 (Section 12.10), which is covered by DD EN ISO/TS 17892-6:2004. Reference has also been made to some of the latest ASTM Standards. Under Eurocode 7, only Class 1 samples may be used for undisturbed shear strength and consolidation testing and this has resulted in a greater diversity in the diameter of samples presented for testing. We have tried to address these issues in Chapters 9 and 13, but would welcome any comments that readers may wish to offer.

Chapter 8, which describes the laboratory equipment needed to perform the tests, has been amended to include summary descriptions of current electronic measuring instruments, which will be discussed in greater detail in Volume 3. However, the actual test procedures are described in terms of manual observation, recording and calculation, so that the principles can be clearly understood. The section on calibration (Section 8.4) has been revised to update references to current British Standards and to take account of the requirements of BS EN ISO 17025:2005. This includes a brief reference to the requirement to estimate the uncertainty of measurement of testing equipment, which will be addressed in greater detail in Volume 3.

A basic knowledge of mathematics and physics has been assumed and the chapters relating to test procedures include some general background information and basic theory. We hope that this will provide an appreciation of the significance and limitations of the tests and an appreciation of certain basic principles which are sometimes difficult to grasp.

We hope that this book will continue provide a useful resource that is well used in the laboratory. We would welcome any comments and criticisms from those who use it.

K. H. Head
Cobham, Surrey

R. J. Epps
Alton, Hampshire

Acknowledgements

We would like to thank ELE International for providing many of the photographs included in this Volume. Thanks are also due to the British Standards Institution, Controls Testing Limited, DH Budenberg, Fugro Engineering Services Limited, Geolabs Limited, Geonor AS, Newton Technology Geomechanics Laboratory, Soil Mechanics Limited, Structural Soils Limited, the University of Hertfordshire, the University of Sheffield, the University of the West of England and John Ashworth of Fugro Engineering Services, for permission to reproduce photographs, data and drawings, as acknowledged in the figure captions. We would also like to thank Geo-Design Consulting Engineers for assistance in providing drawings where needed.

We are also grateful for assistance provided by individuals working for these organisations, especially Ian Bushell, Tim Gardiner, John Masters, Chris Wallace, Paul Kent, John Ashworth and Peter Keeton. In particular, we would like to thank John Masters for his review of the completed draft.

Finally, may we express our thanks to Dr Keith Whittles for providing us with the opportunity of revising and publishing this edition.

Contents

Summary of procedures described in Volume 2

Procedure or Test	Section	Standard or Reference*
Chapter 8		
Calibration	8.4	BS Part 1:4.4
Chapter 9		
Preparation of undisturbed specimens		
From tubes: Shearbox and oedometer	9.2.3	BS Part 1:8.6
Compression (38 mm)	9.2.4	BS Part 1:8.3, 8.4
Set of 3 (38 mm)	9.2.5	BS Part 1:8.4
Compression (100 mm)	9.2.6	BS Part 1:8.3
From blocks: Shearbox and oedometer	9.3.1	BS Part 1:8.7
Compression	9.3.2	BS Part 1:8.5.3
Large diameter	9.3.3	BS Part 1:8.5.3
Large shearbox	9.3.4	BS Part 7:5.4
Encapsulation	9.3.5	BS Part 1:8.5.4
Soil lathe	9.4	BS Part 1:8.5.2
Recompacted: Shearbox and oedometer	9.5.3	BS Part 1:7.7
Compression (38 mm)	9.5.4	BS Part 1:7.7
Compaction mould	9.5.5	BS Part 1:7.7.4
Large diameter	9.5.6	BS Part 1:7.7.5
Chapter 10		
Permeability—Constant head:		
Standard permeameter	10.6.3	BS Part 5:5, ASTM D2434
Axially loaded sample	10.6.4	—
Large permeameter	10.6.5	(Original author)
Filter materials	10.6.6	Lund (1949)
Horizontal permeameter	10.6.7	Dept. of Transport (1990)
Permeability—Falling head:		
Standard permeameter	10.7.2	—
Sample tube	10.7.3	—
Consolidation cell	10.7.4	—
Recompacted sample	10.7.5	—
Erodibility:		
Pinhole	10.8.2	BS Part 5:6.2, ASTM D4647
Crumb	10.8.3	BS Part 5:6.3

continued on next page

Procedure or Test	Section	Standard or Reference*
Dispersion	10.8.4	BS Part 5:6.4, ASTM D4221
Water extract for analysis	10.8.5	Sherard et al (1972)
Cylinder dispersion	10.8.6	Atkinson, Charles & Mhach (1990)
Chapter 11		
California bearing ratio	11.7.2	BS Part 4:7, ASTM D1883
CBR soaking procedure	11.6.9	BS Part 4:7.3, ASTM D1883
Chapter 12		
Direct shear:		
Small shearbox	12.5.6	BS Part 7:4, ASTM D3080
Large shearbox	12.6.4	BS Part 7:5
Drained strength	12.7.4	BS Part 7:4, ASTM D3080
Residual strength	12.7.5	BS Part 7:4
Cut plane	12.7.6	(Original author) et al
Vane shear	12.8.4	BS Part 7:3, ASTM D4648
Pocket shearmeter	12.8.5	(Supplier)
Ring shear	12.9	BS Part 7:6
Fall cone	12.10	BS DD CEN ISO/TS 17892-6
Chapter 13		
Unconfined compression:		
Load frame	13.5.1	BS Part 7:7.2, ASTM D2166
Autographic	13.5.2	BS Part 7:7.3
Remoulded	13.5.3	Terzaghi & Peck (1967)
Triaxial compression:		
Definitive	13.6.3	BS Part 7:8, ASTM D2850
Large diameter	13.6.4	BS Part 7:8, ASTM D2850
Multistage	13.6.5	BS Part 7:9
'Free' ends	13.6.6	Rowe & Barden (1964)
High pressure	13.6.7	—
Special orientation	13.6.8	(Original author)
Reconstituted specimens	13.6.9	Bishop & Henkel (1962) et al
Chapter 14		
Oedometer consolidation:		
BS procedure	14.5.5	BS Part 5:3
ASTM procedure	14.5.8	ASTM D2435
Swelling pressure	14.6.1	BS Part 5:4.3, ASTM D4546
Swelling	14.6.2	BS Part 5:4.4
Settlement on saturation	14.6.3	BS Part 5:4.5, ASTM D4546
Overconsolidated clays	14.6.5	—
Peats	14.7	Hobbs (1987)
Expansion index	14.6.4	ASTM D4829
Expansion of ferrous slags	14.6.4	(Emery (1979)), ASTM D4792

* BS implies BS 1377:1990 unless otherwise stated
ASTM refers to Annual Book of ASTM Standards (2010) Vol 04.08

Chapter 8

Scope, equipment and laboratory practice

8.1 Introduction

8.1.1 Scope of volume 2

It was stated in Volume 1 (third edition), Section 1.1.3, that the laboratory tests normally used for the determination of the physical properties of soils (as defined in the engineering sense) can be divided into two main categories:

1. Classification tests, which are used to identify the general type of soil and to indicate the engineering category to which it belongs
2. Tests for the assessment of engineering properties, such as permeability, shear strength and compressibility

The usual tests which comprise category 1 are described in Volume 1(third edition). Volume 2 deals with the more straightforward tests for the determination of the engineering properties of soils, namely those covered by the general description of category 2.

In this context the term 'shear strength' refers to the 'immediate' undrained shear strength of soils based on the measurement of total stresses. Determination of effective shear strength parameters requiring the measurement of pore water pressures will be covered in Volume 3, as will other types of test of a specialised nature. The only exception is the drained shearbox test for the measurement of the drained 'peak' and 'residual' shear strength of soils, included in Chapter 12 of this volume, because this test does not involve the measurement of pore water pressures.

The tests presented in this volume are described in terms of conventional instruments requiring manual observation and recording of data. Modern developments have led to the introduction of electronic measuring devices with digital displays and with facilities for connection to data-logging and data-processing systems. Electronic equipment of this kind is referred to in this chapter (Section 8.2.6), but the authors feels that traditional manual procedures are more suitable for instruction and for gaining an understanding of basic principles.

8.1.2 Reference to standards

This volume deals with tests that have been generally accepted as standard tests for the determination of engineering properties of soils. Most are covered by BS 1377:1990, in Parts 4, 5 and 7, and the procedures described here are generally in accordance with this British Standard. (Throughout this volume BS 1377:1990 is referred to as 'the British Standard' or 'the BS'.)

These tests, and some others, are also covered in US Standards which are given by their American Society for Testing and Materials (ASTM) numbers, to which appropriate references are made. Differences in detail between the British and American standards, where they occur, are explained. For those tests for which there are at present no British or American Standards, the procedures given follow generally accepted current practice.

Frequent reference is made to Volume 1 (third edition), both for the test procedures described there, and for details of equipment given there and not repeated in this volume.

8.1.3 Presentation

Suggested approach

The general arrangement of this volume is outlined below. Section 8.1.4 gives an outline of good general laboratory practices, with references to Volume 1(third edition) and to other parts of this volume. It is suggested that the technician should study these before proceeding with laboratory tests. Attention should also be given to safety aspects, outlined in Section 8.5.

Equipment

Section 8.2 consists of a review of those items of equipment and tools which are common to several of the tests described in this volume. It is divided into sections relating to measuring instruments; apparatus for preparation of specimens; load frames; constant pressure systems; and general laboratory equipment. Special items required exclusively for a test are listed in the appropriate chapter. An outline description is included of electronic accessories, which can be substituted for conventional methods of observation in many tests.

The proper use and care of measuring instruments are described in Section 8.3, which should be studied carefully before using the equipment. Calibration is covered in Section 8.4.

Specimen preparation

Chapter 9 covers the usual methods for the preparation of undisturbed specimens for testing, from undisturbed or recompacted soil samples. In this context the word 'specimen' is used to signify the portion of material actually used for test, which is usually cut and trimmed from a larger 'sample' (see Section 1.1.7 in Volume 1, third edition). Preparation of other types of test specimen is dealt with in the appropriate chapter.

Tests and their background

Chapters 10–14 are each devoted to a particular type of test or test principle. They start with a general introduction to the topic, followed by a list of definitions of terms as used in this book. Sections on theory present the theoretical background, which should be sufficient to enable the tests and associated calculations and graphical plotting to be understood. This is followed by a brief outline of some of the more important applications of the tests, and of the test results to engineering practice.

The main emphasis of the book is on the detailed procedures to be followed in preparing test specimens, and in carrying out the tests, in the laboratory. For each test a list of apparatus and a list of the procedural stages are given first, followed by step-by-step procedures and practical details. Calculation of results from laboratory data, and the plotting of graphs and their use, are described and illustrated by means of typical examples.

Units and terminology

Throughout this volume, metric (SI) units of measurement are used. In those instances where the relevant Standard makes use of Imperial units (notably ASTM Standards), these are stated with the SI equivalent shown in brackets. All calculations are given in SI units.

In tests which are covered by British Standards the notation, terminology and symbols used here are the same as those used in the BS, unless specifically defined. Otherwise the notation is as listed in the Appendix.

The Appendix provides a summary of the metric (SI) units used in this volume, and conversion factors relating SI to Imperial, US and CGS metric units. A summary of symbols, and a quick reference to useful general data, are included.

8.1.4 Laboratory practice

General

The general laboratory practices and techniques recommended in Section 1.3 of Volume 1 (third edition) apply equally to the procedures described in this volume. Some further points to observe are as follows.

Test equipment, especially motorised items, should be correctly installed and used, and the manufacturer's instructions should be studied and followed. Measuring instruments should be carefully handled, and guidance on the use and care of instruments required for tests described in this volume is given in Section 8.3. Calibration of instruments, covered in Section 8.4, is an essential part of their use, and should be verified at the outset and re-checked subsequently at appropriate regular intervals.

The value of any test depends very much upon the quality of the specimen tested. Procedures for the preparation of good undisturbed test specimens from larger undisturbed samples of soil are given in Chapter 9.

Safety

The importance of safety in the laboratory is emphasised in Section 8.5, where aspects of safety which supplement those given in Section 1.6 of Volume 1(third edition) are provided.

Test data and results

Laboratory tests are usually carried out to the instructions of an engineer, who requires the results as part of the data to be used for the solution of an engineering problem.

Observed test data must always be recorded faithfully and accurately: the objective is to record what is observed, not merely what one thinks ought to be observed. Descriptive comments on what happens during a test should be recorded as part of the test data, and be made available for the engineer when required.

During the course of a test the technician should look for any inaccuracy or source of error which may arise. Appropriate comments should be written against data that are suspect, and wherever possible the readings should be repeated.

When calculating and plotting data, any value which is incompatible with the general trend (e.g. one or more points on a graph remote from a sensible relationship shown by the remainder) should be suspect. If a re-check indicates that the error does not lie in the calculations the test should be repeated, if practicable, for the vagrant points.

At the end of a test the technician should examine the final results critically and consider whether they look reasonable for the type of material and the conditions of the test. If there is any doubt after re-checking the calculations and plotting, it might be necessary to repeat the test after consulting the engineer.

Test results should be reported to a degree of accuracy that is appropriate for the type of test. Any rounding off should be made only on the final results, not at an intermediate stage. Recommended degrees of accuracy for reporting results of classification tests are shown in Table 1.8 of Volume 1 (third edition). For the tests given in this volume they are indicated at the end of each test description, and are summarised in Table 8.1.

8.2 Laboratory equipment

8.2.1 Measuring instruments

Conventional and electronic instruments

The test procedures described in this book relate to the use of 'conventional' mechanical instruments for making measurements of displacement, load and pressure. Operators should be familiar with the use of these instruments because, together with manual recording and evaluation of data, they probably provide the best way to become fully conversant with the test procedures. The types of instruments commonly used are summarised below; their care and use is described in some detail in Section 8.3, and their calibration in Section 8.4.

However, it is recognised that in many laboratories, many of these measurements are now made by using electronic instrumentation, with data capture and storage by means of data loggers and personal desktop or laptop computers. This type of instrumentation may be substituted for conventional instruments in performing most of the tests described in this book. Electronic measuring and monitoring systems are described briefly in Section 8.2.6, but more detailed information will be given in Volume 3.

Displacement gauges

Displacement dial-indicator gauges (referred to here as 'dial gauges') provide the simplest means of measuring displacements to accuracies of 0.01 mm or 0.002 mm. Dial gauges for this purpose were referred to in Volume 1 (third edition), Table 1.1, and two types were shown in Figure 1.1. Of the many types available, those listed in Table 8.2 are the most commonly used in a soil laboratory, and are referred to in this volume. The gauge of type (a) is perhaps the most generally versatile. Further details with illustrations are given in Section 8.3.2.

Useful accessories for dial gauges include extension stems and various types of anvil, which are discussed in Section 8.3.2. A bench stand enables a dial gauge to be used as an independent measuring device, as shown in Figure 8.1(a). A stand fitted with a magnetic base (see Figure. 8.1(b)) provides a secure fixing to a steel surface.

Digital dial gauges are also manufactured (see Section 8.3.2 and Figure 8.21). Electric displacement transducers (see Section 8.2.6) can be used in the same way as dial gauges.

Force measuring devices

The most usual means of measuring force in soil testing is a steel force measuring ring, referred to here as a 'load ring'. The range of load rings described in this volume is summarised in Table 8.3, and some typical rings are shown in Figure 8.2. Load rings are discussed in greater

Table 8.1 Recommended accuracy for reporting laboratory test results

Location	Item	Symbol	Accuracy	Unit
General	Specimen dimension		0.1	mm
	Density	ρ, ρ_D	0.01	Mg/m^3
	Moisture content	w	<10 : 0.1	%
			>10 : 1	%
	Porosity	n	1	%
Chapter 10	Permeability	k	2 significant figures $\times 10^{-n}$ where n is an integer	m/s
Chapter 11	CBR value	CBR	2 significant figures	%
Chapter 12	Angle of shear resistance	ϕ', ϕ'_r	0.5	degree
	Apparent cohesion	c', c'_r, c_u	2 significant figures	kPa
	Vane shear strength	τ_v	2 significant figures	kPa
Chapter 13	Unconfined compressive strength	q_u	2 significant figures	kPa
	ditto	q_u	< 50 : 2	kPa
	(autographic test)		50–100 : 5	kPa
			> 100 : 10	kPa
	Strain at failure	ε_f	0.2	%
	Rate of strain		2 significant figures	% per min
	Sensitivity to remoulding	S_t	1 decimal place	—
	Undrained cohesion	c_u	1	kPa
	Voids ratio	e	0.01	—
Chapter 14	Degree of saturation	S_r	1	%
	Swelling pressure	p_s	1	kPa
	Applied pressures	p	1	kPa
	Coefficient of volume compressibility	m_v	2 significant figures	m^2/MN
	Coefficient of consolidation	c_v	2 significant figures	m^2/year
	Compression ratio: initial	r_0		
	primary	r_p	1	%
	secondary	r_s		
	Coefficient of secondary compression	C_{sec}	2 significant figures	—
	Compression index	C_c	2 significant figures	—
	Swell index	C_s	2 significant figures	—

Table 8.2 Dial gauges for linear measurement

Ref. in Figure 8.20	Face diameter (mm)	Travel (mm)	Graduation (1 div) (mm)	Dial markings and direction	Travel per revolution (mm)	Main application
(a)	57	25	0.01	0–100 (C)*	1	General:shear displacement, axial strain, CBR penetration
(b)	75	50	0.01	0–100 (C)	1	Axial strain (large specimens)
(c)	57	5	0.002	0–100 (C)	0.2	Load rings
(d)	57	12.7	0.002	0–20 (C) or (AC)**	0.2	General (C)
						Consolidation settlement (AC)
(e)	57	12.7	0.002	0–0.2 (AC)	0.2	Consolidation settlement

*(C) = clockwise rotation
**(AC) = anti-clockwise rotation (backwards-reading)
CBR = California bearing ratio

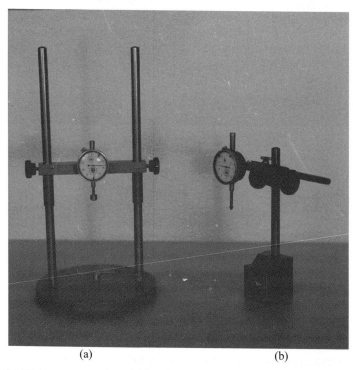

(a)   (b)

Figure 8.1 Dial gauge mountings: (a) bench standard comparator, (b) magnetic base stand

Table 8.3 Load rings for soil testing

Capacity (kN)	Typical sensitivity (N/div)	Reading at maximum working load (divisions)
2	1.3	1500
4.5	3.0	1500
10	7.7	1300
20	18	1100
28	25	1100
50	45	1100
100	100	1000

Figure 8.2 Some typical load rings

detail in Section 8.3.3, and their calibration is described in Section 8.4.4. Other relevant types of force measuring devices include strain gauge load cells and submersible load cells (see Section 8.2.6).

Pressure gauges

The usual types and sizes of pressure gauges used in soil testing for the measurement of fluid (water) pressure are summarised in Table 8.4.

7

Table 8.4 Typical pressure gauges used for soil testing

Type of gauge (see Section 8.3.4)	Diameter (mm)	Working range*	Scale markings (1 division)	Figure reference
Commercial	150 or 200	0–600 kPa	20 kPa	
		0–1000 kPa	20 kPa	8.3(a)
Standard test	200 or 250	0–1000 kPa	10 or 5 kPa	
		0–1200 kPa	10 kPa	
		0–1600 kPa	10 kPa	8.3(b)
Vacuum	80–150	−100–0 kPa	5 or 2 kPa	
		0–760 torr absolute	10 torr	
		of water to 0	0.2 m	8.3(c)

*Pressures (except torr) relate to atmosphere as zero

Ordinary commercial type gauges, such as those shown in Figure 8.3(a), are usually adequate for monitoring a constant pressure system. 'Test' grade gauges, shown in Figure 8.3(b), are more accurate and reliable, and this type is necessary as a reference gauge for calibration purposes. Details are given in Section 8.3.4.

For calibration of these gauges a dead-weight gauge tester is required. Calibration is described in Section 8.4.5.

Water manometers are used and historically, mercury manometers have been used, for the accurate measurement of low pressures. Notes on reading manometers are given in Section 8.3.5.

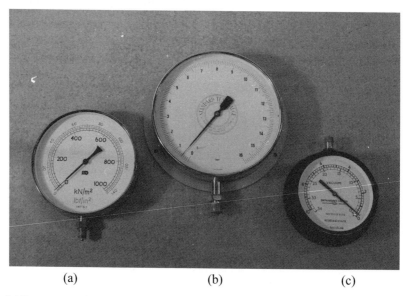

(a)   (b)   (c)

Figure 8.3 Pressure gauges: (a) commercial gauge, 200 mm diameter, 0–1000 kPa; (b) 'test' gauge, 250 mm diameter, 0–1600 kPa; (c) vacuum gauge, 150 mm diameter, −100–0 kPa

Electric pressure transducers (Section 8.2.6) provide an alternative to pressure gauges for routine test measurements, especially when multiple sources of pressure are used. Digital pressure gauges are now manufactured by coupling a pressure transducer to a read-out unit (Section 8.3.4 and Figure 8.25).

8.2.2 Specimen preparation equipment

Equipment which is required for the preparation of undisturbed and recompacted specimens for the tests described in this volume is detailed in Chapter 9, Section 9.1.2. Items are listed under the following headings:

Sample extruders
Sample tubes and formers
Extruder accessories
Small tools
Miscellaneous items
Soil lathe
Small vibrator

8.2.3 Load frames

Three types of equipment come under the general description of load frames.
1.   Frame with a hand-operated or motorised loading device used for compression tests
2.   Frame used for direct shear tests, usually motorised
3.   Dead-weight lever-arm frame used for consolidation tests

In this chapter the term 'load frame' applies to the first type. The second is referred to as a shearbox machine, and is described in Chapter 12. The third is normally called an oedometer press or consolidation press, and is described in Chapter 14.

General purpose load frames used for soil testing range from a 10 kN bench-mounted unit to a large floor-standing machine of 500 kN capacity. Typical machines referred to in this volume are listed in Table 8.5.

Motorised load frames are usually available with steplessly variable speed control, providing platen speeds from about 0.00001 mm/min to 10 mm/min. Modern machines incorporate stepper-motor drive, microprocessor control, LCD display and a connecting port for external computer control. Other features might include a rapid approach and

Table 8.5 Load frames for compression tests on soils

Capacity (kN)	Mounting	Drive	Application
10	Bench	Hand or motor	Uniaxial or triaxial compression up to 50 mm diameter
50	Bench	Hand or motor: multispeed or 5 speeds	Uniaxial or triaxial compression up to 100 mm diameter. CBR; soft rocks; soil cement
100	Floor	Motor; multispeed or stepless	Soils and soft rocks up to 150 mm diameter
500	Floor	Motor; multispeed or stepless	Large soil and rock specimens

unload facility, and over-travel limit switches. Earlier versions of motorised frames were fitted with a 5-speed gear lever and were adjustable to 36 or 42 speeds using gear wheels of varying numbers of cogs to cover a wide range of speeds. Hand-operated machines usually include a 2-speed gearbox, the higher gear being intended for rapid adjustment or rewinding.

8.2.4 Constant pressure systems

Five types of system are possible for applying and maintaining constant water pressures in the laboratory for triaxial compression tests.
1. Self-contained air–water system operated by a foot pump
2. Air–water system operated by an air compressor, driven by electric motor (though a petrol or diesel engine could be used instead)
3. Mercury pot system
4. Motorised oil–water system
5. Air–water system utilising compressed air bottles

Item 1 is suitable in a small laboratory where only a few triaxial machines are in use, or where a motorised unit is impracticable. A separate unit is required for each pressurised test cell. Item 2 is more versatile and is much easier to operate. One installation can serve numerous test machines, and is capable of virtually unlimited extension as long as the capacity of the compressor and air reservoir are sufficient. The continuously available working pressure is usually about 70% of the maximum rated output pressure of the compressor. Item 3 was a very stable and accurate system, but is now obsolete in commercial laboratories as a result of health and safety considerations and is listed here as a historical method for completeness. Item 4 provides an accurate source of pressure up to 1700 kPa, but one unit can provide only one level of pressure at a time. These units are convenient in a small laboratory or for special applications. Item 5 can be used where the installation of an air compressor is not possible, but is not very practicable. Further details are given below.

Air–water system with foot pump

The simplest air-water system consists of a free standing pressure cylinder of gun-metal or other corrosion-resistant metal, fitted with a pressure gauge, a valve and a flexible connection to the triaxial cell, pressurised by an ordinary car-tyre foot pump. The cylinder could alternatively be connected to a motorised pressure system via a regulator valve, as described under item 2. The foot pump system can provide an emergency back-up in case of failure of a motorised compressor.

The principle is shown in Figure 8.4. The cylinder in which air pressure is transmitted to water pressure is about half filled with water. The lower end is connected to the triaxial cell, and the upper end to the foot pump or air pressure system. A photograph of a typical unit is shown in Figure 8.5.

A small triaxial cell can be filled directly from the cylinder if it contains more than enough water initially, but it is usually more convenient to fill the cell first from a separate water supply line. A large cell can be filled more quickly in this way than from the pressure cylinder, which would also require an excessive amount of pumping. When the cell is completely filled, the pressure is built up to the required value by using the foot pump. The volume of air in the cylinder provides a 'cushion' which maintains the pressure constant even if there is a small flow of water into or out of the triaxial cell.

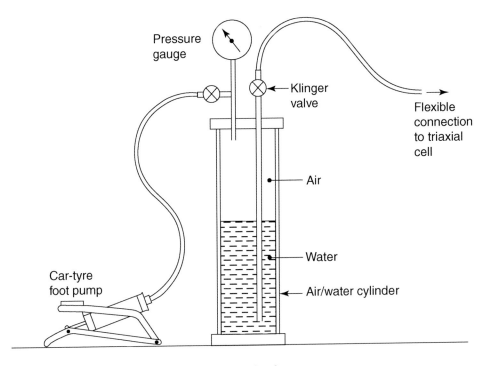

Figure 8.4 Principle of air–water pressure system using foot pump

Motorised air compressor system

An electrically driven air-water constant pressure system comprises the following main items:

Air compressor unit:
    motor compressor
    air receiver
    safety valve
    primary pressure regulator
    pressure switch
    starter
    air filter
    valves
    outlet connector
    drain
Compressed air supply line
Constant pressure reducing valves (regulator valves)
Water trap and secondary filters
Bladder type air-water cylinders
Pressure gauges

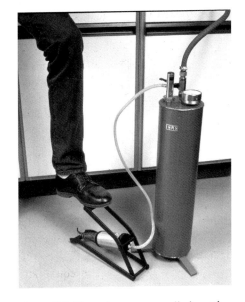

Figure 8.5 Air–water pressure cylinder and foot pump

The principle of the system, to which numerous pressure regulators can be connected, is shown diagrammatically in Figure 8.6. A typical electrically driven air compressor unit, mounted on an air-receiving chamber, is shown in Figure 8.7.

The primary pressure regulator is pre-set to give the required distribution line pressure, which should be set slightly higher than the maximum cell pressure to be used. This regulator controls the starting and stopping of the compressor motor by means of a pressure switch, so

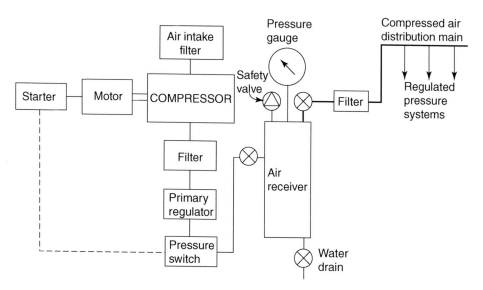

Figure 8.6 Principle of motorised air compressor system

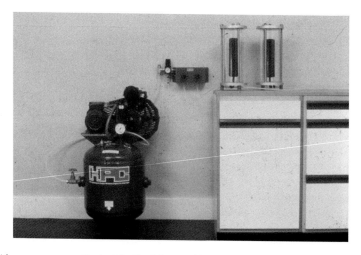

Figure 8.7 Air compressor unit, electrically driven, with air receiver, filter, pressure regulator valves and bladder-type air–water interchange cylinders

that pressure is maintained between an upper and a lower limit, in the same way as a thermostat controls the heater of a constant temperature bath.

The air receiver helps to smooth out transient fluctuations in pressure. If the supply line to the laboratory is of considerable length, a secondary air receiver with safety valve is desirable in the laboratory for the same reason. Alternatively, the presence of a receiver inside the laboratory can be avoided by the use of a compressed air ring main of slightly larger diameter than the required flow rates demand, to act as a slight reservoir.

Filters are necessary to remove oil, dirt, and water. A water trap and secondary filter are essential immediately before each laboratory pressure regulator to remove condensation and any remaining traces of dirt or oil. Some precautions in the operation of compressed air systems are given in Section 8.5.3.

The pressure reducing regulator valve on each pressure line in the laboratory enables the distribution line pressure to be reduced to the pressure required for the test. The regulator operates on a bleed principle, so there is a continual loss of air which can be heard as a quiet hiss. Once adjusted, the regulator should maintain the pressure constant even though the distribution line pressure may vary slightly. Air at the controlled pressure is admitted to the bladder in the air–water cylinder, thus pressurising the water without direct contact between air and water. Pressurised water is led to the triaxial cell, and its pressure is observed on a pressure gauge (Figure 8.8). The close-up view in Figure 8.9 shows the air pressure gauge, water trap and filter, and one of the air pressure regulator valves. Panels with multiple regulator valves are available.

For rapid filling and emptying of a triaxial cell, and to ensure that the air-water cylinder always contains sufficient water, the cell may also be connected to the water supply main, and to a waste outlet, as shown in Figure 8.8. Water from the air–water cylinder is then required only to pressurise the cell and to maintain the pressure.

The non-return valve shown on the water supply line is obligatory in Britain on a connection to a pressure vessel, as a safeguard against contaminated water being forced back into the mains supply.

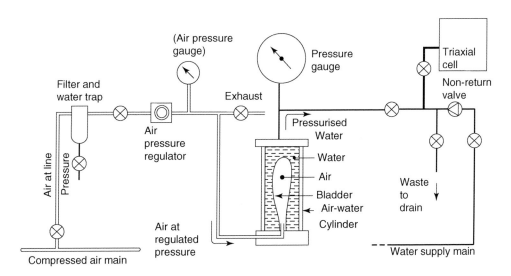

Figure 8.8 Principle of bladder type air–water pressure system

Figure 8.9 Air pressure regulator, air pressure gauge and filter with water trap

Mercury pot system

Details of the self-compensating mercury pot pressure system have been given by Bishop and Henkel (1964). This type of system is no longer used in practice, partly due to the high cost of mercury but mainly because of the health and safety hazards inherent in the handling of mercury.

Motorised oil–water system

The oil and water constant pressure system was originally developed at the Building Research Station by Dr A. D. M. Penman. It provides an alternative to a compressed air system when not more than two sources of equal pressure are needed. The pressure output from an electrically driven pump is continuously variable in the range 0–1700 kPa (although some types can go to 3500 kPa for higher pressure testing). A unit of this kind is shown in Figure 8.10. It is a self-contained system, apart from requiring connection to an elevated supply of de-aired water.

A special type of constant pressure regulating valve, controlled by the front handwheel, regulates the pressure of oil fed to the oil–water interchange vessel, which is fitted with two take-off valves. The electric pump is protected by a thermal overload device. The apparatus should be re-charged with freshly de-aired water at the end of a test. At suitable intervals the system should be drained and the interchange vessel washed, and the apparatus then re-filled with fresh oil and de-aired water. Only the recommended type of oil should be used. These

Figure 8.10 Electrically driven oil–water constant pressure system for pressure up to 1700 kPa

and other checks and maintenance operations should be carried out in accordance with the manufacturer's instructions.

An oil–water system can supply only the pressure which has been selected. Since there is virtually no flow in the supply line it is not possible to use a reducing valve to step down the pressure to a lower value. Multiple units are required if more than one value of pressure is needed at any one time.

Compressed air bottles system

Bottles of compressed air can be used as the source of pressure with the air-water system described under item 2 above, in situations where it is not practicable or permissible to instal an air compressor. The useful life of a bottle is limited, and therefore a large number of bottles may be required. An assured regular replacement service is essential.

An oxygen bottle must *never* be substituted for an air bottle, because of the risk of a serious explosion (see Section 8.5.3).

8.2.5 Other equipment

General items

Many items which were described in Volume 1 (third edition) are necessary for the tests contained in this volume but are not described here, notably the following:

Balances
Ovens and moisture content equipment
Sieves
Apparatus for preparation of disturbed samples
Vacuum pump and vacuum line system
Distilled or de-ionised water supply

Compaction rammers and moulds
Vibrating hammer
Rubber and plastic tubing, pinch clips, etc.

Special items

Many items of equipment are used specifically for certain tests. Those items which are normally used in a soil laboratory, and which have not already been referred to, are described in the appropriate chapter.

Small tools

In addition to the items listed in Section 1.2.9 of Volume 1 (third edition), the following tools are required.

Open-ended spanners of various sizes, for general use and for union nuts on pressure supply lines and gauges (two of each size)
Adjustable spanners (two of each size range)
Stillson wrench (two)
Screwdrivers (various sizes) for slotted head screws
Screwdrivers for Philips head screws.
Wrenches for hexagon-socketed screws (Allen keys)
Pliers and electrical pliers
Ball-pein hammer
Copper- or brass-headed hammer
Pointer lifting and punching tools for use with pressure gauges (see Figure 8.26, Section 8.3.4)

Ideally, there should be two spanners of the correct size to fit every size of nut or union. Unfortunately, because of the numerous screw-thread systems in current use, this is rarely practicable unless a completely new laboratory is being set up and installed by one manufacturer who standardises the fittings to comply with a single system throughout. The best compromise is to have available several sizes of adjustable spanners, two of each size.

The jaws of an adjustable spanner should be closed firmly on to the flats of the nut or bolt-head before exerting any leverage. If the correct size of spanner is not available, use of a well-fitted adjustable spanner is preferable to attempting to use a slightly over-size open-ended spanner, which will very quickly wear the corners off the hexagon.

Typical spanners (one of each size) and screwdrivers are shown in Figure 8.11.

Materials

The following consumable materials are frequently required for the tests described in this volume (see also Volume 1 (third edition), Appendix 5, Sections A5.5 and A5.6).

Polytetrafluoroethylene (PTFE) sealing tape for screw threads
Silicone grease
Petroleum jelly (Vaseline)
French chalk
Viscose sponges
Wiper cloths
Industrial wiper tissue

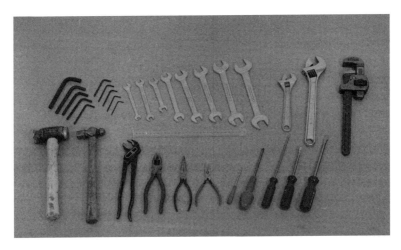

Figure 8.11 Miscellaneous small tools. Top: Allen keys; set of spanners; adjustable spanners; Stilson wrench. Bottom: soft hammer; ball-pein hammer; pliers, screwdrivers

8.2.6 Electronic measuring and monitoring systems

Scope

During the past three decades the application of electronics has brought about the introduction of various types of transducers for the measurement of displacement, load, and pressure in soil testing. Electronic signal conditioning equipment can automatically process the voltage output from these transducers and show the readings as a digital display in engineering units (e.g. mm, N, kPa), thereby greatly easing the task of observing readings and reducing the likelihood of errors. A system of this kind can be extended a stage further by connecting to a compatible data-logging system, to give numerical and/or graphical printouts of data. The data may also be fed into a computer programmed to perform analysis and calculations and to print the required test results both in tabular form and graphically.

Displacement transducers

A typical displacement transducer, known as a linear variable differential transducer (LVDT), is shown in Figure 8.12. A metal rod (the armature) can slide through the axis of an electrical coil, and the resulting changes in inductance are measured and converted to a digital display in units of displacement (mm or μm).

There are a number of alternative linear displacement transducers which are equally reliable and show good linearity with a wide range of resolution. These include potentiometric transducers, linear conversion strain transducers and transducers with internal strain gauges. Their use is increasing and they will be described in more detail in Volume 3.

These instruments may be used for measuring either displacement (axial strain) in a compression test, or vertical movement in an oedometer consolidation test.

Load measurement

A load ring fitted with a displacement transducer instead of a dial gauge is shown in Figure 8.13. The device is calibrated as a unit in order to give an output reading directly in force units (e.g. newtons).

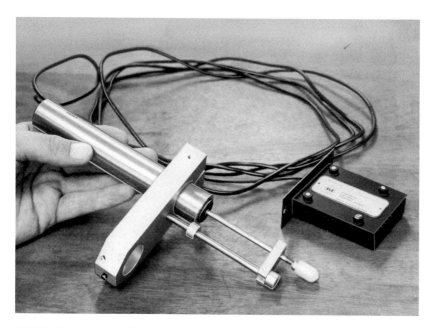

Figure 8.12 Displacement transducer

An alternative to a load ring is a strain gauge force transducer (see Figure 8.14), in which deformation due to load is measured by electrical resistance strain gauges.

Another type of load measuring device, developed at Imperial College, London, is the submersible load cell (see Figure 8.15). This is designed to be accommodated inside a triaxial cell, and readings are not affected by changes in the cell confining pressure. Both positive and negative loads can be measured if appropriate attachments are fitted to the

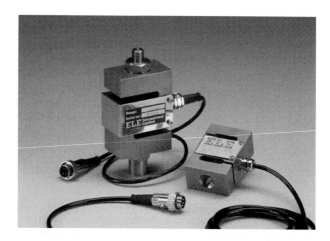

Figure 8.13 Transducer load ring   **Figure 8.14** Strain-gauge load cells

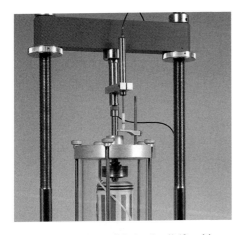

Figure 8.15 Submersible load cell (fitted in a triaxial cell)

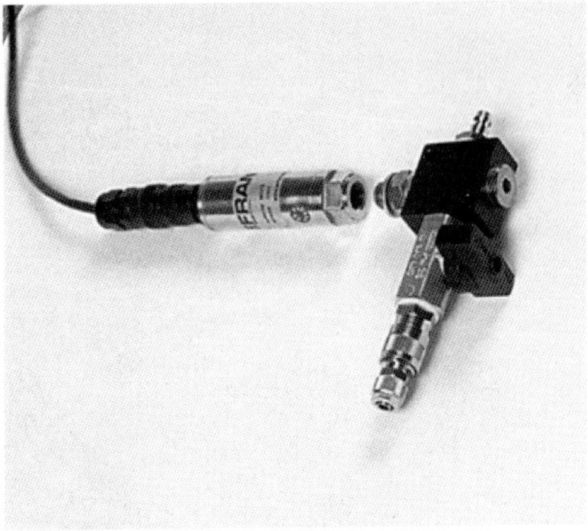

Figure 8.16 Pressure transducer

loading piston. It has the advantage that the applied force is measured within the cell, thereby eliminating the effects of piston friction on the readings. The cell is calibrated to give output readings in engineering units, usually newtons.

Pressure transducers
A typical pressure transducer is shown in Figure 8.16. A thin diaphragm on which strain gauge circuits are bonded or etched is mounted in a rigid cylindrical housing, and protected by a porous filter. A change in pressure causes an extremely small deflection giving rise to an out-of-balance voltage which is amplified and converted to a digital display in pressure units, usually kPa. The pressure range required for most test applications is 0–1000 kPa, but transducers covering higher pressure ranges are available. A pressure transducer must not be used to measure pressures below atmospheric unless it has been specifically designed for that purpose.

Data-processing system
A logical development based on the instrumentation referred to above is an automatic data acquisition and monitoring system which incorporates a computer to control the acquisition of data, to analyse the data and to produce final test results in tabular and graphical form. A system of this kind is shown in Figure 8.17. The computer can be programmed to initiate the test sequence, and to monitor and process numerous tests simultaneously. But provision should always be made for the intervention of the operator when important decisions have to be taken. It can be applied to most soil laboratory tests, including non-standard variations, and to many other types of test.

Essential requirements for electronic systems
Use of electronic equipment requires care and attention to certain details which may not always be appreciated, otherwise erratic and unreliable data may be obtained. Some of the essential factors are listed below.

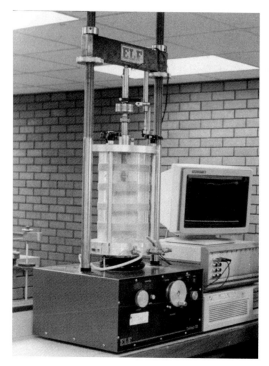

Figure 8.17 Data acquisition system providing automatic graphical presentation, linked to triaxial compression test

1. Stabilised mains voltage and transducer energising current
2. Stand-by generator with automatic start, to provide uninterrupted electricity supply in case of mains failure
3. Screening of signal leads from all electrical fields, especially AC circuits and supply lines, both inside and outside the building
4. Adequate ventilation and reliable temperature control and protection from excessive humidity
5. Suitable interface units for use with computers
6. Adequate calibration facilities for instruments, circuits, and programs
7. 'Fail-safe' data storage facilities

The design and assembly of electronic data-logging and processing systems should be entrusted only to electronics specialists who are experienced in the application to soil testing.

8.3 Use and care of instruments

8.3.1 General comments

This section enlarges on the proper use and care of the measuring instruments referred to in this book. Procedures for their calibration are given in Section 8.4.

Measuring devices are precision-made instruments, and should be treated with respect and protected against damage, dirt, dust and damp. They should be installed and used strictly in accordance with manufacturers' instructions.

For non-standard applications the manufacturer's advice should be sought. Most manufacturers are willing to give their advice free of charge, and to provide additional literature including technical guidance and details of available accessories. Difficult or unusual installation problems can usually be simplified by the use of the appropriate components and accessories.

8.3.2 Dial gauges

Types of gauge

The dial gauges listed in Table 8.2 are all of the continuous reading type, i.e. the main pointer describes numerous revolutions during the full travel of the stem, the secondary pointer indicating the number of revolutions. Standard gauges show an increasing clockwise reading as the stem is compressed, but the 'backwards-reading' gauge of type (e) has an anti-clockwise scale to facilitate recording of readings of settlement in the oedometer consolidation test.

The main features of a typical dial gauge are shown in Figure 8.18(a). Dial gauges can be fitted with various types of backplate, the most useful having an integral fixing lug offset from centre, giving four possible fixing positions (see Figure 8.18(b)). Dial gauges mounted in load rings are sometimes held in a fixing which grips the bush around the stem, but this type of fixing should be used only if it makes an accurately machined fit, otherwise pinching of the stem housing may interfere with free movement of the stem.

The mechanism of dial gauges can be the source of large periodic errors. Systematic calibration of dial gauges is therefore necessary (see Section 8.4.5).

Use of dial gauges

A dial gauge should provide more than enough travel for the overall movement to be measured, and should be sensitive enough for its purpose. It should be securely clamped

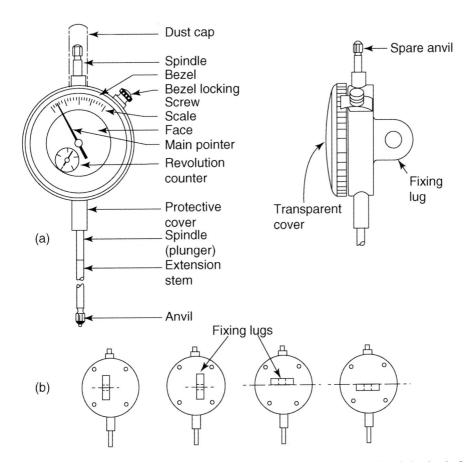

Figure 8.18 Typical dial gauge details: (a) main features; (b) backplate with offset fixing lug in four positions

in place by means of the fixing lug provided, as close as possible to the axis of movement, and should not be mounted at the end of a slender bracket or on a flimsy support. The gauge should be set up so that its axis is parallel to the direction of movement to be measured.

When setting up a dial gauge for continuous reading it is convenient to adjust it to read zero at the initial position, if possible, or to read an exact number of millimetres or multiples of 100 μm. After tightening the clamping screw, fine adjustment can be made if necessary by rotating the bezel (see Figure 8.18(a)) which can then be locked in position. It is advisable to restrict this adjustment to within about one-tenth of a revolution (10 divisions on a 100 divisions dial) either side of the normal zero position, otherwise the revolution counter will not be in phase with the main pointer which may lead to mistakes in subsequent readings.

Accessories

Extension stems of various lengths can be fitted to the plunger.
Several different types of anvil are available for fitting to the end of the stem, and the correct type should be selected. The types most often used for soil testing are shown in Figure 8.19 their main applications being as follows:

(a) Standard anvil: steel ball: for general purposes when bearing on a flat machined surface
(b) Rounded or button anvil: uses similar to (a); also on non-machined surfaces or where high precision is less important
(c) Flat anvil: for curved (convex) surfaces, and for direct measurements on soil samples when using a bench stand
(d) Chisel (offset) anvil: for resting on a narrow ledge such as the end of a California bearing ratio (CBR) mould (see Chapter 11, Figure 11.27)

An extension stem or anvil should be screwed home securely, but not over-tightened, with thumb and finger, while gripping the plunger between thumb and finger of the other hand.

Dial faces and readings

The types of dial face used on gauges described in this volume are shown in Figure 8.20, and essential details are given in Table 8.2. Before starting a test in which dial gauges are to be observed it is essential to understand clearly how to read the main pointer and the

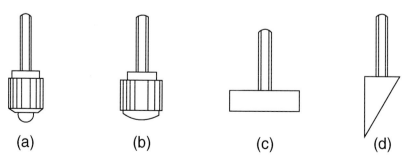

Figure 8.19 Anvils for dial gauges: (a) standard, steel ball; (b) rounded (button); (c) flat; (d) chisel (offset)

revolution counter, and the reading of both should always be recorded. The interval between each graduation mark is usually printed on the dial.

Gauges of type (a) in Figure 8.20 present no difficulty because the main pointer indicates hundredths of a millimetre and the revolution counter (small pointer) indicates whole millimetres. The reading of gauge (a) is therefore 7.38 mm. On the extended travel type, gauge (b), the revolution counter numbered up to 25 describes two revolutions over the full range of travel (50 mm), but it is usually evident whether the reading is greater or less than 25 mm. Gauge (b) could be indicating either 13.82 or 38.82 mm.

Gauge faces graduated with 0.002 mm divisions are usually less easy to read than the above types. The numbered intervals may indicate the number of such divisions (as in Figure 8.20(c)), or multiples of 0.01 mm (d), or millimetres (e). In all cases the graduation marks indicate intervals of 0.002 mm (2 μm).

In Figure 8.20(c) the numbered intervals refer to multiples of 10 divisions, therefore,

10 represents 0.02 mm
20 represents 0.04 mm
50 represents 0.10 mm
100 represents 0.2 mm (1 revolution)

One revolution is indicated by one division of the revolution counter, five divisions of which represents 1 mm. The gauge reading in example (c) is obtained as follows:

Main pointer:	18 divs. × 0.002 =	0.036 mm
Revolution counter:	8 divs. × 0.2 =	1.6 mm
	Reading	1.636 mm
	or	1636 μm

Always read the lower division next to the pointer on the revolution counter.

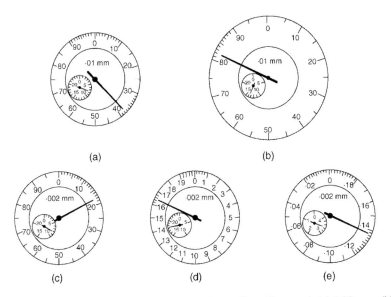

(a) (b) (c) (d) (e)

Figure 8.20 Dial gauge faces (for details see Table 8.2). Readings illustrated: (a) 7.38 mm; (b) 13.82 mm or 38.82 mm; (c) 1636 μm (1.636 mm); (d) 3565 μm (3.565 mm); (e) 3.537 mm

In Figure 8.20(d), each numbered interval is of five divisions and therefore represents multiples of 0.01 mm, therefore

1 represents 0.01 mm
2 represents 0.02 mm
5 represents 0.05 mm
10 represents 0.1 mm
20 represents 0.2 mm (1 revolution)

One revolution is indicated by one division of the revolution counter, as for (c). The gauge reading shown in example (d) is obtained as follows, reading to the nearest half division:

Main pointer:	$16.5 \times 0.01 =$	0.165 mm
Revolution counter:	$17 \times 0.2 =$	3.4 mm
	Reading	3.565 mm

This gauge has a total travel in excess of one revolution of the counter (5 mm), so on the second time round the reading would be 8.565 mm. An anti-clockwise reading gauge of this type is usually used for oedometer consolidation measurements.

The gauge face shown in Figure 8.20(e) is similar to (d) except that it is numbered directly in millimetres and is therefore easier to read. The example shown (which is anti-clockwise rotation) gives a reading of $(0.137 + 3.4) = 3.537$ mm or 8.537 mm on the second time round.

A dial gauge should be read from directly in front of the face and normal to it, so as to avoid parallax errors.

When the stem of a dial gauge is extending, ensure that the anvil maintains contact with the surface on which it bears and that the gauge does not stick. A light tap with a pencil on the dial face immediately before taking a reading may be needed to overcome local sticking, but excessive knocking should be avoided.

Figure 8.21 Example of a digital dial gauge

Digital dial gauges

Electronic dial gauges with a digital display, as shown in Figure 8.21 can be used in place of mechanical gauges and are gradual replacing analogue dial gauges as they can be read directly by data-logging systems. The power supply is provided by a lithium battery and they operate on the same principle as a linear transducer, using an integral analogue to digital converter and digital display in 0.001 mm divisions.

General care

Dial gauges are fitted with a delicate mechanism similar to that of a watch, and should be treated with similar care. If a gauge is dropped or subjected to a sharp blow or to vibration the pivots may be damaged or the alignment of bearings upset. A damaged gauge should be inspected and repaired only by a competent instrument maker, or returned to the manufacturer.

Never push the stem rapidly in and out, nor release the stem when compressed so that it extends with a jerk.

The spindle is intended to work dry, and neither it nor the mechanism should be oiled or greased. Even thin oil can pick up dirt which may cause sticking. Any oil, grease or moisture on the spindle should be carefully wiped off with a clean cloth or tissue.

If the spindle tends to stick, remove the anvil from the end of the stem, slide off the protective cover (see Figure 8.18) and wipe the spindle and the inside of the cover with a soft dry cloth. Replace the cover, and check that the spindle moves freely before setting up for use.

Replace a gauge in its box when not in use. Gauges permanently attached to test machines or other equipment should be covered with a small polythene bag to protect the working parts from dust.

Do not lay a dial gauge down on the bench on its back with it partly resting on the spindle. This may cause bending of the spindle, and consequent jamming. The safest way is to lay the gauge face downwards on a cloth or sheet of clean paper as a protection against scratching, with the spindle clear of the bench and any object which might knock against it.

8.3.3 Force measuring devices

Principle

Steel load measuring rings are a traditional means of measuring applied forces in the laboratory. They are robust, easily calibrated, and combine good stability and repeatability with sufficient accuracy. Each ring is normally supplied with the manufacturer's calibration certificate relating dial gauge reading to the applied force.

The principle of a load ring is the same as that of a calibrated spring. The deflection of the ring under the application of an applied force is measured by means of a dial indicator (dial gauge) which is mounted in the ring in the manner shown in Figure 8.22. The dial gauge

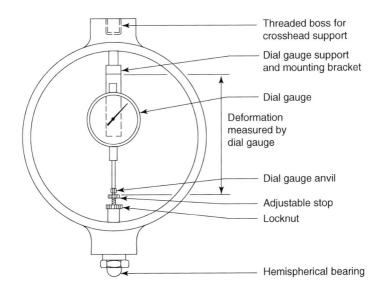

Figure 8.22 Mounting of dial gauge in load ring

is an integral part of the force measuring instrument. Within its working range the observed deflection enables the force to be determined to within a known degree of accuracy.

Load rings of this type are inherently non-linear, and the dial gauge mechanism can also be a source of error. Calibration against known forces is therefore essential, and a second-order analysis is necessary to determine errors and repeatability of the indicated force (see Section 8.4.4). Where the departure from linearity is not significant in comparison with the order of accuracy required from a test, a mean linear calibration can be used by assuming that the applied force is directly proportional to the measured deflection. Clause 4.4.4.6 of BS 1377:Part 1:1990 permits a deviation of up to ±2% from linearity as the range within which the mid-range calibration factor is acceptable. A properly conducted calibration is necessary in order to assess the validity of this assumption.

A development of this principle is to fit a displacement transducer in place of a dial gauge which can be connected to a signal conditioning module, as described in Section 8.2.6, to provide a digital display.

Types of ring

Load rings may be either of the clamped boss type, shown in Figure 8.2 (left), or of the integral boss type, shown in Figure 8.2 (right). Integral boss rings are the more expensive to produce, but offer better mechanical stability by eliminating the possibility of boss movement. Clamped boss rings are more prone to changes in calibration characteristics over a period of time, but remain serviceable if regularly calibrated and carefully handled. For both types, high tensile steel gives a greater load capacity for a given size of ring compared to mild steel.

The deflection characteristics of a load ring are not the same in tension as in compression, therefore separate calibrations are necessary for the two modes of use. Integral boss rings can be calibrated and used in both compression and tension, but clamped boss rings must be specially designed if they are to be used in both modes. For most soil testing applications it is necessary to measure only compressive forces.

A soil laboratory should hold a range of load rings of differing capacities and sensitivities, so that a load ring appropriate to the strength properties of the soil, and the test conditions, can be selected. All significant readings of applied force taken during a test, which are used for determining soil strength parameters, should lie within the calibration range. The range of load rings referred to in this volume is summarised in Table 8.3. The sensitivities (N/div.) shown in Table 8.3 are typical nominal values; actual working values are obtained from the calibration data supplied with each ring.

Care and use

The dial gauge or transducer fitted to a load ring should not be removed unless absolutely necessary. If a replacement gauge is fitted, the ring must be recalibrated, because of the non-linearity tolerance permitted in commercial dial gauges. A load ring should never be loaded beyond its maximum capacity. Load rings should be kept in their case when not in use. Additionally, the dial gauge of a ring which remains fitted to a test machine should be protected from dust.

A load ring fitted in a compression test loading frame should be fixed tight up against the crosshead. If there is any slackness the test specimen will carry the weight of the ring in addition to the indicated load.

When setting up a load ring immediately before a test, ensure that the anvil of the dial gauge makes contact with the adjustable stop of the ring. Adjust the stop if necessary so that the gauge reading corresponds to the calibration reading for zero load. For a compression ring, this reading will normally be zero. For a tension ring it will be a value which will allow the ring to extend to its working limit without the stop separating from the gauge anvil. Tighten the stop in position after adjustment.

Check that the spindle of the dial gauge moves freely, and returns to the zero position smoothly and without sticking when released. If there is a tendency to stick, clean the spindle and its cover as described under the general care of dial gauges, after first screwing down the adjustable stop to its lowest position. Never oil the dial gauge spindle.

8.3.4 Pressure gauges

Types of gauge

The range of commercial and standard test gauges referred to in this volume is summarised in Table 8.4.

Commercial gauges of good manufacture should be correct to within 1% of the maximum scale reading over the range 10–90% of that value. Readings outside this range, especially at the low end of the scale, may be unreliable. Pressures from zero (atmospheric) up to about 100 kPa can be measured accurately by using sensitive pressure transducers (see Section 8.2.6).

Gauges of 'test' grade are supplied as secondary pressure standards which have been calibrated against a primary pressure standard such as a dead-weight tester. Each gauge is issued with a calibration certificate, and is guaranteed to an accuracy of within 0.25% of the scale range at any point of the scale.

Measurement of pressures below atmospheric, such as on a vacuum line, requires a special vacuum gauge. Usually an indication of the degree of vacuum is all that is required, and not an accurate measurement, therefore a gauge of 80–150 mm diameter, of the type shown in Figure 8.3, is adequate. Vacuum gauges may be calibrated in negative pressure units (kPa, bars, or metres of water), or in millimetres of mercury (torr) of absolute pressure.

Scale markings

Pressure gauges supplied with soil testing equipment for use in Britain are usually purpose-graduated in SI pressure units, i.e. kPa. Some have dual scales in kPa and lb/in². Gauge manufacturers normally supply gauges graduated in bars (1 bar = 100 kPa exactly). A scale marked in bars may be converted to kPa by adding two zeros to each whole number on the scale, and replacing the word 'bar' by 'kPa' or 'kN/m²'.

Principle

The Bourdon tube principle, on which most pressure gauges are based, is illustrated in Figure 8.23. An elliptical-section phosphor bronze tube (the Bourdon tube), closed at its end and bent into the arc of a circle, acts as a pressure spring, and uncoils slightly under pressure. The free end lifts by an amount which is proportional to the applied pressure, and transmits the motion to the rotation of the pointer through a connecting link and quadrant gear. A hairspring takes up the small amount of backlash in the gears and pivots.

The following three types of gauge casing are available to allow for three methods of fixing:
- Direct mounting: supported only by the connection to the pipework
- Flush mounting: with a front flange and clamp arrangement
- Surface mounting: with a back flange for fixing to a panel

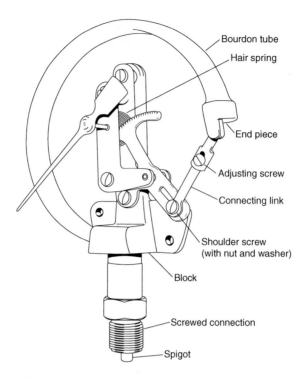

Bourdon tube

Hair spring

End piece

Adjusting screw

Connecting link

Shoulder screw
(with nut and washer)

Block

Screwed connection

Spigot

Figure 8.23 Principle of Bourdon tube type of pressure gauge (diagram courtesy of Budenberg Gauge Co. Ltd.)

Component parts of the surface mounting type, which is the most convenient in a soil laboratory, are shown in Figure 8.24.

Digital pressure gauges

Electronic pressure gauges, as shown in Figure 8.25, can be used in place of mechanical gauges. These are mounted on a pressure transducer block and use an analogue to digital converter to display pressure in kPa. Power is provided by a lithium cell, for which the manufacturers usually guarantee a life of one year.

Use and care of gauges

Pressure gauges should normally be mounted with the dial face vertical, at about the same level as the point at which pressure is to be measured.

When used for measuring water pressure, the connecting tubing should be completely filled with water. A bleed valve fitted next to the gauge will facilitate removal of air. The joint between the gauge and connecting pipe should be made tight to avoid leakage.

It is not normally necessary to replace all the air in the Bourdon tube with liquid, because of its relatively small volume. However, if it is essential to read very rapid response to pressure

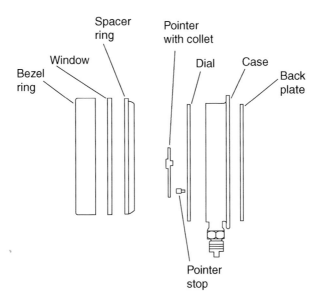

Figure 8.24 Component parts of pressure gauge casing (diagram courtesy of Budenberg Gauge Co. Ltd.)

Figure 8.25 Example of a digital pressure gauge

changes, this can be arranged by the manufacturers who would then describe it as a 'filled' gauge. A vacuum should never be applied to the Bourdon tube in an attempt to de-air it.

A gauge should be observed from a position normal to its face, to avoid parallax errors. A light tap just before reading may be needed to overcome local sticking, but excessive knocking may upset the delicate mechanism.

A gauge should not be subjected to vibration, or to pressure pulsations, or to sudden large changes in pressure. Pressures should be increased or decreased gradually.

Never pressurise a gauge beyond the end of its scale. Manufacturers recommend that gauges should not be subjected to steady continuous pressures exceeding about 75% of the maximum scale reading.

Never apply negative pressure (suction) to a gauge not designed for that purpose.

A pressure gauge must *never* be used with oxygen unless it is of the safety type, compliant with BS EN 837-1:1998 (see Section 8.5.3).

Gauges which are shut off when not being used for a length of time should be left under a small pressure (say 5% of the maximum reading) as a safeguard against changes in temperature or atmospheric pressures causing internal suction. Alternatively, the gauge may be left open to atmosphere.

If the bearings and pivots of a gauge require lubrication, use only a drop or two of thin oil and remove any excess very carefully. The teeth of the quadrant gear should never be oiled.

A gauge which develops a fault, or inaccuracies which cannot be rectified by the calibration procedures described below, should be returned to the manufacturer or supplier for overhaul.

Adjustment

Adjustment to pressure gauges should be carried out by a competent instrument mechanic.

If calibration of a gauge reveals a constant error throughout the scale range, this can be corrected by adjusting the pointer on its spindle. First remove the bezel ring and window. Lift off the pointer by using a pointer remover (see Figure 8.26(a)), taking great care not to bend the tapered spindle. Apply a pressure exactly equal to that indicated by the first main division on the dial, and push the pointer back on to the spindle with the fingers, as closely as possible to that reading. Check the readings at several other pressures up to the maximum, and ensure that the mechanism is working freely. Then secure the pointer by tapping it on to the tapered spindle using a pointer punch (see Figure 8.26(b)) and a light hammer. After re-assembly, calibrate the gauge again.

If the error is not constant, but progressively increases or decreases linearly as the pressure increases, the magnification ratio requires adjustment. Either the back of the gauge, or the dial, must first be removed. The shoulder screw (see Figure 8.23) is slackened, and the end of the connecting link is moved slightly towards the centre if readings become progressively

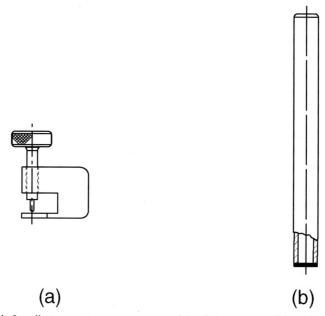

(a)   (b)

Figure 8.26 Tools for adjustments to pressure gauges: (a) pointer remover; (b) pointer punch (diagram courtesy of Budenberg Gauge Co. Ltd.)

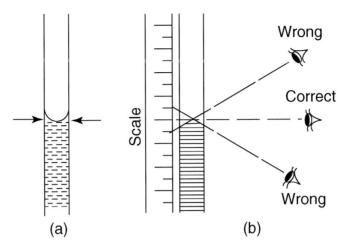

Figure 8.27 Reading water manometers: (a) water in standpipe or manometer, observe bottom of meniscus; (b) observe in horizontal direction to obtain true reading on scale

lower, or away from the centre if they become higher, with increasing pressure, Re-tighten the screw, re-check the calibration, and make a further adjustment if necessary.

If non-linear errors are present, the gauge should be returned to the supplier for adjustment or replacement.

8.3.5 Piezometers and manometers

Water pressures of very low values, such as are used in permeability tests, can be measured as a head of water in a piezometer standpipe. A typical arrangement is shown in Figure 10.24, Chapter 10. If the head of water is denoted by h mm, the pressure is equal to

$$\frac{h}{1000} \times 9.81 \text{ kPa}$$

The bottom of the curved surface (meniscus) of the water in a manometer tube should be observed (see Figure 8.27(a)), and should be viewed horizontally (see Figure 8.27(b)).

Historically, small differential pressures have been measured with a mercury manometer, but this practice has been discontinued, due to the health and safety hazards associated with the handling of mercury. Low pressures and small differential pressures are generally measured using a sensitive pressure transducer.

8.3.6 Hanger weights

Although they are not strictly measuring instruments, hanger weights are used for the application of a known force which remains constant over a period of time, either during a test or for calibration purposes. They can therefore be regarded as components of a measuring system, and should be treated and calibrated as such.

Hanger weights usually consist of discs or squares of cast iron (see Figure 8.28), slotted so that they can be suspended from a hanger in the manner shown in Figure 14.23 (Chapter 14). They are provided with recessed lead plugs to facilitate fine adjustment of their mass to nominal values, to within a reasonable degree of accuracy. Typical sizes which are available

Figure 8.28 Slotted weights

are 1 kg, 2 kg, 5 kg, 10 kg, and these may be supplemented by ordinary balance weights if necessary to obtain an exact load of any given magnitude.

Weights are often marked with their mass in the casting, but these values should not be accepted without verification. Each piece should be weighed to an accuracy of within 0.1%, and the true mass (kg) painted on. For some applications it is convenient to have the weight (i.e. force, in kN or N) painted on, preferably with a distinctive colour. The units used, whether kg, kN or N, should be included. The hanger and loading yoke from which the weights are suspended should also be weighed and marked.

Conversions from mass to force units are as follows:

$$\text{force (newtons)} = \text{kilograms (kg)} \times 9.807$$
$$= \text{pounds (lb)} \times 4.448$$

When not in use, slotted weights should be stored in a rack, or stacked carefully with only a small number to a pile, the largest at the bottom. They should be safeguarded against being knocked over, and protected from splashing, dirt and dust. Weights of the type used for balances should be stored in a box.

8.4 Calibration

8.4.1 General

This section covers the calibration of measuring instruments (additional to those referred to in Volume 1 (third edition)) which are required for performing the tests described in this volume. Routine checks and calibrations of other items of test equipment are also given (see Section 8.4.7).

Calibration of measuring instruments used for tests ('working instruments') may be carried out either by an external organisation, or in-house using the laboratory's own standards of reference. For a small laboratory it might be difficult to justify the expense of obtaining and maintaining its own reference standards, in which case all calibrations would be done externally. In a medium-sized or large laboratory it is usually more convenient to hold certain reference standards, against which some or most of their measuring instruments can be calibrated in-house.

An external organisation entrusted to provide calibrations should hold suitable qualifications (such as the United Kingdom Accreditation Service (UKAS) (Calibration) accreditation) for performing the relevant measurements, and must show that their calibrations are traceable to national standards of measurement. If the laboratory holds its own reference standards, these must be calibrated at regular intervals by a similarly accredited organisation. Recommendations for the use and calibration of reference standards are given in Section 8.4.8.

Reference should be made to Section 1.7 of Volume 1 (third edition) for general requirements and comments.

8.4.2 Principles

The principles of calibration of measuring instruments and the need for calibration to be traceable to national standards of measurement were outlined in Volume 1 (third edition), Section 1.7.1. A regular programme of calibration, together with properly maintained records, is an essential requirement for UKAS accreditation of a testing laboratory.

Instruments which are used for making measurements as part of test procedures ('working instruments') must be calibrated against reference standards for which a valid certificate of calibration, traceable to national standards, is applicable. The precision of the reference standard should generally be one order higher than that of the instrument being calibrated. This means that the range of uncertainty of measurement of the reference standard should be one-tenth, or at most one-fifth, of that of the working instrument. In some cases a ratio of one-half could be acceptable.

Manufacturers usually issue a calibration certificate with measuring devices such as load rings and pressure gauges. If the certificate includes confirmation that the calibration is traceable to national standards, and identifies the traceability route, the calibration data can be accepted as the initial calibration and can be used as the basis for preparing a calibration chart or graph. If there is no such confirmation, initial calibration against a certificated reference standard by an accredited organisation is necessary.

The performance of most instruments changes over a period of time, especially with intensive use, and can also vary with changes of environmental conditions such as temperature. Therefore it is essential to re-calibrate instruments at regular intervals, and to ensure that up-to-date calibration data are readily available for reference when tests are being carried out and results analysed. Calibration should be carried out in an appropriate environment, and the calibration temperature should always be recorded. The maximum intervals between calibrations of the types of instruments referred to here, as specified in BS 1377:Part 1:1990, are summarised in Table 8.6.

The following information should be displayed on or adjacent to measuring instruments, wherever practicable, as well as being entered in the calibration records:

Identification number of instrument
Date calibrated
By whom calibrated
Date when next calibration is due

Table 8.6 Calibration and checking of laboratory working instruments

Item	Maximum intervals between calibrations	Routine checks	Section reference
Displacement dial gauges	1 year	Check free movement of stem over full range. Anvil tight Gauge securely held	8.3.2 8.4.4
Displacement transducers	1 year (using working read-out unit)	As above	8.2.6
Load rings	1 year	Dial gauge or transducer as above. Contact made on anvil stop under zero load. Securely mounted. Anvil stop tight.	8.3.3 8.4.3
Pressure gauges	6 months	Reading zero at atmospheric pressure. Leave under small pressure when not in use	8.3.4 8.4.4
Pressure transducers	6 months (using working read-out unit)		8.2.6
Linear and torsion springs	1 year	Securely fixed in apparatus	8.4.4

The relevant working data (e.g. mean calibration factors; tabulated corrections; graphical relationships, as appropriate) should also be clearly displayed for easy reference.

8.4.3 Uncertainty of measurement

It is a requirement of BS EN ISO 17025:2005 that laboratories shall have procedures in place to estimate the uncertainty of measurement for all calibrations and tests. The UKAS sets out the procedures to follow for the determination of uncertainty of measurement in their guidance document M3003 and this is covered in greater detail in Vol 3.

8.4.4 Calibration of force measuring devices

General requirements

Calibration of devices used for measuring force is outlined in Clause 4.4.4.6 of BS 1377:Part 1:1990. The term 'load ring' used here (see Section 8.3.3) refers to a conventional steel force measuring ring fitted with either a dial indicator or a displacement transducer for measurement of its deflection under the application of an applied force. The dial indicator or transducer forms an integral part of the load ring. The same principles of calibration apply to other types of force measuring device.

Load rings should be calibrated before first use, and then recalibrated at least once every 12 months or more often if subjected to intensive use or if frequently loaded near to maximum capacity. A working load ring used for test measurements is calibrated against a suitable calibrated proving device, which may be a proving ring or an electrical resistance strain gauge load cell or similar instrument. The proving device must be certificated to show traceable calibration, and should meet the requirements of Grade 1 of BS EN ISO 376:2002 (Table 2) for repeatability and interpolation. It should be of a range and sensitivity appropriate to the load ring being calibrated.

Alternatively a suitably calibrated dead-weight pressure gauge tester fitted in a reaction frame can be used as a self-contained calibration device (see Figure 8.31(b) in Section 8.4.5). The known pressure resulting from calibrated weights acting on a piston acts on a second piston to generate a known force, which is applied to the load ring being calibrated.

In a soil laboratory load rings are usually interchangeable between testing machines, and it is therefore not practicable to allocate each ring to a particular load frame. Clause 4.4.4.6.1 of BS 1377:Part 1:1990 states that under these circumstances each ring should be calibrated in a 'dedicated' load frame. It is seldom practicable to reserve one load frame for calibration purposes only, so as a compromise one of the working load frames, which should be clearly identified, should always be used for calibration.

Load rings are inherently non-linear devices and significant deviations from a smooth relationship between force and deflection can occur, sometimes in the form of a 'spike'. These deviations are often not evident from a conventional first-order plot of deflection against force, of the kind shown in Figure 8.29(b), but can be identified by the method of analysis of the calibration data described below.

Limits of calibration

The lower limit of verification defined in BS EN ISO 7500-1:2004, Clause 6.4.5 (Note 1), applies to compression testing machines, not to load rings. For the calibration of load rings, the applied forces should cover the range from 10% of the scale maximum upwards, which provides a sensible lower limit for the calibration procedure. Alternatively the force corresponding to a reading of 200 divisions or digits on the load ring gauge or transducer could be taken as the lower limit of verification. The upper limit of verification should be as close as possible to, but not greater than, the working capacity of the load ring, i.e. its maximum scale reading.

Calibration procedures

The calibration procedure presented below is generally as described in BS EN ISO 7500-1:2004 (Clauses 6.4 and 6.5).

1. Mount the load ring above the proving device in the load frame so that the forces are applied along the axis of the frame. A spherical seating arrangement facilitates this condition.
2. Allow enough time for both devices to attain a steady temperature, and record that temperature at the beginning and end of calibration to the nearest 1°C.
3. Load and unload the two devices three times, up to the maximum capacity of the load ring and down to zero load, without recording any readings.
4. If necessary, re-set the gauge and the load ring to read zero under zero force.
5. Apply the initial force (as determined by the proving device) and record the reading of the load ring indicator when conditions are steady. The initial force should be equal to about 20% of the nominal capacity of the load ring (maximum scale reading), or the lower limit of verification as determined above, whichever is greater.
6. Increase the force in four approximately equal additional increments up to the maximum scale reading of the load ring, and record the load ring indicator reading under each force (giving five forces in all).
7. Reduce the force in decrements, and record the load ring indicator reading at the same values of force as during loading. Remove the force completely after taking the reading at the initial force and record the reading of the load ring indicator.

Calibration of force measuring ring

Ring number	987–6–543		Date of calibration	1.4.93
Gauge number	XY 210		Calibration temperature	20° C

Calibrated in compression

Maximum gauge reading must not exceed　　1900　　divisions

Applied force KN	Dial indicator reading (divs)				Factor	Repeatability	
	Test 1	Test 2	Test 3	Average	N/div	Spread divs	%
2	182.4	179.9	180.0	180.8	11.062	2.5	1.36
4	367.0	362.0	365.0	364.7	10.968	5.0	1.36
6	531.6	530.0	530.0	530.5	11.310	1.6	0.29
8	735.2	729.0	732.2	732.1	10.927	6.2	0.84
10	921.3	916.5	919.0	918.9	10.883	4.8	0.52
12	1109.8	1104.0	1106.0	1106.6	10.844	5.8	0.53
14	1298.0	1295.0	1290.0	1294.3	10.817	8.0	0.62
16	1487.9	1485.5	1485.0	1486.1	10.766	2.9	0.20
18	1680.0	1676.4	1679.5	1678.6	10.723	3.6	0.22
20	1872.0	1868.5	1870.0	1870.2	10.694	3.5	0.19

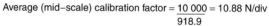

Average (mid–scale) calibration factor = $\dfrac{10\ 000}{918.9}$ = 10.88 N/div

(a)

(b)

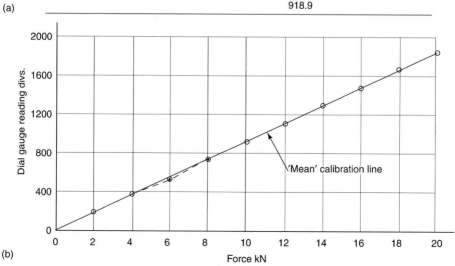

Figure 8.29 Calibration of typical load ring: (a) calibration data, (b) conventional first-order plot of readings

8. Repeat steps 5, 6 and 7 twice more to give three cycles of loading and unloading.
9. If the lower limit of verification is less than 20% of the scale maximum, and if appropriate, apply additional forces down to the lower limit of verification, in steps of about 5% of the maximum scale reading, otherwise following the above procedure.
10. From the three sets of readings at each force, calculate the following:
    Average load ring reading (R_a)
    Spread of each set of readings (R_s) (difference between highest and lowest)
    Repeatability (%) (R_r) $R_r = \dfrac{R_s}{R_a} \times 100\%$

Analysis

In one accepted method of analysis, the calculated values of R_a, R_s and R_r are tabulated with the sets of force and gauge readings. Errors in the load ring are then calculated in terms of departures from linearity, using a calculated linear relationship derived from the mid-scale force readings.

A more practical method of analysis and graphical presentation for a soil laboratory is given in BS 1377. This method, together with an example, is as follows:

1. For each set of readings divide the applied force (newtons) by the average load ring reading (R_a) to obtain the calibration factor C_R (in N/div or N/digit) for that force. (If the ring calibration were exactly linear, these factors would all be the same.)
2. Plot each value of C_R, to a suitably enlarged scale, against the corresponding average reading R_a, as shown in Figure 8.30. This type of plot (a second-order relationship) emphasises the deviations from linearity. The example used shows a pronounced 'spike' at about 500 divisions (6 kN).
3. Draw a horizontal line corresponding to the mid-scale factor, or the closest to mid-scale where readings have been taken. Draw horizontal lines corresponding to factors 2% higher and 2% lower than the mid-scale factor.
4. Where the plot of calibration factor lies within the ± 2% limits, the mid-scale calibration factor may be used for calculating forces in a test, provided that the calculated repeatability (R_r) does not exceed 2%. Outside these limits the calibration factor should be obtained from the graph.

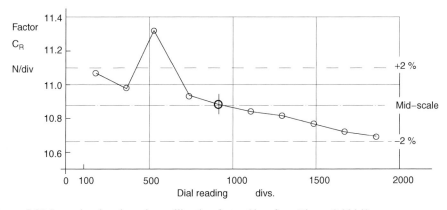

Figure 8.30 Second-order plot using calibration factor (data from Figure 8.29(a))

In the example shown in Figure 8.30, the mid-scale calibration factor would be applicable for readings from 180 to 370 divisions, and from 730 to 1870 divisions. The factor in the region of the 'spike' is less certain, except at the point where readings were taken, which may or may not be the highest point of the spike.

Calibration near the lower verification limit of a load ring of low capacity can be carried out by using calibrated weights suspended from a loading yoke.

Temperature correction

A temperature correction is given in Appendix B of BS EN ISO 376:2002 for correcting the deflection of a steel load ring when calibrated at a temperature other than 20°C. The same correction would be applicable when a load ring calibrated at 20°C is used in a test at a significantly different temperature.

The correction equation is

$$d_{20} = d_t[1 - K(t - 20)]$$

where d_{20} is the reading of the load ring at 20°C, d_t is the observed reading at t°C, t is the working temperature and K is the temperature coefficient, which for a steel ring is taken to be 0.00027/°C. By way of illustration, at a working temperature of 30°C (10°C above calibration temperature) the observed readings would be reduced by 0.27%, which is well within the 2% range referred to above.

For load rings made of other materials, and for strain gauge load cells, the value of the temperature coefficient K should be obtained from the manufacturer. Electric strain gauge devices are likely to be much more sensitive to temperature changes than load rings.

8.4.5 Calibration of other instruments

Pressure gauges

Pressure gauges should be recalibrated at least once every six months, preferably at the start of the summer and winter seasons. Gauges subjected to intensive use should also be checked and recalibrated at intermediate intervals.

A calibrated dead-weight gauge tester of the type shown in Figure 8.31(a) provides a suitable reference standard for a soil laboratory. Accurate weights resting on a precision-made piston (the critical component) fitting into a matching vertical cylinder generate a pressure in the oil of the system which is known to a high degree of accuracy. When calibrating working gauges which are used with water, an oil–water interchange unit (available from the manufacturer of the gauge tester) should be incorporated to avoid contamination with oil.

(a)

(b)

Figure 8.31 (a) Dead-weight pressure gauge tester; (b) dead-weight tester with reaction frame for calibration of load rings and transducers

A gauge tester of the type shown in Figure 8.31(a) typically covers a pressure range up to 1500 kPa, which can be extended by using an additional set of weights. The instrument must be on a firm level surface in a dust-free atmosphere. It should be covered when not in use and protected from dust and damp.

Alternatively a calibrated pressure gauge of 'test' grade may be used as a laboratory reference standard, provided that it is reserved for calibration purposes only. This enables working gauges to be calibrated in-situ, but the working gauge and the reference gauge must be at the same level during calibration. The reference gauge should initially be calibrated on a dead-weight tester, or by an accredited external organisation, and then recalibrated at the prescribed intervals.

The procedure for calibrating a pressure gauge, in accordance with BS 1377:Part 1:1990: Clause 4.4.4.7 is as follows: (It should be noted that Clause 4.4.4.7 makes reference to Clause 1 of BS 1780:1985, which has been superseded by BS EN 837-1:1998, but the procedure described here remains valid, as specified in BS 1377:Part 1:1990.)

1. Mount the gauge vertically, either in the dead-weight tester or connected to a pressure source in parallel with the reference gauge.
2. Allow enough time for a steady temperature to be attained, and record that temperature, which should be 20°C, ± 3°C.
3. Pressurise the gauge to its maximum scale value, and release the pressure slowly without making any adjustments.
4. Raise the pressure in convenient increments (e.g. 100 kPa) as determined by the weights on the dead-weight tester, or measured by the reference gauge, up to the maximum scale reading, and record gauge readings and true pressures.
5. Reduce the pressure in decrements and record gauge readings at the same pressure levels as in step 4.
6. Repeat steps 4 and 5 to give two cycles of readings.
7. Calculate the average gauge reading corresponding to each level of pressure.
8. Calculate the error at each pressure as the difference between true pressure (P) and gauge reading (G); i.e.

$$\text{Error} = P - G$$

   Tabulate the error (positive or negative) alongside the average readings as shown in Figure 8.32.
9. Plot the error (positive upwards, negative downwards) to a suitably enlarged scale against the corresponding reading of the gauge being calibrated (see Figure 8.32), and join the points to give a correction graph.
10. Display the graph, or table (or both) alongside the gauge to which it relates. Any gauge reading is then increased (if the correction is positive) or decreased (if negative) to give the correct pressure.

Displacement gauges

Dial gauges and electric transducers used for measuring displacements should be calibrated at least once a year. Electric transducers must be calibrated together with the same read-out unit or data logger as used for tests.

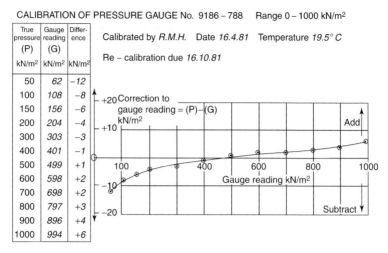

Figure 8.32 Pressure gauge calibration chart and correction graph

Table 8.7 Tolerances of non-linearity of dial gauges (scale divisions of 0.01 mm) Taken from BS 907:2008

Interval of reading	Limit of error in reading over stated interval (mm)	Repeatability of reading	Discrimination
Any 0.1 mm	0.005		
Any half revolution	0.0075		Within 0.003 mm
Any one revolution	0.01	Within 0.002 mm	during gradual change
Any two revolutions	0.015		of ≈0.025 mm
Any larger interval	0.020		

A dial gauge or displacement transducer can be calibrated against a calibrated micrometer device by observing readings at regular intervals of displacement. Alternatively a gauge can be mounted in a comparator frame so that readings corresponding to thicknesses of calibrated gauge blocks or length bars can be observed.

The tolerances of non-linearity specified in BS 907:2008 for commercial gauges are summarised in Table 8.7.

Springs

Torsion springs for the laboratory vane apparatus should be recalibrated at least once a year, each to an accuracy within 2% of its indicated torque over its working range. Springs would normally be returned to the manufacturer for this service. Alternatively, a method for in-house calibration is outlined in ASTM D 4648. A pulley wheel of about 115 mm diameter (which is accurately measured) is attached to the stem of the vane shear apparatus in place of the vane, to act as a lever arm. The apparatus is secured with its axis horizontal, and known weights suspended from a string attached to the wheel rim provide a known torque. A range of weights is used, and from the corresponding readings of the spring deflection a calibration graph can be obtained.

Extension springs for the autographic unconfined compression test apparatus should be recalibrated once a year. A loading yoke with a suitable arrangement for measuring extension of the spring could be set up in the laboratory, but it is probably more convenient to return the springs to the manufacturer for calibration. Calibration of each spring should be to an accuracy within 5% of the indicated force for the upper 90% of the working range.

8.4.6 Calibration of electronic instrumentation

Electrical measuring instruments such as transducers should be calibrated at least once a year, using methods similar in principle to those described above for conventional instruments. Electrical devices should always be calibrated with the same read-out units as are used in normal testing procedure.

The environmental temperature during calibration should be maintained constant and recorded. Electronic instrumentation is more sensitive than conventional instruments to temperature variations. Electrical circuits should be energised several hours in advance of calibration to allow for warming up so that readings become stable.

Further details on calibration of electronic instruments will be found in Volume 3.

8.4.7 Checking and calibration of test apparatus

In addition to measuring instruments, numerous other items of test apparatus need to be checked or calibrated before use and at regular intervals thereafter. Items required for the tests described in this volume (additional to the items referred to in Table 1.14, Volume 1 (third edition)) are listed in Table 8.8, together with a summary of the relevant checks that are needed. Further details are given in the sections referred to in Table 8.8.

Initial checks or calibrations should be made on each item before it is first put into use. Nominal dimensions or manufacturer's data (except for certified traceable calibrations) should not be accepted without verification. Subsequent re-checks are necessary at regular intervals to make allowance for wear due to use. According to Clause 4.1.3 of BS 1377:Part 1:1990, when critical dimensions differ from the specified dimensions by more than twice the manufacturing tolerance the item no longer complies with the Standard, and should be withdrawn from use. The intervals between re-checks suggested in Table 8.8 provide a general guide, but for items subjected to intensive use additional intermediate checks should be made.

It is particularly important to check the height of cutting rings for shearbox and oedometer specimens before every use. Cutting edges wear easily, and a small difference in specimen height, which is not allowed for, can give rise to significant errors in calculated voids ratio.

Initial checks and re-checking should be carried out in accordance with a planned schedule, and all observations should be properly documented and the records retained.

8.4.8 Reference standards

Reference standards of measurement which are held in the laboratory for calibration of working instruments should be used for calibration only and for no other purpose (see Section 1.7.2 of Volume 1 (third edition)). Their use should be restricted to authorised personnel who have been suitably trained in calibration procedures.

Reference standards (additional to those referred to in Section 1.7.2 of Volume 1 (third edition)) appropriate for calibrating measuring instruments used in tests described in this

Table 8.8 Checking and calibration of test apparatus

Item	Measurement or checking procedure	Maximum interval between re-checks	Section reference
Permeameter cells	Mean internal diameter Overall internal length Distance between manometer glands Manometer tube diameters	1 year (Initial check sufficient)	10.6.3 10.7.2
CBR moulds	Internal diameter and height Mass	1 year	11.6.3
Sampling tubes	Internal diameter and length Condition of cutting edge	1 year Before each use	9.1.3
Split formers	Internal diameter and length when assembled	1 year	9.1.3
Shearbox	Internal dimensions Mass	1 year	12.5.3
Cutting rings	Internal diameter Mass	1 year	14.5.5(2)
	Internal height Condition of cutting edge	Before each use	
Hanger weights	Mass (to 0.1%) Clean; no corrosion or damage Mass or weight (force) clearly marked	2 years Before each use	8.3.6
Oedometer load frame	Deformation characteristics under load	2 years or when new porous discs fitted.	14.8.1
	Beam ratio	(Initial check sufficient)	

Table 8.9 Calibration intervals for laboratory reference standards (see also Table 1.14, Volume 1 (third edition))

Reference standard	Maximum interval between calibrations
Proving device (Grade 1.0)	5 years
Pressure gauge tester	5 years
Reference pressure gauge	1 year

volume are listed in Table 8.9. These instruments should be calibrated initially, and recalibrated within the stated periods, by a suitably qualified organization such as a laboratory having UKAS (Calibration) accreditation. Calibration certificates must include confirmation that the calibration is traceable to recognised standards of measurement. Essential details are as follows:

- Name of the calibrating organisation
- Name and location of organisation for whom the calibration was made
- Description and identification number of item calibrated
- Method of calibration and equipment used
- Calibration certificate number of the reference device used for calibration, and the traceability route if it was not calibrated by a UKAS organisation
- Calibration data and results, including calibration temperature

- Date of calibration
- Signature of person responsible for the calibration

8.5 Safety

The notes on safety given below supplement those regarding the laboratory generally, given in Volume 1 (third edition), Section 1.6.

8.5.1 Machinery

Moving parts of machinery should always be covered with a protective casing or cage when in operation. This applies particularly to gears, belt drives and chain drives.

When changing gears or drive sprockets on a machine, first make sure that the motor is switched off *and disconnected from the mains*.

When motor drive is to be used on a test machine fitted with a hand wheel, disengage the hand winding handle before switching on. If it cannot be disengaged, keep loose clothing and other items well clear.

The vertical alignment of a sample or cell in a load frame must be maintained during a compression test, especially when high loads are applied. Deviation from correct alignment indicates inherent instability, which can be very dangerous.

Do not interfere with over-run cut-out switches if fitted to a machine.

8.5.2 General

When testing hard or brittle samples for unconfined compressive strength, they should be surrounded by a protective cage as a precaution against flying fragments.

The importance of the stability and strength of bench supports for oedometer presses is dealt with in Chapter 14, Section 14.5.3, Item 10. The same considerations apply to any apparatus in which dead-weight loading is used.

Slotted weights when not in use should be stored tidily, and not stacked so high that they are likely to be unstable. When stacked on a hanger the slots should not all be in alignment but staggered at 90° or 180°.

8.5.3 Compressed air

Compressed air, like electricity, must be treated with respect. All tubing, connections and fittings should be of a type specifically designed for use with compressed air at a pressure 50% in excess of the maximum working pressure to be used. Tubing and fittings supplied for hydraulic systems are not necessarily safe for use with compressed air at the same pressure.

Unless specially designed and certificated for pneumatic use, triaxial cells should be used only with water or oil as the pressurising fluid, never compressed air. Failure of a cell under liquid pressure would be messy and inconvenient; but failure under air pressure could result in a dangerous, if not lethal, explosion.

Never expose any part of the body to a jet of compressed air, and never direct a jet of air towards another person.

Provision should be made in all compressed air pipelines and vessels for the removal of condensate water. Pipelines should be laid to a slight fall in the direction of the flow, with drainage valves at low points. Accumulated water should be drained from pipelines

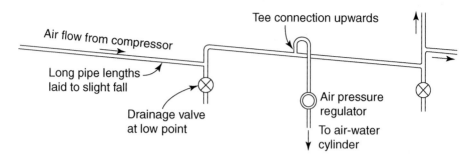

Figure 8.33 Compressed air pipelines: making provision for collection and removal of condensation

and air-receiving vessels regularly. Branches should leave the main airline in an upward direction (see Figure 8.33).

Bladders and their fixing clips used in air–water exchange cylinders should be inspected regularly, and replaced as soon as they show signs of deterioration.

If bottled compressed air is used, an oxygen bottle should *never* be used as a substitute. Oxygen under pressure in contact with oil or grease forms an explosive mixture, and could result in a serious explosion. Thin films and slight deposits of oil or grease, which are unavoidable unless pipework and connections are scrupulously cleaned, are particularly susceptible to ignition. Pressure gauges are always likely to be contaminated with oil, and special safety gauges together with appropriate precautions must be used with oxygen. Further details are given in BS EN 837-1:1998:Clause 9.8.

8.5.4 Mercury

Under EC Regulation No 552/2009, manometers, barometers or thermometers containing mercury may no longer be sold commercially. Such instruments will in the future be replaced by instruments that use alcohol, or digital alternatives such as are widely used in modern soil laboratories.

Laboratories that hold a supply of mercury and continue to use it should appreciate that mercury falls under the Control of Substances Hazardous to Health (COSHH) Regulations. Staff handling mercury must be made aware of the hazards and risks through safety data sheets, appropriate risk assessments and training. Good practice, including the treatment of spilled mercury, is outlined in Section 1.6.7 of Volume 1 (third edition). Reference should also be made to the Health & Safety Executive, Guidance Note MS 12 (rev) (Health and Safety Executive, 1996).

Mercury vapour must not be inhaled, and direct contact with the skin should be avoided. Always be on the lookout for leaks in the system. Mercury under pressure escapes from minute cracks or pinholes as a fine spray, which readily evaporates. A leak is difficult to observe until some quantity has accumulated. Deep trays containing a little water should be placed under pipe joints and connecting tubing to catch any leakage. A free surface of mercury should be covered by a thin layer of water to prevent escape of mercury vapour.

References

Annual Book of ASTM Standards (2010) Section 4 Construction: Volume 04.08 Soil and Rock. American Society for Testing and Materials, Philadelphia, PA, USA

BS 907:2008 *Specification for dial gauges for linear measurement*. British Standards Institution, London

BS 1377:1990 *Methods of test for soils for civil engineering purposes*. British Standards Institution, London

BS EN ISO 7500-1:1998 *Metallic materials verification of static uniaxial testing machines. Tension/compression testing machines. Verification of the force measuring system*. British Standards Institution, London

BS EN 837-1:1998 *Pressure gauges: Bourdon tube pressure gauges. Dimensions, metrology, requirements and testing*. British Standards Institution, London

Bishop, A. W. and Henkel, D. J. (1964) *The Measurement of Soil Properties in the Triaxial Test*. Edward Arnold, London

Health and Safety Executive (1996). Guidance Note MS 12 (rev). Mercury – Medical Surveillance. Health and Safety Executive, London

Two recommended textbooks

Lambe, T. W. and Whitman, R. V. (1979) *Soil Mechanics, SI version*. Wiley, New York

Scott, C. R. (1974) *An Introduction to Soil Mechanics and Foundations*. Applied Science Publishers, Barking, UK

Chapter 9

Preparation of test specimens

9.1 Introduction

9.1.1 Scope

Preparation of disturbed samples for testing was covered in Section 1.5 of Volume 1 (third edition). Preparation of test specimens from undisturbed samples is described in this chapter. These procedures in general are based on Clause 8 of BS 1377:Part 1:1990.

The procedures described below relate to the hand trimming of test specimens of soils that possess some cohesion. The types of specimen covered are:

1. Cylindrical specimens for compression tests
2. Square specimens for standard shearbox tests
3. Disc specimens for oedometer consolidation tests
4. Undisturbed specimens in a 'core-cutter' tube, for falling head permeability tests.

Procedures for specimens of types 2 and 3 are almost identical and are treated as one. Use of a soil lathe, including a motorised lathe, is described, together with comments on its possible mis-use.

The words 'sample' and 'specimen' are used here broadly in the sense defined in Volume 1 (third edition), Section 1.1.7. A 'specimen' refers to the relatively small portion of soil which has been trimmed from a larger 'sample' and on which a certain test is performed. The word 'sample' may also be used for a relatively large mass of soil required for a test (such as a large diameter triaxial test) if it consists of all, or the greater part of, the original sample. Material recompacted into a mould, and from which smaller specimens may be trimmed, is also referred to as a sample.

Test specimens prepared by hand trimming are generally referred to here as 'undisturbed', although the sample from which they are cut may consist either of undisturbed soil, or of material which has been recompacted or remoulded. Procedures are given for the preparation of recompacted samples.

Some aspects of sample preparation not included here are those requiring the placement of granular (non-cohesive) soils, and those for which special procedures have been developed for a particular test. Procedures of these types are described in the appropriate chapters, as follows:

> Chapter 10: Setting up sandy soils at various porosities in the constant head permeameter cell (Section 10.6.3, Procedural stage 4)
> Chapter 11: Several different methods for the preparation of samples for CBR test (Section 11.6)

Chapter 12: Formation of cohesionless specimens at various porosities for standard shearbox tests (Sections 12.5.4 and 12.6.3)

Chapter 13: Preparation of cylindrical specimens of cohesionless material in a special former for triaxial compression tests (Section 13.6.9)

Preparation of orientated cylindrical specimens, requiring a specially developed apparatus (Section 13.6.8)

Procedures for the handling and sub-division of disturbed (cohesionless) samples were described in Volume 1, Section 1.5.5.

9.1.2 Equipment

Items of equipment required for the sample preparation procedures described in this chapter are listed below, and their use is outlined in Section 9.1.3. Special items required for procedures given elsewhere are described in the appropriate chapter.

Extruders

Sample extruders for the removal of undisturbed samples from various kinds of sampling tubes may be either hand-operated or motorised. Perhaps the most versatile is the hand-operated hydraulic vertical extruder shown in Figure 9.1, which has an integral double-acting pump. It is designed to accept standard U-100 sample tubes, and adaptors can be obtained for other sizes of tubes and moulds, from 38 to 150 mm diameter. A motorised vertical extruder, which incorporates an oil pressure gauge and provides a pressurised quick-return movement, is shown in Figure 9.2.

Figure 9.1 Hand-operated hydraulic vertical sample extruder

Figure 9.2 Motorised hydraulic vertical extruder

A hand-operated horizontal screw type extruder, designed for extruding samples from U-100 tubes, is shown in Figure 9.3. Horizontal extrusion is more satisfactory for soft or friable soils, and for samples that need to be examined closely before test specimens are prepared. The sample must be supported by a semi-circular section trough along its whole length as it emerges from the tube.

A hydraulic motorised extruder which was specially designed for use with piston sampling tubes up to 1 m length is shown in Figure 9.4. Horizontal extrusion is essential for long samples of this type. The motorised pump unit enables a steady and controlled force to be applied to the sample, resulting in minimum disturbance as it is pushed out.

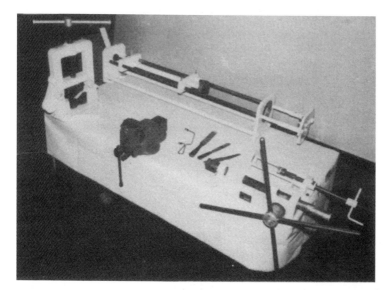

Figure 9.3 General equipment for sample extrusion (photograph courtesy of Department of Civil and Structural Engineering, University of Sheffield)

Figure 9.4 Motorised hydraulic horizontal extruder for piston samples up to 1 m long (photograph courtesy of Soil Mechanics Ltd.)

Smaller hand-operated extruders are available for extruding specimens from 38 mm diameter tubes. These are the vertical hydraulic type (see Figure 9.5), and the rack and pinion type shown in Figure 9.3. Extruders making use of a hand-operated hydraulic jack, similar to a car jack, for removing compacted samples from a compaction mould and a CBR mould, are shown in Figures 9.6(a) and 9.6(b), respectively.

Sample tubes and formers

Moulds into which undisturbed samples are extruded, or into which disturbed material is placed, for preparing test specimens comprise several kinds of tubes, rings and formers, as follows:

1. Sample tubes with cutting edge. The standard British size is 38 mm internal diameter and 230 mm long. Other diameters are 35 mm, 50 mm, 70 mm and 2.8 in. Notes on their use are given in Section 9.1.3.

Figure 9.5 Vertical hydraulic type extruder for 38 mm diameter sample tubes

2. Cutting rings for oedometer test specimens, typically 75 mm diameter and 20 mm high in Britain, but there are several other sizes (see Chapter 14, Section 14.5.3).

3. Cutting 'rings' for shear box specimens, usually 60 mm square or 100 mm square and 20 mm high (see Chapter 12, Section 12.7.3).

4. Split formers, consisting of either two or three segments, of internal dimensions exactly corresponding to cylindrical test specimen sizes. A two-segment former 38 mm diameter, fitted with wing-nut clamps, is shown mounted on the extruder in Figure 9.5. The use of split formers is outlined in Section 9.1.3.

5. Split formers, of two or three segments, for preparing triaxial test specimens on a triaxial pedestal, as described in Section 13.6.9. This type has a recessed extension to fit around membrane O-ring seals, and is usually fitted with a suction connection. Formers of several sizes are shown in Figure 9.7.

Extruder accessories

Accessories designed for holding sample tubes and formers for cutting test specimens as a sample is extruded from its tube or mould, mounted on a vertical extruder, are listed below.

1. Adaptor for preparing and extruding a single 38 mm diameter specimen (see Figure 9.8(a)).

2. Triple-tube holder (see Figure 9.8(b)) for cutting three 38 mm diameter specimens simultaneously all from the same horizon, as shown in Figure 9.8(c).

3. Adaptor for extruding the whole 100 mm diameter sample from a sample tube or thin-walled piston sampling tube.

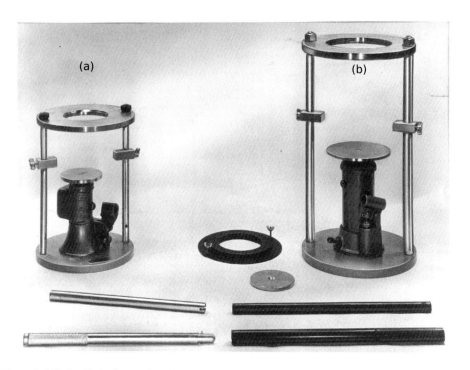

Figure 9.6 Hydraulic jack extruders for removing compacted samples from: (a) BS compaction mould; (b) CBR mould

Figure 9.7 Split formers (two segments) for triaxial test specimens of 38, 70, 100, 150 mm diameter

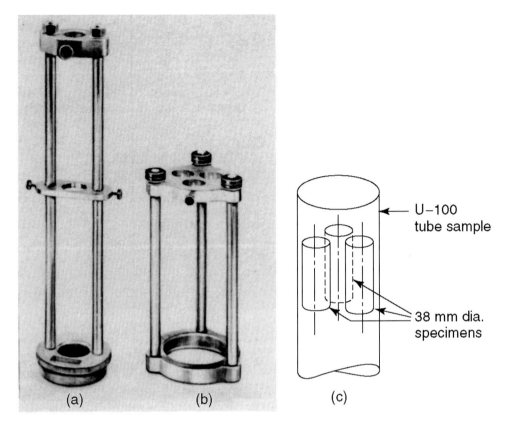

Figure 9.8 (a) Single-tube adaptor for 38 mm diameter specimen; (b) triple-tube adaptor for preparing three 38 mm diameter specimens; (c) set of three 38 mm diameter specimens from one horizon in U-100 tube

    4.   Adaptors for extruding from a compaction mould (see Figure 9.9(a)) or a CBR mould (see Figure 9.9(b)).

    5.   Jig fitted with interchangeable locating plates for holding either a square cutter for obtaining a shearbox specimen, or a circular oedometer consolidation ring cutter (see Figure 9.16 in Section 9.1.3).

For removing specimens from cutting tubes and rings, simple 'pushers' of metal or wood may be used. A cylindrical pusher of the same length as a tube, graduated in millimetres to indicate the exact length remaining in the tube after trimming the ends, is shown diagrammatically in Figure 9.10(a). A square pusher for transferring a specimen from cutting 'ring' to shearbox is shown in Figure 9.10(b).

Small tools

Tools required for the preparation of test specimens are listed below (see Figure 9.11).
- Trimming knife (cobbler's knife)
- Scalpel (craft tool to which several shapes of blade can be fitted)
- Wire saws (piano wire about 0.4 mm diameter, spiral wire)

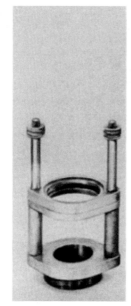

(a)

(b)

Figure 9.9 Adaptors to vertical extruder for extruding from: (a) BS compaction mould; (b) CBR mould

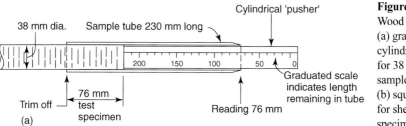

Cylindrical 'pusher'

38 mm dia.   Sample tube 230 mm long

200   150   100   50   0

Graduated scale indicates length remaining in tube

76 mm test specimen

Trim off

(a)

Reading 76 mm

Figure 9.10
Wood 'pushers':
(a) graduated cylindrical pusher for 38 mm diameter sample tube;
(b) square pusher for shearbox specimen

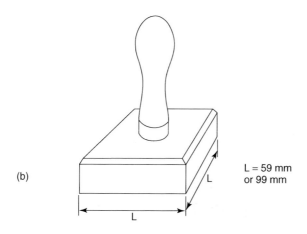

(b)

L = 59 mm or 99 mm

L

L

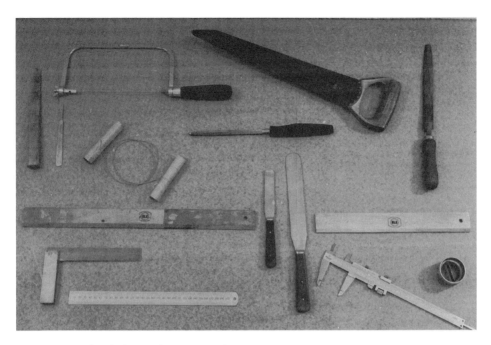

Figure 9.11 Hand tools for specimen preparation

- Cheese wire
- Toothed saw
- Half-round file
- Knife sharpener
- Steel straight-edge trimmer
- Spatulas (small and large)
- Steel straight-edge
- Steel try-square
- Steel rule, graduated to 0.5 mm
- Vernier callipers, readable to 0.1 mm
- End trimmer for tube specimen
- Clinometer (see Figure 13.40)
- Metalworking vice (included in Figure 9.3)
- Pipe clamp, 150 mm capacity (see Figure 9.3)
- Chain wrenches (see Figure 9.12)

Figure 9.12 Chain wrench, suitable for gripping U-100 tube and end-caps

The last three items are sometimes necessary for removing seized-up end caps from sample tubes used in the field.

Miscellaneous items

Other items frequently required for specimen preparation are as follows (some of these were referred to in Volume 1 (third edition)):

• Flat glass plate	• Burette stands
• Metal tray	• Polythene sheet and bags
• Watch glass	• Cloths
• Funnel	• Wax and wax pot
• Scoop	• Labels and markers
• Tamping rod	• PTFE spray
• Half-round plastic guttering	• Thin lubricating oil
• Moisture content tins	• Penetrating oil (e.g. WD40)

Soil lathe

A hand-operated soil lathe is shown in Figure 9.30 (see Section 9.4.1). A motorised lathe (no longer in production) is shown in Figure 9.13. Use of a soil lathe is described in Section 9.4.

Small vibrator

For the preparation of test specimens of sand in a dense state, an engraving tool (see Figure 9.14) can be used as a small vibrator, if fitted with a square or round tamping foot for preparing shearbox or triaxial specimens, or with an extension foot for use in a permeameter cell.

9.1.3 General principles

Figure 9.13 Motorised soil lathe for specimens up to 100 mm diameter

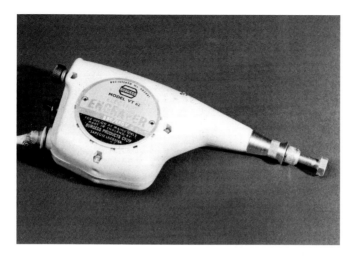

Figure 9.14 Hand engraving tool, for use as a small vibrator (photograph courtesy of University of West of England)

Preparation of test specimens from soil samples requires care, a certain degree of skill, and a great deal of patience. Undisturbed samples are expensive to obtain, and it is very easy to spoil a sample through carelessness or excessive haste. Some of the basic skills to be acquired for sample trimming and measurement were indicated in Volume 1 (third edition), Sections 1.5 and 3.5.2. Further comments are given below, together with notes on the use of the equipment listed in Section 9.1.2.

Tools

Hand tools should be maintained in good condition, and cleaned and dried after use. Trimming knives should be sharpened before use and re-sharpened frequently. Worn scalpel blades should be discarded (well wrapped). Straight-edges and rulers used for checking flatness should not be used as scraping tools, otherwise the edges will become worn and quickly lose their straightness. Scrapers, which are inexpensive, should always be used for this purpose.

Tubes and cutting rings

Sample tubes for preparing compression test specimens may be either of plain bore (see Figure 9.15(a)), or with a 'relieved' bore, i.e. the internal diameter of the tube is very slightly greater than that of the cutting edge (see Figure 9.15(b)). The latter type offers less resistance to entry and removal of the sample than the plain type, and is therefore less likely to cause disturbance of sensitive soils. When extruding from the relieved bore type, the specimen must leave the tube in the same direction as it entered, i.e. it must be pushed out at the non-cutting end (see Figure 9.15(c) and Section 9.2.2). Plastic protective end caps are available for both types of tube.

The cutting edges of sample tubes and cutting rings should be inspected before use to ensure that they are sharp, in good condition and truly circular. Any burrs on the inside of the cutting edge should be removed by careful use of a half-round file. Inside faces generally

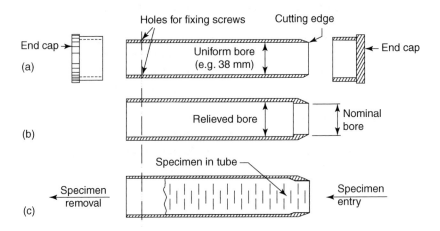

Figure 9.15 Tubes for compression test specimens: (a) plain bore tube; (b) relieved bore tube; (c) extrusion of specimen from relieved bore tube

should be smooth, and may be very lightly coated before use with thin oil, or a PTFE spray. Checks should be made periodically to ensure that a regular shape is maintained. Tubes and cutters should be cleaned and dried after use, and put away with their cutting edges protected against possible damage.

Split formers

A split former may consist of either two or three segments. The mating faces of a three-part former are usually marked with reference letters or numbers to ensure that the segments are always assembled in the correct order.

A split former of type 4 in Section 9.1.2 is used to hold a specimen which has been extruded from a tube while it is trimmed exactly to length. This ensures that the ends are made flat and square to the axis of the cylinder. A former of type 5 is used as a mould into which soil is compacted or moulded for the preparation of a recompacted or remoulded specimen. The segments must be securely clamped together; if a clamping device is not provided with the former, two or three Jubilee clips of the appropriate size can be used. Some formers incorporate a connection to which a vacuum tube can be fitted, to enable a rubber membrane to be held in contact with the internal face while a specimen is being built up inside. If this is used the mating faces of the former should be made airtight by coating with silicone grease.

Extruders and accessories

Extruders and components to which specimen tubes and rings are attached should be maintained in good condition, and cleaned after use. In particular, screw threads and bearing surfaces should be cleared of dirt and protected against damage. A smear of thin oil on screw threads will prevent them from binding, provided that they are protected from dirt.

Check that all separate securing screws and nuts are present after each use; nevertheless, it is desirable to keep a few spares.

When setting up tubes and cutters, and various accessories for extrusion, it is essential to fit all components in correct alignment, and to secure them tightly so that alignment is maintained under the large forces applied during extrusion. When attaching a U-100 tube to the extruder, and fittings to the end of the tube, avoid cross-threading, and screw on securely by giving several full turns.

Some extruder top plates may not be threaded as shown in Figure 9.16, but are machined with a bevel to allow differing types of sample tubes or liners to be accommodated.machines adapted in this way should be fitted with appropriate guards to prevent accidental trapping of fingers between the top plate and sample tube as the two engage.

Samples

When preparing an undisturbed specimen, it is essential to avoid disturbing the portion of the soil that will form the test specimen, and to protect the sample against drying out. Avoid making too large a cut, which may disturb the soil within the tube or ring. Cutting with a knife tends to produce a 'smear' effect on the surface, which could impede drainage in some

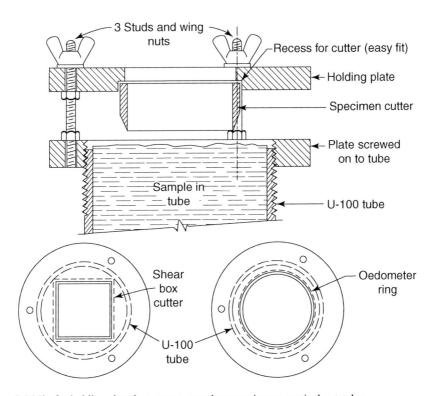

Figure 9.16 Jig for holding shearbox cutter or oedometer ring on vertical extruder

tests. Excess 'wiping' of surfaces with any kind of blade should be avoided. Smearing can be offset by lightly scarifying the exposed surfaces with a fine brass-bristle brush.

The presence of small stones or hard nodules in the soil is one of the commonest causes of difficulty in specimen trimming. A stone which protrudes from the specimen surface may be carefully removed and the cavity filled with 'matrix' material pressed well in. Stones are the most frequent cause of damage to cutting edges of moulds and tubes. Soil which is suspected of containing stones should be observed carefully as the sample is extruded, so that a stone lodged against a cutting edge can be removed before it creates grooving or scoring of the specimen.

Before extruding a sample from the tube in which it was taken, its identification details should be recorded on the appropriate laboratory test sheet. The end caps, any packing materials, and wax seals are then carefully removed. The face against which the extruder piston will bear should be trimmed to give a reasonably flat surface.

Samples must always be protected against drying out while exposed during preparation. For this purpose a sample preparation room maintained at a relative humidity of about 95% is desirable. Where this is not practicable, a local area of humidity can be provided on the workbench by constructing a small 'tent' of dampened cloths supported by burette stands.

A sample or specimen should never be left exposed to the atmosphere even for a few minutes, but should be covered with a piece of polythene or similar impervious material. This also applies to the end face of a portion remaining in a sample tube. Before returning to stores, tube samples should be protected with wax as described in Volume 1 (third edition), Section 1.4.4.

9.2 Undisturbed specimens from sample tubes

9.2.1 Open drive samples

BS EN ISO 1997-2:2007 stipulates that samples for strength and compressibility testing shall be of sample quality class 1. Samples taken in open drive thick walled samplers such as the U-100, which have traditionally been used in the United Kingdom for shear strength and compressibility testing, are deemed by BS EN ISO 22475-1 to be at best sample quality class 2. Only rotary core samples, or samples taken by pushing thin walled tubes into the ground under steady pressure, designated OS-TW in BS EN ISO 22475-1, are recognised as meeting the requirements for quality class 1 samples.

An open drive thin-walled sampler has been developed commercially (Gosling and Baldwin, 2010) which meets the requirements of the standard in terms of geometry (wall thickness 3.0 mm, area ratio <15%, inside clearance ratio < 0.5%, edge taper angle < 5°). This has been shown by field trials to be robust and a limited programme of shear strength testing suggests that for stiff clays, the undrained cohesions measured on samples recovered with the open drive thin walled sampler are noticeably higher than those from conventional open drive samples. So although this new sampler, designated UT-100 by Gosling and Baldwin (2010), may not produce class 1 samples, it provides a significant improvement on the conventional U-100 sampler.

Increased use of samples from both pushed and open drive thin-walled tubes may be anticipated. Sample extruders will require minor modification to accommodate their use as the reduced wall thickness of the open drive UT-100 tube cannot accommodate a standard 4 inch British Standard pipe thread and a thread with a square profile has been adopted.

9.2.2 Extrusion from sample tubes taken in-situ

The following remarks relate especially to U-100 samples taken from boreholes, but can apply to other types of tube sample which have been obtained in-situ.

Before mounting on the extruder the top and base ends of the tube must be identified. The method of labelling should clearly indicate which end is the top, and it is better to rely on markings on the tube itself rather than on the end caps, which could inadvertently be interchanged.

When the sample is extruded from the base end of the tube, i.e. with the tube mounted base uppermost in a vertical extruder, tests are carried out on the least disturbed soil from the lowest horizon in the tube. However, it is common practice to extrude the sample in the same direction as the soil entered the tube, to minimise the possibility of inducing complex stresses and stress relief. The top end of the sample may include an indeterminate length of disturbed material, resulting from boring operations immediately preceding the driving of the sample tube. If it is necessary to use material for testing from near the top of the tube, any soil which appears to be disturbed should first be discarded.

The wax sealing disc may be left in place if its surface is reasonably flat and level. It will then help to distribute the load from the extruder piston (which is slightly smaller than the tube bore) more uniformly across the area of the sample.

In the descriptions of specimen preparation which follow, the words 'top' and 'upper' relate to the sample as mounted in the extruder, irrespective of which end was the top when in-situ.

9.2.3 Shearbox and oedometer specimens from U-100 tube (BS 1377:Part 1:1990:8.6)

Extrusion of undisturbed samples from a U-100 tube into specimen cutting rings (e.g. 60 mm square shearbox specimens, or 75 mm diameter oedometer specimens) is best carried out by mounting the ring in a specially designed jig which can be attached to the sample tube. The arrangement for use with a vertical extruder is shown in Figure 9.16, and ensures that the soil enters the ring squarely and with the minimum disturbance. Each shape and size of ring requires a purpose-made holder which can be bolted or clamped to the jig.

The method of specimen preparation is as follows. The maximum size of particle in a specimen 20 mm thick should not exceed 4 mm (see Sections 12.5.4 and 14.5.1.)

1. Attach the sample tube to the extruder, normally with the bottom end uppermost. Remove the protective covering and extrude any loose or disturbed material, then extrude a short length (20–30 mm) of sample for examination, and moisture content determination if required.
2. Cut off the extruded portion with a wire saw or knife, and trim the end of the remaining sample flat and flush with the end of the tube.
3. Fit the jig onto the end of the sample tube, and secure the specimen ring in position so that there is a gap of about 5 mm between the cutting edge and the surface of the soil in the tube. If the ring cannot be attached to the holding plate, place the ring

on the surface of the sample and guide it so that it seats correctly as the sample is jacked upwards.

4. Extrude the sample steadily, cutting away surplus soil from the outside of the ring as the sample enters it (see Figure 9.17(a)), until the top surface projects a few millimetres above the top of the ring.

5. Cut off the sample close to the sampling tube with the cheese wire saw (see Figure 9.17(a)), working inwards towards the centre.

6. Remove the jig from the tube, and take the ring out of the jig. Take care to prevent the specimen from falling out of the cutting ring.

7. Place the sample and ring on the flat glass plate, cut off the soil projecting above the ring (see Figure 9.17(b)) and trim flat and flush with the end of the ring.

8. Turn the ring over and progressively cut away the surplus soil to within about 1 mm of the final surface, as indicated in Figure 9.17(c), using a sharp blade, and starting with the annulus of the excess diameter.

9. Carefully trim flat and flush with the cutting edge of the ring, avoiding damage to the cutting edge (see Figure 9.18).

During trimming operations, avoid excessive remoulding or smearing of the soil surface. Check for flatness with the reference straight-edge. Observe the general directions given in Section 9.1.3.

If the holding jig is not available, extrude about 50 mm of sample after stage 2 above, position the cutting ring (see Figure 9.19) and cut away the soil from outside the ring while pushing it squarely into the sample (see Figure 9.20). Cut off with the wire saw (see Figure 9.21) and proceed as described above (stage 7 onwards).

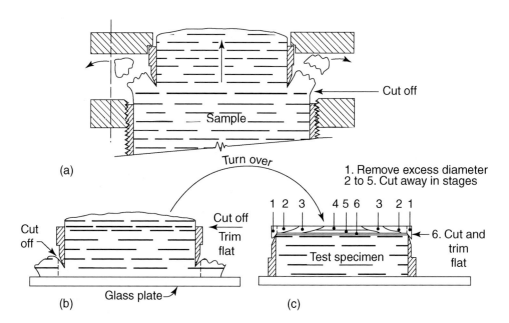

Figure 9.17 Preparing a test specimen in cutter from sample in U-100 tube: (a) extrusion from tube into cutter; (b) trimming bottom face; (c) trimming top face

Figure 9.18 Trimming surface of specimen in oedometer ring

9.2.4 Compression specimen from U-38 tube (BS 1377:Part 1:1990:8.3)

The following procedure is for the preparation of a 38 mm specimen, about 80 mm long, from a sample contained in a 38 mm diameter sample tube. The sample may have been taken in the tube either in-situ, or in the laboratory from a larger tube sample or a block sample.

Figure 9.19 Sample partly extruded from U-100 tube for trimming by hand into cutting ring

Figure 9.20 Cutting soil from around ring as it is pushed into sample

Figure 9.21 Cutting off specimen in ring using wire saw

The procedure is similar for samples in tubes of other sizes up to about 70 mm diameter, but the method given in Section 9.2.5 is usually preferable for larger diameters.

The specimen should normally be trimmed to a right cylinder having a height:diameter ratio of 2:1, or a little greater. The largest particle should not exceed 6.3 mm for a 38 mm diameter specimen or one-fifth of the specimen diameter for larger specimens (see Table 13.3, Section 13.6.1).

1. Remove the protective covering and trim off any surplus material so that the specimen ends are flush with the tube.
2. Fit the tube, and the split mould into which the specimen is to be transferred, on to the extruder. A tube with relieved bore should be placed with its cutting edge furthest from the split mould, so that the sample leaves the tube in the same direction as it entered (see Figure 9.15(c)). Check that the tube and mould are correctly aligned. Insert the clamping plate which secures the split mould in position (see Figure 9.22). Place a disc of oiled paper between the sample and the extruder ram, or lightly oil the face of the ram, to prevent the soil adhering to it.
3. Extrude the specimen into the split mould by winding the extruder steadily, and then retract the extruder head. Any disturbed soil from the top end of the tube should be pushed out beyond the end of the split mould so that it can be cut off and discarded (see Figure 9.23).
4. Cut off surplus material with the wire cutter, and trim the ends flat and flush with the ends of the mould using the end trimmer (see Figure 9.24) or a sharp knife.
5. Protect the ends of the specimen from loss of moisture, e.g. by using polythene sheet, until required for testing. Wrap and seal the surplus soil, and seal any soil

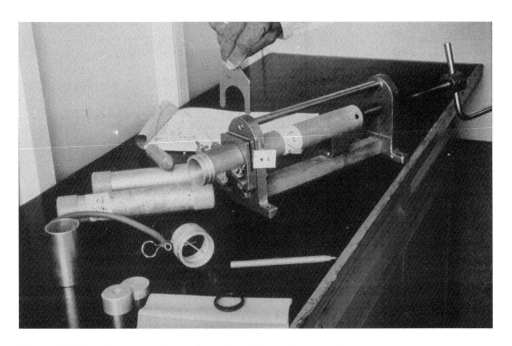

Figure 9.22 Preparing to extrude specimen from 38 mm diameter tube

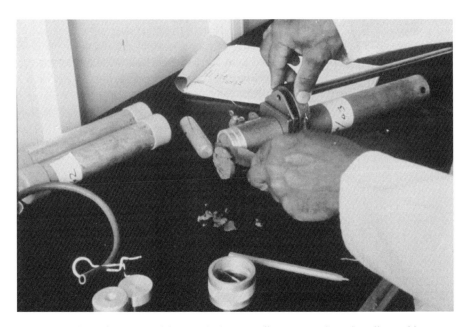

Figure 9.23 Cutting off surplus soil from end of 38 mm diameter specimen in split mould

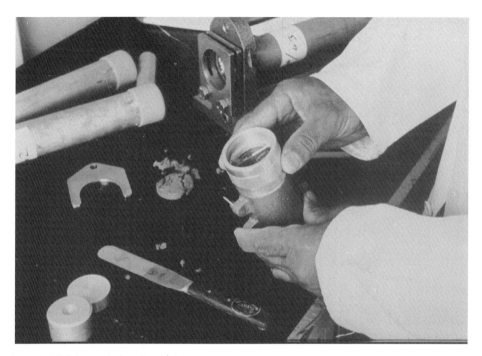

Figure 9.24 Using end trimming tool

remaining in the sampling tube. Ensure that both the specimen and the surplus soil are adequately labelled.

6. When ready to test the specimen, stand it on a flat surface and carefully remove the split mould. Patch up any surface defects.

If a screw-type extruder and split mould are not available, a graduated pusher can be used instead. Trim the specimen flush with the cutting edge of the tube. Apply the pusher from this end until the required length (nominally 76 mm) is read on the scale, which indicates the length of the specimen remaining in a standard 230 mm long sampling tube (see Figure 9.10(a)). If a tube of different length is used, the equivalent scale reading must be calculated. Trim off the extruded material flush with the top end of the tube. When ready for the test, extrude the specimen by applying the pusher in the same direction.

9.2.5 Set of three compression specimens from U-100 tube (BS 1377:Part 1:1990:8.4)

The procedure for preparing a set of three 38 mm diameter specimens from one horizon in a U-100 sample tube, which is common practice for triaxial testing in the UK, is as follows. A similar procedure is used for preparing a single specimen of smaller diameter than the sampling tube.

1. Select three identical 38 mm diameter tubes which are in good condition and very lightly oiled (see Section 9.1.3). Mount them in the three-tube adaptor (see Figure 9.8(b)) and lock them securely in place with the screw which engages one of the holes at the top end of each tube

2. Attach the sample tube securely to the extruder.

3. Remove the protective covering and extrude any loose or disturbed material.

4. Extrude the sample a little at a time until the upper face of the horizon to be tested is exposed. Cut off the surplus soil, and trim flush with the end of the tube.

5. Attach the three-tube adaptor securely to the end of the sample tube so that the cutting edges of the tubes are about 10 mm above the end of the sampling tube. Ensure that the tubes are held in the correct alignment.

6. Extrude the sample until its upper edge touches the three tubes, and again check that the tubes are vertical and parallel.

7. Continue extruding at a steady rate, making sure that the tubes do not splay apart, and cut away from around the tubes any surplus soil which may cause obstruction (see Figure 9.25). Stop extruding when the tubes are about half to two-thirds full.

8. Trim away the excess soil from around and between the tubes. The material can be used for natural moisture content measurements, and for Atterberg limits or other classification tests, but some should be retained for inspection later. Any variation in soil type within the extruded length should be recorded.

9. Using a cheese wire, sever the clay adjacent to the top of the sampling tube.

10. Unscrew and remove the three-tube attachment. Protect the soil remaining in the sample tube by re-sealing with wax and replacing the cap.

11. Cut off the soil from the ends of the tubes and trim flush.

12. Remove the tubes from the adaptor, and clean off any soil adhering to the outsides. Fit the end caps for protection against loss of moisture.

13. Clean the tube adaptor, paying particular attention to the screw-threads.

Each tube specimen is subsequently trimmed and extruded as described in Section 9.2.3.

9.2.6 Single 100 mm specimen from U-100 tube (BS 1377: Part 1:1990:8.4)

Preparation of a 'whole core' specimen from the standard U-100 or U-4 sampling tube used in the UK is described below. The procedure is similar for other sizes of tube exceeding 70 mm diameter. Samples of this size require very careful handling, and may need two people when extruding, transporting and setting up.

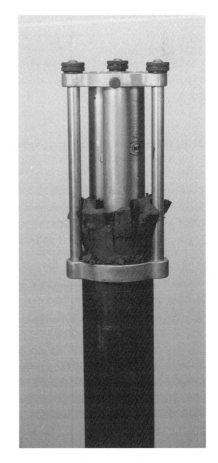

Figure 9.25 Extruding sample from U-100 tube into three 38 mm diameter tubes

Extrude the whole sample from the U-100 tube and lay it out for inspection, either on a purpose-made trough or on a 450 mm length of 100 mm diameter plastic guttering. If the sample extruder operates horizontally, the sample can be extruded directly onto the trough or guttering. If the sample is extruded vertically, it should be supported between two lengths of guttering while being removed and laid down on the inspection bench.

The outer 'skin' or 'smear' of soil which was adjacent to the sample tube should be lightly scraped away to reveal the undisturbed material. After inspection and description, the length of sample which is required for testing can be selected.

Usually only one specimen can be prepared. But it is sometimes possible to obtain two specimens of about 200 mm length from a nearly full sample tube.

A sample taken in an aluminium alloy liner tube inside a standard U-100 tube is about 100 mm diameter and can be trimmed to length by enclosing it in the split former. A sample taken in the steel U-100 tube itself is approximately 106 mm diameter, and would have to be trimmed down to 100 mm diameter to fit inside the split mould. It is usually better to test the sample without trimming, especially if it is friable or contains stony material. It can be held between two pieces of guttering 200 mm long for trimming the ends, which should then be checked for flatness and squareness and re-trimmed if necessary. The small excess in diameter can be accommodated in the standard 100 mm diameter cell and membrane without trimming.

If a specimen length is less than 200 mm, one end is trimmed first in the split former or guttering, and the sample is then moved along the tube (or the guttering is moved) to expose the other end for trimming.

9.3 Undisturbed specimens from block samples

9.3.1 Shearbox and oedometer specimens (BS 1377:Part 1:1990:8.7)

The following instructions relate to the preparation of specimens up to 100 mm square for shearbox tests, and of disc-shaped specimens for oedometer consolidation tests, trimmed by hand into cutting rings from undisturbed block samples of cohesive soil, or from a sample already extruded from a sampling tube.

1. Cut away enough material from the outside of the block sample to enable a roughly square or disc-shaped piece somewhat larger than the test specimen to be cut from inside the block (see Figure 9.26(a)). The plane of the disc should normally be parallel to the plane which was horizontal in-situ unless the specimen has to be orientated in any other direction.
2. Trim one face flat, using a sharp blade and checking with a straight-edge. Place the sample, trimmed face downwards, on a flat surface such as a glass plate. Trim the upper surface roughly flat and parallel to the glass plate.
3. Using the cutting ring as a template, trim the sample to 1 or 2 mm larger than the final specimen size for a short distance ahead of the cutting edge (see Figure 9.26(b)).
4. Push the ring down slowly and steadily, keeping its axis vertical, ensuring always that it is pushed squarely to avoid disturbance of the clay, and allowing the cutting edge to pare away the last slither of soil (see Figure 9.26(c)). Ensure that the specimen is a close fit in the ring, and that no voids are formed against the inside face of the

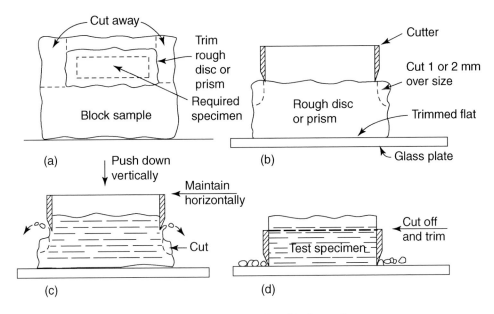

Figure 9.26 Hand trimming test specimen into cutter from block sample

ring, especially at the corners of the square ring. Continue the above process until the cutting edge reaches the glass plate (see Figure 9.26(d)). This procedure is easier if a second ring (a 'driving ring') is placed on top of the cutting ring with the internal surfaces in alignment. A flat piece of wood can be used to push the driving ring.

5. Cut off and trim the remaining surplus soil flush with the end of the ring, as in Figure 9.17(c).

An alternative method is to use the apparatus shown diagrammatically in Figure 9.27. This is based on a device given in the *Earth Manual* (US Department of the Interior, 1990) for guiding the cutting ring so as to ensure that its axis remains vertical as it is advanced onto the sample.

9.3.2 Compression test specimens (BS 1377:Part 1:1990:8.5.3)

A cylindrical compression test specimen 38 mm diameter may be obtained from a block sample of soft or fairly firm clay by pushing in a thin-walled 38 mm sampling tube which has a sharp cutting edge. The block sample should be firmly supported on a flat surface, but the sides around the sampling location should not be laterally restrained. The tube should be pushed in squarely with a steady pressure, for a distance of about 90 mm. The apparatus shown in Figure 9.27, fitted with a suitable tube adaptor, will ensure that the axis of the tube remains vertical. Before withdrawing the tube it should be rotated one complete turn to shear off the soil at the end.

For a firm or stiff clay a satisfactory cylindrical specimen can be prepared easily by using a soil lathe (described in Section 9.4). In the absence of a lathe, a procedure similar to that described in Section 9.3.1 can be used, as follows:

1. Cut a rough cylindrical shape that is somewhat larger than the required sample.

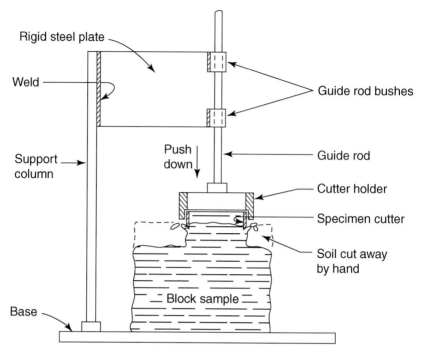

Figure 9.27 Apparatus for trimming specimens from block samples (based on Earth Manual (US Department of the Interior, 1990)

2. Trim one end flat so that it can stand on a glass plate with its axis vertical. Trim the other end flat and parallel to the first.
3. Using the sample tube as a template, trim the specimen slightly larger than the final diameter just ahead of the cutting edge.
4. Allow the cutting edge to pare away the final 1 mm or so as the tube is gradually advanced, ensuring that the axis remains vertical. The apparatus shown in Figure 9.27 could be used for this purpose.
5. When the tube contains the required length of specimen plus a small excess, cut off any surplus material from the end.
6. Final preparation can be done as described in Section 9.2.3.

9.3.3 Large diameter specimens

Cylindrical compression test specimens of 100 mm diameter and larger are more difficult to prepare, and require greater care in handling, than the conventional specimens referred to above. If a soil lathe of large enough capacity is available, the procedure described in Section 9.4 is perhaps the easiest method for firm or stiff clays.

The principle of hand-trimming from a block is similar to that described in Section 9.3.2, but final trimming to the correct diameter should be done with a U-100 cutting shoe (see Figure 9.28). A full-length sample tube of this diameter would be too awkward to handle.

This procedure can also be used for obtaining an undisturbed sample in a core-cutter for a falling head permeability test (see Section 10.7.2). The same principle can be used for obtaining an undisturbed sample in a CBR mould, either from a block sample or from the ground in-situ, if the mould is fitted with a cutting shoe (see Section 11.6.8).

9.3.4 Block sample for large shearbox

Block samples of materials from which it is difficult to obtain small test specimens, or which cannot be properly represented by small specimens, are sometimes required for large shearbox tests.

Block samples of soils such as stiff fissured clays, and soft rocks, require very careful handling and preparation. Some materials, especially those that are highly fissured, can

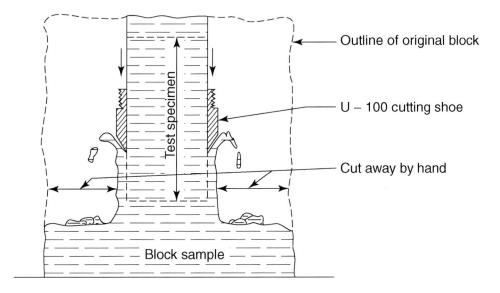

Figure 9.28 Trimming 100 mm diameter specimen from block sample using U-100 cutting shoe

only be satisfactorily sampled on site directly into the shearbox, for which a duplicate box is desirable. Other materials which can be prepared from block samples and tested in the relatively undisturbed state include soils of the 'boulder clay' type, colliery spoil materials, and some cohesive industrial waste products.

A suggested procedure for preparing block samples for the large shearbox is as follows:
1. Trim the lower face of the block sample, if necessary, to enable it to stand firmly on a flat surface such as the workbench top.
2. Trim the upper face to provide a surface which is approximately level and reasonably flat.

3. Place one half of the shearbox on the top face, and mark the outline of the inside face of the box with a knife blade.
4. Remove the half box, and carefully trim a square section about 5 mm outside the marked outline, to a depth about equal to the depth of the half box (see Figure 9.29(a)).
5. Cut to a depth of about 20 mm from the upper surface to the exact outline of the box.
6. Replace the half box and ease it over the portion trimmed to size (see Figure 9.29(b)).
7. Keeping the half box horizontal, ease it down gradually by cutting away the remaining soil just ahead of it, so that the box makes a close fit around the sample, until the half box is almost filled (see Figure 9.29(c)).
8. Place the other half of the box in position, and bolt or clamp the two halves together in correct alignment.
9. Cut the block to a depth slightly in excess of the required sample thickness to an outline about 5 mm outside the box size (see Figure 9.29(d)).
10. Allow the box to advance downwards by trimming ahead of it, as in stage 7, until the surface of the sample just projects above the top flange of the box (see Figure 9.29(e)).
11. Cut off the surplus soil so that the top surface of the block is level and flush with the end of the box.
12. Trim to a level surface below the flange of the box a distance equal to the thickness of the bottom grid plate, which is then fitted in position (see Figure 9.29(f)).
13. Turn the block sample over, and repeat stage 12 to fit the upper grid plate (see Figure 9.29(g)). If practicable, first separate the sample from the remainder of the block by undercutting about 20 mm below the box.

The whole sample is then ready for lifting and placing in the carriage of the shearbox machine.

During the trimming process, any large particles dislodged from a prepared face may be replaced by fine 'matrix' material well pressed into the resulting cavity. It is important to obtain a tight fit of the sample against the sides of the box, and if necessary additional 'matrix' material should be packed down the sides using a long spatula blade.

9.3.5 Encapsulation (BS 1377:Part 1:1990:8.5.4)

The procedures described in Sections 9.3.1 and 9.3.2 are not practicable for some brittle soils, and for these the following method may be satisfactory. It can be used for obtaining a test specimen from an irregular lump of soil, as well as from cylindrical or rectangular samples.

1. Protect the lump of soil with a waterproof coating, such as thin clinging film or several coats of paraffin wax.
2. Place the sample in a container that can be held in place in a sample extruder, such as a one-litre compaction mould.
3. Surround the sample with a suitable plaster (e.g. Polyfilla), mixed with water to a workable paste. Ensure that the sample is completely encapsulated. Alternatively, damp sand can be packed around the sample.

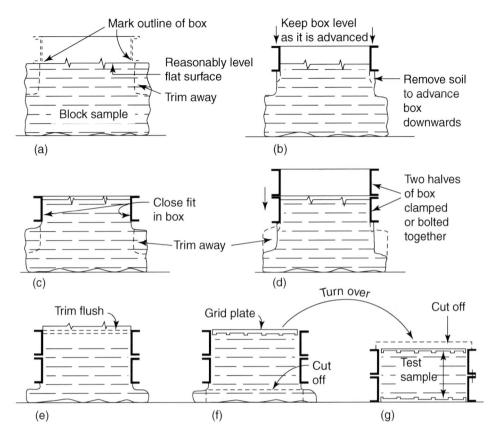

Figure 9.29 Hand trimming a block sample for large shearbox

4. Allow the plaster to set, but not to harden completely. An overnight period is usually sufficient.
5. Clamp the mould containing the encapsulated sample to the frame or crosshead of a hydraulic extruder. Place a cutting ring, with a suitable driving ring, on the ram and jack the ring into the sample with a continuous steady movement. This operation is shown in Figure 9.30(a).
6. An alternative method is to invert the arrangement described in step 5, as shown in Figure 9.30(b). The ring, with a driving ring, is restrained by the crosshead while the sample is jacked upwards against it. This arrangement is necessary when jacking a sample tube instead of a ring, because it enables some lateral restraint to be provided for the tube by means of a floating guide plate.
7. Trim and prepare the test specimen as described in Section 9.2.3.

 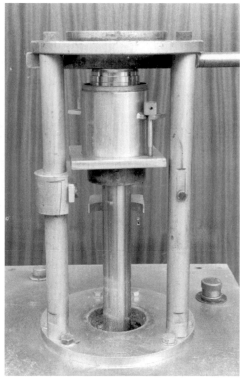

Figure 9.30 Arrangements for taking test specimen from encapsulated sample: (a) jacking cutting ring into sample; (b) jacking sample against cutting ring

9.4 Use of soil lathe

9.4.1 Hand-operated soil lathe (BS 1377:Part 1:1990:8.5.2)

The small hand-operated soil lathe (see Figure 9.31(a)–(c)) is designed for the preparation of 38 mm or 50 mm diameter specimens of clays in the firm to stiff, or very stiff, consistency range. Softer clays may be subjected to excessive disturbance by this procedure, but satisfactory specimens of this type of material can usually be obtained by means of sample tubes.

Before mounting in the lathe a sample should first be trimmed roughly to shape, using a sharp blade or wire saw (see Figure 9.31(a)). The length should be somewhat longer than the required specimen length, and the end faces trimmed flat and parallel. These faces are mounted between the platens of the lathe, and the upper platen is brought firmly into contact with the upper surface and locked into position so that the sample is securely held. Surplus material is removed progressively from the sample by means of a series of fine vertical cuts, rotating it slightly between each cut (see Figure 9.31(b)). For firm clays a wire saw is suitable, but for stiffer clays a sharp knife may be more effective. Avoid any tendency to cause distortion by dragging at the soil. Any stones or

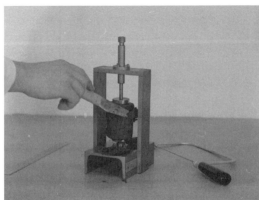

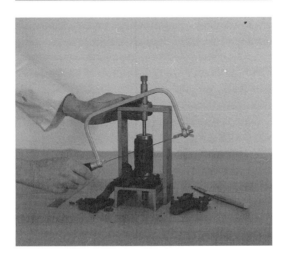

Figure 9.31 Use of soil lathe: (a) sample roughly trimmed and mounted in lathe; (b) removing surplus soil by progressive vertical cuts; (c) trimming to final diameter using wire saw against vertical guide plates

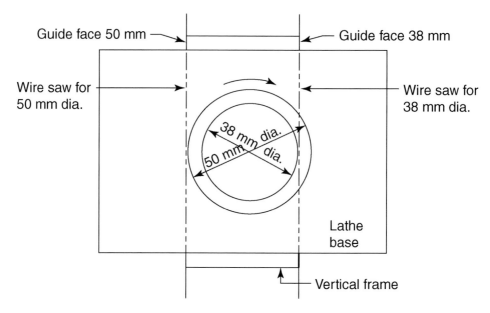

Figure 9.32 Use of soil lathe guide plates for preparing either 38 mm or 50 mm diameter specimens

hard nodules should be removed carefully, and resulting cavities should be filled with fine material from the parings.

The specimen is trimmed accurately to the final diameter by using the frame of the lathe as a guide for the wire saw when making the last few cuts (see Figure 9.31(c)). One type of lathe has the rotating platen mounted eccentrically so that specimens of either 38 mm or 50 mm diameter may be accurately prepared (see Figure 9.32). The specimen is rotated slightly between each cut, until a smooth cylindrical surface is obtained. It is then placed in a split former and the ends are trimmed to give the correct length, using the wire saw or a sharp blade.

9.4.2 Motorised soil lathe

The procedure for using a motorised soil lathe is similar to that given above. The lathe shown in Figure 9.13 can be used for the preparation of cylindrical specimens up to 100 mm diameter and 350 mm long, which is its main advantage. Trimming should be done by using a wire saw, or a sharp knife, rather than the cutting tool mounted on the machine. Only a fine cut should be taken with each stroke, because too deep a cut may cause distortion or tearing of the sample. The specimen should be rotated a few degrees between each cut, and should remain stationary while cutting vertically downwards, against the firm support of the machine platen, as indicated in Figure 9.33(a). The conventional 'turning' process, as used when turning a metal bar in a workshop lathe (see Figure 9.33(b)) can induce torsional stresses in the soil which may lead to excessive disturbance, and should be avoided. It may be possible to finally trim the specimen while it is being rotated at a very low speed, provided that a very fine cut is taken with a wire or sharp blade.

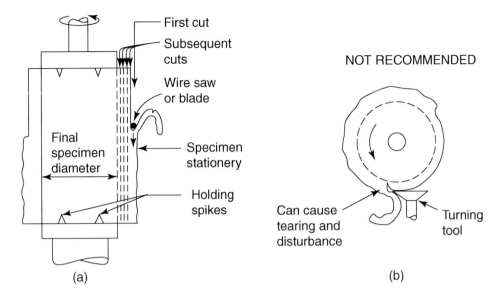

Figure 9.33 Use of soil lathe: (a) recommended procedure applying fine vertical cuts to stationary specimen; (b) rotational cutting ('turning') not recommended

9.5 Recompacted specimens (BS 1377:Part 1:1990:7.7)

9.5.1 Compaction criteria

Standard compaction procedures for soils were described in Volume 1 (third edition), Chapter 6. It is sometimes necessary to carry out shear strength, compressibility or permeability tests on specimens of soil that have been recompacted in the laboratory according to a specified procedure. Methods of compaction are described below. (The preparation of recompacted samples for CBR tests is described separately in Chapter 11, Section 11.6.)

Before a compacted sample can be prepared, the criterion for compaction must be ascertained. Compaction is used for either of the following two objectives:

- To bring the soil to a specified dry density or voids ratio
- To apply to the soil a known compactive effort

The first condition is not difficult to achieve in a specimen mould of any size because the mass of soil required to fill the known volume can be calculated. However, it is important to obtain a uniform distribution of density throughout the volume and some preliminary trials may be necessary.

The second condition is easy to obtain in a standard compaction mould, by using recognised standard compaction procedures such as those summarised in Volume 1 (third edition), Section 6.3.3, Table 6.3. In a larger mould the degree of compaction can be increased in proportion to the volume of soil, for which Table 9.1 provides some guidance. When compacting into moulds or tubes that are too small for the standard compaction equipment, the degree of compaction to be applied by small hand rammers should first be ascertained by trial. Small compaction devices such as the Harvard compactor (Vol. 1 (third edition), Section 6.5.10) provide some measure

Table 9.1 Compaction of large specimens

Specimen size	Volume	BS 'light' compaction (2.5 kg rammer)		BS 'heavy' compaction (4.5 kg rammer)		Maximum size of particles
(mm)	*(cm³)*	*layers*	*blows/layer*	*layers*	*blows/layer*	*(mm)*
100 × 200	1571	5	25	8	27	20
105 × 210	1818	5	29	8	31	20
150 × 300	5301	8	54	13	54	28

of control for hand compaction, as well as a basis for comparison with standard compaction procedures.

Whenever sufficient material is available, compaction into a standard compaction mould (see Section 9.5.5) is preferable to compacting cohesive soil directly into small tubes and rings (see Sections 9.5.3 and 9.5.4). The compacted sample can then be extruded and trimmed in the same way as for an undisturbed sample. Larger samples, such as 100 mm diameter cylindrical specimens which are 200 mm long, can be compacted directly into sample formers.

9.5.2 Soil preparation

The soil used for preparation of test specimens should be prepared as described in Volume 1 (third edition), Section 6.5.2. The soil should not be dried initially.

Remove over-size particles by use of appropriate sieves. The maximum size of particles depends on the type of test and the test specimen dimensions, i.e. height (H) and diameter (D), as summarised below.

Type of test	Maximum size of particles
Shearbox	$H/10$
Consolidation	$H/5$
Compressive strength ($H \approx 2D$)	$D/5$
Permeability	$D/12$

Sub-divide the soil by riffling, to provide a representative sample of appropriate size.

Adjust the moisture content to the desired value by thoroughly mixing in additional water, or by partial air drying under close control to prevent local over-drying (see Section 6.5.2, 5) of Volume 1 (third edition). Take representative samples for determination of the moisture content. Place the prepared soil in a sealed container and store for at least 24 hours before compaction. After compaction, seal the soil in the mould and allow to mature for at least 24 hours before trimming test specimens.

9.5.3 Shearbox and oedometer specimens

If the quantity of soil available is not enough to fill a compaction mould (see Section 9.5.5), a small shearbox or oedometer specimen may be prepared by compacting or remoulding the soil directly into the cutting ring. The soil is first brought to the required moisture content and allowed to mature. The ring is placed on the flat glass plate, and soil is tamped into the ring in two or three layers, keeping the ring firmly in contact with the glass plate. Tamping may be done either with a metal flat-ended rod, or by using the Harvard compaction device (see Section 6.5.10, Volume 1 (third edition)), with which a controlled amount of compaction can be given. The degree of tamping necessary should be determined previously by trial. The specimen end faces are trimmed flat, and the specimen should be sealed and allowed to mature for at least 24 hours before proceeding with a test.

9.5.4 Compression specimens 38 mm diameter

If the quantity of soil available is not enough to fill a compaction mould (Section 9.5.5), the soil can be compacted directly into a 38 mm diameter sample tube or split former. A similar procedure applies to other sizes up to about 70 mm diameter. Compaction is done by hand tamping with a steel rod, or by using a compression machine to apply static compaction if a specified density is to be obtained. The appropriate mass of soil is weighed out and compacted into the known volume of tube or mould.

Compaction to a specified compactive effort is difficult to achieve consistently by hand tamping, and for this purpose the Harvard compaction device (Volume 1 (third edition), Section 6.5.10) is useful. However, this device should first be calibrated against standard procedures using similar material. After compaction the specimens should be sealed and stored for at least 24 hours before extruding and testing.

9.5.5 Use of compaction mould

The most satisfactory method for the preparation of test specimens of compacted soil is to start by compacting the soil into a compaction mould, provided that there is enough material available to fill it. If the moisture content and degree of compaction are specified, the procedure given in Sections 6.5.3, 6.5.4 or 6.5.5 of Volume 1 (third edition) is followed, after bringing the soil to the required moisture content. If the soil is to be compacted at a given moisture content to a specified dry density, a static compaction procedure similar to that described in Section 11.6.3 is usually the most suitable. Alternatively, dynamic compaction could be used, after carrying out preliminary trials, following the principle given in Section 11.6.5. The density achieved should be within 2% of the specified density.

After compaction the soil should be sealed in the mould for a maturing period of at least 24 hours, to allow any developed pore pressures to dissipate. The sample can then be treated as a U-100 tube sample, using a special adaptor to attach the compaction mould to the extruder.

For the preparation of a set of three 38 mm diameter compression test specimens, the triple-tube holder is attached to the compaction mould with a special adaptor, and extrusion and specimen preparation can proceed as described in Section 9.2.4. Shearbox and oedometer specimens can be extruded into the appropriate cutting ring as described in Section 9.2.2, taking each specimen from near the mid-height of a compacted layer (see Figure 9.34(a)).

Alternatively, the whole compacted sample can be extruded from the mould, and individual specimens hand-trimmed. For the preparation of a set of three compression test specimens, the sample is cut longitudinally into three equal sectors (see Figure 9.34 (b)), each of which is then hand trimmed or formed on the soil lathe into a cylinder (see Figure 9.34(c)). For the

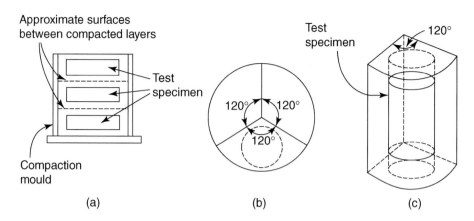

Figure 9.34 Preparation of test specimens from soil compacted into mould: (a) three oedometer or shearbox specimens; (b) vertical division into three sectors; (c) cylindrical specimen prepared from one sector

preparation of shearbox and oedometer specimens, the compacted soil is divided horizontally at the boundaries between each compacted layer, and each disc can be used to form a specimen as shown in Figure 9.34(a).

9.5.6 Compaction of large specimens

The following procedure relates to the preparation of recompacted cylindrical specimens of 100 mm diameter and over, for compression tests.

The specimen is prepared in an extended split former (see Figure 9.7) of the appropriate diameter, with the lower end cap acting as the base of the mould. The former must be rigidly clamped together. For non-cohesive or friable soils a rubber membrane should be fitted inside the former before placing the soil, holding it against the walls by a continuously applied vacuum. The membrane will reduce the risk of collapse when the former is removed. Cohesive soils can also be compacted into the membrane in this way instead of fitting the membrane later. If stony material is present a second membrane should be fitted in case the inner one is punctured by sharp-edged fragments.

To obtain a given density the amount of soil required to fill the known volume is weighed out and placed in the mould in layers, either under static pressure or by dynamic compaction, as described in Section 11.6.4 or 11.6.5. If a specified compactive effort is to be applied, the amount of compaction corresponding to BS 'light' or BS 'heavy' compaction, using hand rammers for specimens 100 mm, 105 mm and 150 mm diameter, may be obtained from Table 9.1 (see section 9.5.1). This is based on compactive energy per unit volume being the same as that used in the BS compaction tests.

The compacted sample should be sealed and allowed to mature for at least 24 hours before testing. The segments of the split former should then be removed carefully. Any cavities may be patched up with surplus soil, and the end surface should be trimmed flat and square with the sample axis. The sample should be handled with care when setting up to test, especially if it is of non-cohesive or friable material.

References

BS EN ISO 1997-2:2007. *Eurocode 7—Geotechnical design—Part 2: Ground investigation and testing*. British Standards Institution, London

BS EN ISO 22475-1:2006. *Geotechnical investigation and testing—sampling methods and groundwater measurements—Part 1: Technical principles for execution*. British Standards Institution, London

Gosling, R. and Baldwin, M. (2010) Development of a thin wall open drive tube sampler (UT-100). *Ground Engineering*, March 2010, pp. 37–39

Norbury, D.R. (2010) *Soil and Rock Description in Engineering Practice*. Whittles Publishing, Caithness, Scotland.

US Department of the Interior, Bureau of Reclamation (1990) *Earth Manual* (third edition). US Government Printing Office, Washington, D.C.

Chapter 10

Permeability and erodibility tests

10.1 Introduction

10.1.1 Scope

The permeability of a soil is a measure of its capacity to allow the flow of a fluid through it. The fluid may be either a liquid or a gas, but soils engineers are concerned only with liquid permeability, and the liquid is usually understood to be water.

Procedures for the direct measurement of permeability of soils in the laboratory, using two different types of test, are described in this chapter. Indirect methods for the assessment of permeability by means of calculations based on other soil properties are also given. The application of laboratory test procedures to the problems of suffusion and erosion, and to the design of filters, is outlined.

Only inorganic soils are considered here. The permeability characteristics of highly organic soils such as peats are more complex and require field tests to obtain realistic values.

Field measurements of permeability, which can take into account features such as the soil fabric, are for that reason often more satisfactory than laboratory tests. Permeability tests on natural undisturbed soil are probably carried out more frequently in-situ than in the laboratory, but field inspection and testing are beyond the scope of this book.

10.1.2 Principle

Soils consist of solid particles with voids between them. In general, the voids are interconnected, which enables water to pass through them: that is, soils are 'permeable' to water. The degree of permeability is determined by applying a hydraulic pressure difference across a sample of soil, which is fully saturated, and measuring the consequent rate of flow of water (see Sections 10.3.1 and 10.3.2). The 'coefficient of permeability' is expressed in terms of a velocity.

The flow of water through soils of all types, from 'free-draining' gravels and sands to 'impervious' clays, are governed by the same physical laws. The difference between the permeability characteristics of extreme types of soil is merely one of degree, even though a clay can be ten million times less permeable than a sand. Clays, and some other materials such as concrete, contrary to casual observation, are not completely impermeable, although they may appear to be so if the rate of flow through them is no greater than the rate of evaporation loss. The method used for measuring permeability depends upon the characteristics of the material.

the permeability. If the degree of saturation is less than about 85%, air is likely to be continuous, instead of being in isolated bubbles, which invalidates Darcy's law.

In permeability testing efforts are made to eliminate air so that the soil can be assumed to be fully saturated. This can be difficult in partially saturated fine-grained soils.

6. Soil fabric

Many soils in their natural state are not homogeneous but are anisotropic, often due to stratification, i.e. they consist of layers or laminations of different soil types. The permeability (k_H) of a stratified deposit in the direction parallel to stratification (often near to the horizontal) is usually several times greater than the permeability (k_v) normal to stratification (vertically). In repeatedly layered soils, such as varved clays, the ratio of the two permeabilities k_H/k_v, known as the permeability ratio, can be 100 or more.

Other features of anisotropy which can affect permeability are discontinuities such as fissures; lenses or intrusions of silt or sand; or pockets of organic material. The bulk permeability of cohesive soils in the field is greatly affected by the presence of discontinuities, and can be several orders of magnitude greater than the permeability measured on small samples in the laboratory. The permeability of fine-grained soils is also influenced by their microstructure, which includes the state of flocculation or dispersion of the particles, their orientation, and their state of packing.

7. Nature of fluid

The 'absolute' or 'specific' permeability, K (see Section 10.3.3), is a constant for a given soil in a particular state. The coefficient of permeability, k, depends upon the properties of the permeating fluid; this is generally assumed to be water and the value of k would be different for other fluids.

The properties of a fluid which are relevant to permeability are density and dynamic viscosity. For water the density, ρ_w, varies little over the range of temperatures normally experienced (0–40°C), but viscosity, η_w, decreases by a factor of about 3 over this range. The test temperature is therefore significant, as discussed under item 9 below.

The origin of water used for tests, and its treatment, can also be significant, and these factors are discussed in Section 10.6.2.

8. Type of flow

One of the assumptions on which Darcy's law is based (see Section 10.3.2) is that the flow of water is 'laminar' or streamline, which occurs when the velocity is relatively low. Above a certain critical velocity the flow becomes turbulent and Darcy's law on which permeability calculations depend is no longer valid. This can apply in sand and especially in materials of medium gravel size and coarser in which little or no fine material is present. In the large void spaces the velocity of flow may be high enough to give turbulence.

The effect of the 'critical' rate of flow on the measure permeability is discussed in Section 10.3.8.

9. Temperature

An increase in temperature causes a decrease in the viscosity of water, i.e. the water becomes more 'fluid', which affects the value of the measured permeability. The measured value can be corrected to a standard temperature, as described in Section 10.3.4. For a laboratory test the standard temperature is usually 20°C, while the temperature for a typical field permeability test in the UK may be about 10°C.

10.2 Definitions

Permeability The ability of a porous material to allow the passage of a fluid.

Pressure head (h_p) Pressure in water at a point expressed in terms of height of a column of water above that point.

Elevation head (h_e) The height of a point above a fixed datum level.

Total head (h) Water pressure expressed in terms of the height of a column of water above the datum level.

$$h = h_p + h_e$$

Hydraulic gradient (i) The ratio of the difference in total head on either side of a soil layer, to the thickness of the layer measured in the direction of flow. It is a dimensionless number.

Darcy's law The velocity of the flow of water through a soil is directly proportional to the hydraulic gradient, i.e. $v = k\,i$, where k is the coefficient of permeability.

Coefficient of permeability (k) The mean discharge velocity of flow of water in a soil under the action of a unit hydraulic gradient. It is usually expressed in metres per second (m/s).

Absolute or *specific permeability* (K) The permeability of a soil expressed as an area, and independently of the nature of the permeating fluid. The unit is mm²

$$K = \frac{k\eta_w}{\rho_w g}$$

Laminar flow Flow in which the movement is steady and continuous, and generally parallel.

Flow line Smooth path assumed to be followed by particles of water flowing through soil at the discharge velocity.

Discharge velocity (v) The mean velocity of flow of water related to the full cross-sectional area A of the soil through which it flows

$$v = q/a$$

where q is the rate of flow.

Permeameter Apparatus in which the permeability of soil is measured, consisting of two types, the constant head and the falling head permeameters.

Piezometer tube Open-topped tube containing water, the level of which indicates the pressure at the point to which it is connected.

Constant head test Permeability test in which water is made to flow through a soil sample under a constant difference in head, or hydraulic gradient.

Falling head test Permeability test in which the piezometer tube used for measuring the head also provides the water which passes through the sample, and therefore the level falls during the test.

Piping Movement of water carrying soil particles which progressively erodes channels beneath the surface, leading to sudden and disastrous collapse. The effect can be simulated in a constant head permeameter.

Specific surface (S) The ratio of the surface area of a collection of particles to the volume of the solid particles. It is usually expressed as mm^2/mm^3, or mm^{-1}.

Dynamic viscosity (η) The resistance of a fluid to flow. For water at 20°C the dynamic viscosity η_w is equal to about 1 mPas (1 millipascal second), but it varies with temperature.

Permeability ratio The ratio of the average coefficients of permeability measured in the horizontal and vertical direction on soil in-situ.

$$\text{Permeability ratio} = \frac{k_H}{k_D}$$

Erosion Removal of soil particles by the movement of water.

Suffusion Movement of fine soil particles from a matrix of coarser particles due to the flow of water through the soil. Fine particles may be either carried away, or moved into another soil horizon.

Erodibility (dispersibility) Erosion of fine-grained soils by a process in which individual clay particles go into suspension in practically still water.

Dispersive soils Soils that are erodible in still water. They usually contain a preponderance of sodium cations in the pore water.

Filter A layer of material with a particle size distribution that meets certain requirements, placed adjacent to a soil from which water is flowing in order to prevent suffusion.

10.3 Theory

10.3.1 Flow of water in soil

Flow lines

The flow of water through a column of soil is illustrated in Figure 10.1. An individual particle of water moving from level A to level B follows a tortuous path around the solid particles and through the voids, as shown by the thick line in Figure 10.1(a). The velocity of the particle varies from moment to moment, depending upon the size and configuration of the pore space through which it travels. However, for most practical purposes we can consider the soil mass as a whole and each water particle can be assumed to travel along a smooth path, known as a flow line, as shown in Figure 10.1(b).

Flow lines may be straight lines or smooth curves, and they represent the flow of water at a steady constant velocity known as the discharge velocity (see below).

Piezometric head

Horizontal flow of water through a tube containing a sample of soil is illustrated in Figure 10.2. The flow is initiated because the inlet reservoir (near the point P) is at a higher level than the outlet reservoir (near Q)

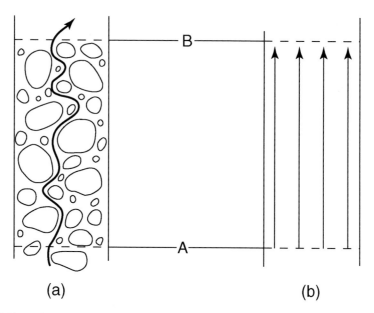

(a)   (b)

Figure 10.1 Flow of water through soil: (a) actual path of water particle; (b) idealised flow lines

If a piezometer tube is inserted in the soil at any point X, water will rise in the tube up to a level which indicates the static pressure of the water at X. The level of water in the pipe is called the piezometric level, and the pressure of water is denoted by the pressure head, i.e. the height of the column of water, h, in the pipe, measured in millimetres or metres above the point X. This can also be expressed in terms of pressure units, by using the relationship

$$p = h\rho_w g \tag{10.1}$$

where p is the pressure, ρ_w is the density of water and g is the acceleration due to gravity. The customary units used are p in kPa (kN/m²), h in mm, $\rho w = 1.00$ Mg/m³ and $g = 9.81$ m/s². Equation (10.1) then becomes

$$p = \frac{9.81}{1000} h \text{ kPa}$$

Therefore 1 m or 1000 mm of water = 9.81 kPa and 1 kPa = 101.9 mm of water.

In most permeability tests, pressures are measured as the total head of water (in mm in laboratory work) above an arbitrary but fixed datum level.

In order to induce a flow of water through the soil from point P to point Q in Figure 10.2, there must be a difference in pressure between P and Q, i.e. the piezometric head at P must be greater than at Q. This is analogous to the flow of an electric current along a wire, in which a potential difference is necessary to induce a current to flow against its resistance.

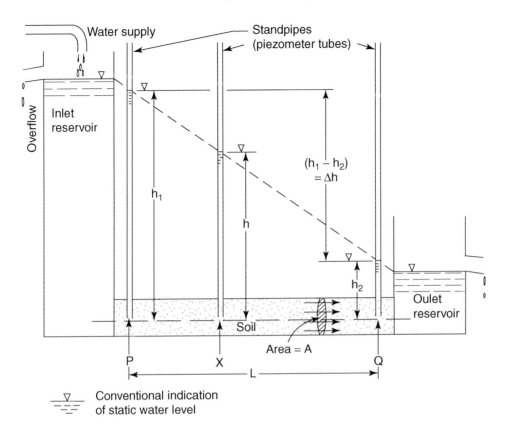

Figure 10.2 Horizontal flow of water through soil under a hydraulic gradient

Hydraulic gradient

The difference in piezometric level between P and Q is the hydraulic head between these two points.

If the piezometric heights (in mm) above P and Q are denoted by h_1 and h_2 mm respectively, the head difference (Δh) is equal to $(h_1 - h_2)$ mm. The ratio of the difference in head to the distance L (mm) between P and Q is the hydraulic gradient, denoted by i

$$i = \frac{h_1 - h_2}{L} = \frac{\Delta h}{L} \quad (10.2)$$

The piezometric heights and the horizontal distance must be measured in the same units, and the hydraulic gradient is a dimensionless number.

If the flow of water is not horizontal, but is inclined at some angle as indicated in Figure 10.3, the difference in level between the point S and T, i.e. the difference in 'position head', must be taken into account. This is achieved by measuring the height of the piezometric

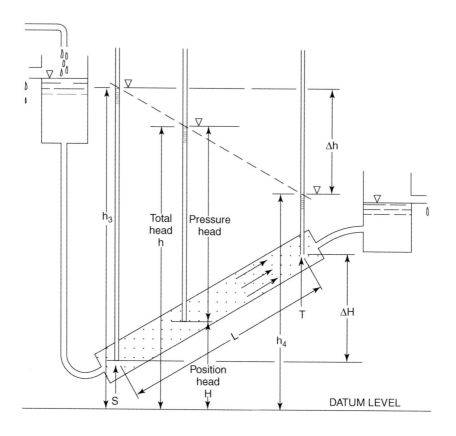

ΔH = difference in 'position head' between S and T
Δh = difference in 'total head' between S and T
hvdraulic aradient i = Δh

Figure 10.3 Flow of water through soil in an inclined direction

levels h_3 and h_4 above a common datum level, which gives the 'total head' at each point. The length L is the distance between points S and T measured along the direction of the flow lines, i.e. along the axis of the cylinder of soil. The calculation of the hydraulic gradient is similar to that given above, i.e.

$$i = \frac{h_3 - h_4}{L} = \frac{\Delta h}{L}$$

Discharge velocity

The water velocity which is considered in connection with flow through soils is almost invariably the 'discharge velocity' and not the actual and erratic velocity of particles of water through the void spaces. If a quantity of water Q (m³) flows through a column of soil of area of cross-section

A (m^2) (measured perpendicular to the flow lines) in a time t seconds, the rate of flow, denoted by q (m^3/s) is equal to Q/t, and the discharge velocity v (m/s) is given by the relationship

$$v = \frac{Q}{At} = \frac{q}{A} \quad \text{m/s} \tag{10.3}$$

That is, the discharge velocity is equal to the rate of flow per unit area of cross-section.

The actual velocity of flow through the voids is known as the 'seepage velocity', and is greater than the discharge velocity, v, to an extent which is inversely proportional to the porosity of the soil. However, the seepage velocity does not enter into laboratory analysis.

10.3.2 Darcy's law

The relationship discovered by Darcy in 1856 concerning the flow of water in sands states that the rate of flow is proportional to the hydraulic gradient. Using the symbols given in Section 10.3.1, this relationship can be written as

$$\frac{Q}{t} \propto i$$

or

$$Q = Akit \tag{10.4}$$

where A is the area of cross-section of the soil, and k is known as the 'coefficient of permeability' of the soil, and has the dimensions of velocity. It is usually expressed in metres per second (m/s).

Equation (10.4) is the basic equation for permeability calculations. It is based on the assumption that the flow of water is laminar, or streamline, and not turbulent. This assumption is generally valid for soils ranging from clays to coarse sands, but may not be so for coarser materials.

The units normally used in laboratory work are as follows:

Quantity of water	Q	ml
Area	A	mm^2
Time	t	min
Rate of flow	q	ml/min
Hydraulic gradient	i	(dimensionless)
Coefficient of permeability	k	m/s
Temperature	T	°C

Equation (10.4) can be written in terms of the above units as follows:

$$k = \frac{Q}{60\,Ait} \quad \text{m/s}$$

or

$$k = \frac{q}{60\,Ai} \,\text{m/s} \tag{10.5}$$

10.3.3 Permeability coefficients

A more generalised version of Darcy's law states that the discharge velocity of flow of fluid through a porous granular medium, under steady conditions, is proportional to the excess hydrostatic pressure causing the flow, and inversely proportional to the viscosity of the fluid. This can be expressed by the proportionality

$$v \propto \frac{i_p}{\eta} \tag{10.6}$$

where v is the discharge velocity, η is the dynamic viscosity of the fluid, ip is the pressure gradient, equal to $\Delta p/L$, i.e. the pressure difference per unit length along the flow lines. The above relationship can be expressed by the equation

$$v = \frac{K\rho_w g i}{\eta} \tag{10.7}$$

since $i_p = \dfrac{\Delta p}{L} = \dfrac{(\Delta h)\rho_w g}{L} = i\rho_w g$ (using Equation (10.1))

K is an empirical constant known as the 'absolute' or 'specific' permeability, having the dimensions of area.

The 'absolute' permeability is constant for a given soil at a given porosity and is independent of the properties of the fluid. In most engineering applications we are concerned with the flow of water, therefore the following substitution is made:

$$v = \frac{K\rho_w g}{\eta_w} \tag{10.8}$$

where k is the coefficient of permeability referred to in Section 10.3.2 and η_w is the dynamic viscosity of water. Therefore Equation (10.7) can be written as

$$v = ki$$

The 'absolute permeability', K, is not often encountered, and the term 'permeability' as generally used refers to the 'coefficient of permeability', k. The relationship between k and K, in SI units, is derived as follows, where k is measured in m/s and K is expressed in mm². For water, $\rho_W = 1.00 \ \mathrm{Mg/m^2}$

$$\eta_w = \text{about 1.0 mPas (at 20°C)} = 10^{-3} \text{ Pas}$$
$$g = 9.81 \ \mathrm{m/s^2}$$

$$\therefore \ k = \frac{K}{(1000)^2} \times \frac{1000 \times 9.81}{10^{-3}} = 9.81K$$

or $k(\mathrm{m/s}) = 10 \ K \ (\mathrm{mm^2})$ approximately.

Throughout this book the term 'permeability' relates to the coefficient of permeability, k, measured in m/s.

10.3.4 Effect of temperature

From Equation (10.8) it can be seen that the permeability, k, is not constant for a given soil but is related to the dynamic viscosity of the fluid (water), denoted by η_w. Viscosity varies with temperature, and increases by about 30% from 20°C to 10°C. (The relationship between the dynamic viscosity of water and temperature, from 0–40°C, is given in Volume 1 (third edition), Table 4.13.) Therefore the water temperature should always be taken into account when performing permeability tests, whether in the laboratory or in the field.

It is convenient to relate permeability data to a standard temperature of 20°C. The ratio of the dynamic viscosity of water at any temperature $T°C$ (η_T) to that at 20°C (η_{20}) is shown graphically in Figure 10.4. If a permeability test carried out at $T°C$ gives a coefficient of permeability, k_T, the corresponding value at 20°C, k_{20} is calculated from the equation

$$k_{20} = k_T \left(\frac{\eta_T}{\eta_{20}} \right) \tag{10.9}$$

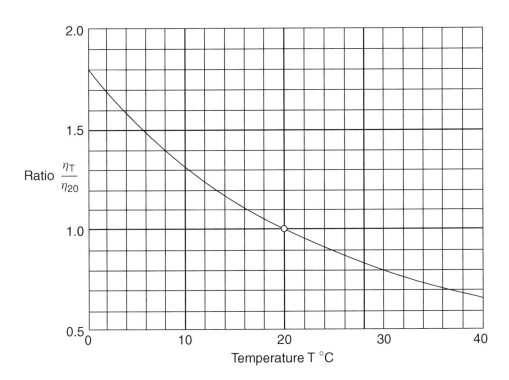

Figure 10.4 Relationship between dynamic viscosity of water and temperature (based on data from Kaye and Laby, 1973)

where (η_T/η_{20}) is read from the graph at $T°C$.

The in-situ permeability corresponding to $T°C$, k_T, can be calculated from the laboratory value measured at 20°C from the equation

$$k_T = \frac{k_{20}}{\left(\dfrac{\eta_T}{\eta_{20}}\right)} \tag{10.10}$$

where (η_T/η_{20}) is read from the graph at $T°C$.

In the UK the average ground temperature from a depth of about 2 m is fairly constant at about 10°C, at which temperature the viscosity ratio is about 1.3. The viscosity of groundwater is therefore greater than that used in a normal laboratory test by about 30%.

10.3.5 Flow in constant head permeameter

Arrangements for carrying out a constant head permeability test on a sample of clean sand are shown diagrammatically in Figure 10.5 with the flow downwards, and in Figure 10.6 with the flow upwards. The principle is the same in either case. The test is described in Section 10.6.3.

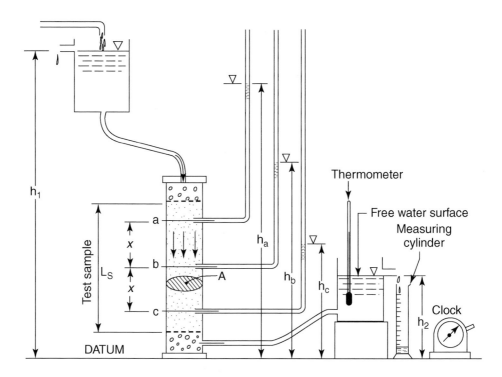

Figure 10.5 Principle of constant head permeability test: downward flow

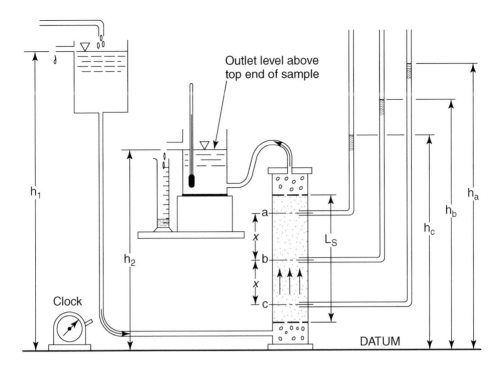

Figure 10.6 Principle of constant head permeability test: upward flow

The hydraulic gradient in the sample is determined by means of piezometer tubes inserted at three different levels, denoted by a, b and c. The water levels in these tubes are at heights h_a, h_b and h_c, respectively above bench level, which is taken as the common datum level. Considering piezometer tubes at a and b only, the distance between the points of insertion into the sample is denoted by x, and the hydraulic gradient i is equal to $(h_a - h_b)/x$. The area of cross-section of the sample is A mm^2.

If a quantity Q ml of water at $T°C$ passes through the sample in a time t min, the coefficient of permeability of the sample, from Equation (10.5), is given by the equation

$$k_T = \frac{Q}{60\,Ait}\,\text{m/s}$$

$$= \frac{Qx}{60\,A(h_a - h_b)t}\,\text{m/s} \qquad (10.11)$$

Equation (10.11) is used for the calculation of permeability from the tests described in Section 10.6.3.

The hydraulic gradient is obtained from piezometer level readings, rather than from the difference between the permeameter inlet and outlet levels ($h_1 - h_2$), in order to measure a difference in hydrostatic head within the sample itself. The difference in head between the

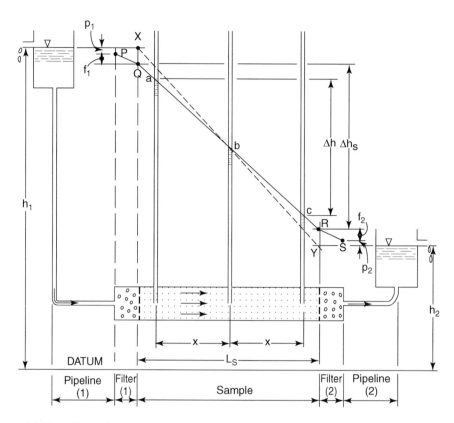

Figure 10.7 Head losses in constant head permeameter cell and connections

inlet and outlet levels includes the small pressure losses which occur within the filter layers, and possibly in the pipeline connections. These losses are illustrated diagrammatically in Figure 10.7, which depicts horizontal flow. The difference in head between piezometer tubes a and c, a distance $2x$ apart, is denoted by Δ_h. Head losses in the filters are denoted by f_1 and f_2, and those in the pipelines by p_1 and p_2. The difference in head across the whole length of the sample L_s and filters is indicated by the line PQRS, assuming that the gradient from Q to R is the same as in the portion between piezometers a and c. The assumption that the head loss along the length of sample L_s is equal to the difference between the inlet and outlet levels $(h_1 - h_2)$ is represented by the line XY, which is steeper than QR and therefore gives a hydraulic gradient which is too large.

However, if the rate of flow of water is small, head losses in pipelines and filters would be very small compared with those in the sample, and the error in using the assumed head loss XY would not be very significant. For higher rates of flow the error would be larger.

10.3.6 Flow in falling head permeameter
The principle of the falling head permeability test on a sample of low permeability such as clay is shown in Figure 10.8. The test is described in Section 10.7.2.

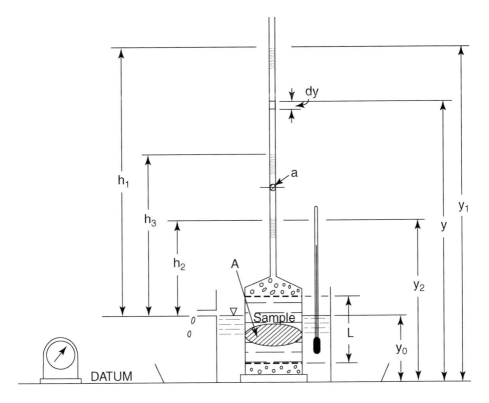

Figure 10.8 Principle of falling head permeability test

The notation used in the analysis given below is as follows:

Length of sample	L
Area of cross-section of sample	A
Area of cross-section of standpipe tube	a
Heights of water above datum in standpipe:	
at time t_1	y_1
at time t_2	y_2
at any intermediate time t	y
fall during small time increment dt	dy
Quantity of water flowing through sample in	
small time increment dt	dQ
Height of outlet level above datum	y_0

At any time t, the difference in height between the inlet and outlet levels is equal to $(y - y_0)$. The hydraulic gradient i at this instant is therefore equal to $(y - y_0 / L)$.

The quantity of water flowing through the sample in time dt is equal to the area of the standpipe multiplied by the drop in height of the water level, i.e.

$$dQ = -a\ dy$$

(the negative sign appears because y is decreasing).

But from Darcy's law, Equation (10.4),

$$dQ = Aki \, dt = \frac{Ak(y - y_0)}{L} \, dt$$

$$\therefore \; -a \, dy = \frac{Ak(y - y_0)}{L} \, dt$$

$$\text{or} \; -\frac{1}{(y - y_0)} \, dy = \frac{kA}{aL} \, dt \tag{10.12}$$

Integrating Equation (10.12) between the limits $y = y_1$ to y_2 and $t = t_1$ to t_2, we obtain

$$-\int_{y_1}^{y_2} \frac{dy}{y - y_0} = \int_{t_1}^{t_2} \frac{kA}{aL} \, dt$$

$$\text{i.e.} -[\log_e(y - y_0)]_{y_1}^{y_2} = \left[\frac{kAt}{aL}\right]_{t_1}^{t_2}$$

$$\therefore \log_e \frac{y_1 - y_0}{y_2 - y_0} = \frac{kA}{aL}(t_2 - t_1) \tag{10.13}$$

Putting $y_1 - y_0 = h_1$ and $y_2 - y_0 = h_2$. Equation (10.13) can be re-written as

$$k = \frac{aL}{A(t_2 - t_1)} \; \log_e\left(\frac{h_1}{h_2}\right) \tag{10.14}$$

The time difference $(t_2 - t_1)$ can be expressed as the elapsed time, t (min). The heights h_1 and h_2 and the length L are expressed in millimetres, and the areas A and a in square millimetres. Equation (10.14) then becomes

$$k \; (\text{mm/s}) = \frac{aL}{A \times 60t} \; \log_e\left(\frac{h_1}{h_2}\right)$$

To convert natural logarithms to ordinary (base 10) logarithms, multiply by 2.303. If k is expressed in m/s, the above equation becomes

$$k \; (\text{m/s}) = \frac{2.303 \, aL}{1000 \times A \times 60t} \; \log_{10}\left(\frac{h_1}{h_2}\right)$$

or

$$k = 3.84 \frac{aL}{At} \; \log_{10}\left(\frac{h_1}{h_2}\right) \times 10^{-5} \; \text{m/s} \tag{10.15}$$

Equation (10.15) is used for the calculation of permeability from the falling head test described in Section 10.7.2.

When carrying out a test it is convenient to divide the drop in standpipe level, h_1 to h_2, into two portions at an intermediate level h_3, such that for a given sample the time from h_1 to h_3 (t_{1-3}) is theoretically equal to the time from h_3 to h_2 (t_{3-2}). Equation (10.15) can be re-arranged to give the relationship of h_1 and h_2 to time, i.e.

$$t = 3.84 \frac{aL}{Ak} \log\left(\frac{h_1}{h_2}\right) \times 10^{-5}$$

or for a given soil

$$t = \text{constant} \times \log\left(\frac{h_1}{h_2}\right)$$

Therefore $t_{1-3} = \text{constant} \times \log\left(\frac{h_1}{h_3}\right)$

and $t_{3-2} = \text{constant} \times \log\left(\frac{h_3}{h_2}\right)$

If these times are equal

$$\log\left(\frac{h_1}{h_3}\right) = \log\left(\frac{h_3}{h_2}\right)$$
$$\therefore \frac{h_1}{h_3} = \frac{h_3}{h_2}$$

$$\text{or } h_3 = \sqrt{(h_1 h_2)} \qquad (10.16)$$

This relationship is used for marking suitable level markings on the standpipes for the test.

It should be noted that all heights denoted by y are measured from the datum level, and those denoted by h are measured from the outlet water level at y_0 above datum.

10.3.7 Empirical relationships

Outline

Several formulae have been published relating the permeability of soils, especially sands, to their particle size characteristics and other classification data. Of these, two which have been widely accepted are given below. The first is that given by Hazen (1982), which is simple but indicates only the order of magnitude of permeability and is based only on particle size data. The second is the Kozeny formula (1927) and its modification by Carman (1939), which agrees better than most other formulae with measured permeabilities (Loudon, 1952) and takes into account the shape of particles and the soil porosity, as well as the particle size distribution.

Both formulae are intended only for clean sands, but their use is sometimes extrapolated to finer soils to obtain an approximate indication of their permeability.

Hazen's formula

This is based on experimental work with fine, uniform sands, and is an attempt to relate permeability to the effective size of the particle, D_{10} (see Volume 1 (third edition), Section 4.2 and Figure 4.1). The formula as given by Terzaghi and Peck in 1948 (see Terzaghi and Peck (1967), (updated as Terzaghi *et al.*, 1996) is

$$k \ (\text{cm/s}) = C_1 \ (D_{10})^2$$

where D_{10} is the effective particle size in centimetres, and C_1 is a factor (dimensions $\text{cm}^{21} \ \text{s}^{21}$) which is stated to be about 100. According to Taylor (1948) the value of C_1 can range from about 40 to 150. Lambe and Whitman (1979) quote results of tests by Lane and Washburn (1946) on a wide range of soils (coarse gravel to silt) which give an average value for this factor of about 16, whereas Hazen's observations were limited to sands of fairly uniform grain size.

Conversion of Hazen's formula to conventional SI units gives the equation

$$k \ (\text{m/s}) = C_1 \ (D_{10})^2 \times 10^{-4}$$

in which D_{10} is expressed in millimetres. If the factor C_1 is assumed to be 100, this equation becomes

$$k = 0.01 \ (D_{10})^2 \ \text{m/s} \tag{10.17a}$$

If $C_1 = 16$ (Lambe and Whitman, 1979)

$$k = 0.0016 \ (D_{10})^2 \ \text{m/s} \tag{10.17b}$$

These alternative equations give calculated values differing by a factor of about 6, but this method does not take into account the considerable variations due to voids ratio. These equations should not be applied to clays.

Kozeny's formula

The equation proposed by Kozeny (1927) relates permeability to particle size, porosity (n), angularity of particles, specific surface (S), and viscosity of water (η_w). The general equation is

$$k = \frac{\rho_w g n^3}{C \eta_w S^2 (1-n)^2} \tag{10.18}$$

This equation was modified by Carman (1939) who replaced porosity by voids ratio, using the substitution $n = e \ / \ 1 + e$. This is known as the Kozeny–Carman equation

$$k = \frac{\rho_w g}{C\eta_w S^2} \cdot \frac{e^3}{1+e} \qquad (10.19)$$

For a collection of spherical particles uniformly distributed in size between diameters d_1 and d_2, the specific surface (i.e. the surface area per unit volume of grains), S, is obtained from the equation

$$S = \frac{6}{\sqrt{(d_1 d_2)}} \qquad (10.20)$$

If d_1 and d_2 are expressed in millimetres, S is expressed in mm²/mm³, or mm⁻¹. The term $\sqrt{(d_1 d_2)}$ is the geometric mean particle diameter. The relationship between specific surface and geometric mean diameter is shown graphically in Figure 10.9.
In Equations (10.18) and (10.19) the constant C is a shape factor, equal to 5 for spherical particles. However, sand particles are rarely spherical, so an angularity factor, f, by which C is multiplied, is introduced into the above equations to allow for the irregular shape of the grains. Equation (10.19) then becomes

$$k = \frac{\rho_w g}{5 f \eta_w S^2} \left(\frac{e^3}{1+e} \right) \qquad (10.21)$$

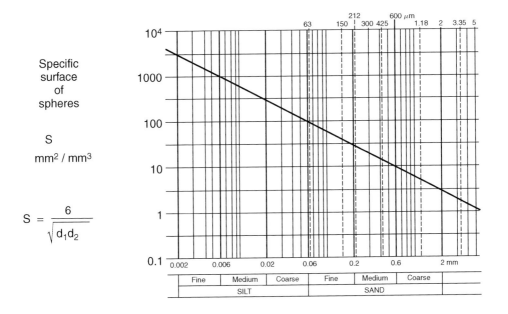

Figure 10.9 Specific surface of sand-size and silt-size spherical particles

Values of f stated by Loudon (1952) are as follows:

Rounded grains $\quad f = 1.1$
Sub-rounded grains $f = 1.25$
Angular grains $\quad f = 1.4$

Classification of grains into the above shape categories can be obtained with the aid of Figure 7.3 of Volume 1 (third edition). For a sand consisting of a range of particle sizes, an angularity factor is first assigned to each sieve fraction by inspection. A factor for the whole can then be obtained by combining these separate factors in proportion to the percentages by mass retained on each sieve, as illustrated by the worked example given in Section 10.5.3.

Equation (10.21) can be expressed in the customary SI units given below

k	=	coefficient of permeability	$= \mathrm{m/s}$
g	=	acceleration due to gravity	$= 9.81 \ \mathrm{m/s^2}$
ρ_w	=	mass density of water	$= 1.00 \ \mathrm{Mg/m^2}$
η_w	=	dynamic viscosity of water, which at about 20°C	$= 1 \ \mathrm{mPas}$
e	=	voids ratio of soil	
S	=	specific surface of grains $(\mathrm{mm^2/mm^3})$	$= \mathrm{mm^{-1}}$

Substituting these values in Equation (10.19) with the appropriate multiplying prefixes, for a temperature of 20°C,

$$k_{20} = \frac{(1 \times 1000) \times 9.81}{5f \times \dfrac{1}{1000} \times (S \times 1000)^2} \left(\frac{e^3}{1+e} \right) \mathrm{m/s}$$

or for practical purposes

$$k_{20} = \frac{2}{fS^2} \left(\frac{e^3}{1+e} \right) \mathrm{m/s} \tag{10.22}$$

10.3.8 Critical hydraulic gradient

Darcy's law is valid only for conditions of 'laminar' or streamline flow (see Section 10.1.5, item 8). When the velocity of flow exceeds a certain critical value, turbulence occurs and Darcy's law no longer applies. If the flow is in a downward direction, frictional drag of the water on the surfaces of the soil particles tends to bring the particles closer together and increases the contact pressure between them. If the water flows upwards, frictional drag tends to lift the particles and force them apart. The hydraulic gradient under which the particles begin to lose contact is known as the 'critical hydraulic gradient', i_c, and is equal to the ratio of the submerged density of the soil, ρ (see Volume 1 (third edition), Section 3.3.2) to the density of water, ρ_w. That is

$$i_c = \frac{\rho'}{\rho_w} \tag{10.23}$$

The submerged density of most soils is about the same as that of water, therefore the critical gradient is usually about equal to 1.

The loss of contact between soil grains results in loss of effective stress, and the soil then has zero shear strength. This gives the 'quick' condition, in which the soil behaves as a liquid having a density about twice that of water. A 'quick' condition can occur in any cohesionless soil which is subjected to an upward hydraulic gradient equal to or greater than the critical gradient. A hydraulic gradient in excess of the critical value is also responsible for the 'boiling' of sand at the bottom of an open excavation below the water table, and for sub-surface erosion known as 'piping'.

The effect of the critical hydraulic gradient can be demonstrated in the constant head permeability cell (see Section 10.6.3). The principle is described below, with reference to Figure 10.10 which is a simplified version of the arrangement shown in Figure 10.6.

Water is allowed to flow upwards through a sample of the sand held between suitable wire mesh screens in a container, and completely submerged. The difference in level between the inlet and outlet reservoirs, at heights h_1 and h_2 above a datum level, denoted by Δh, is

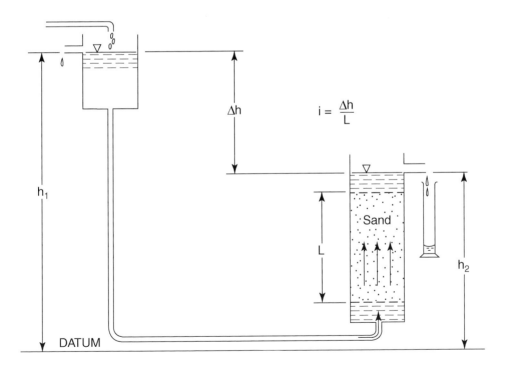

Figure 10.10 Illustration of critical hydraulic gradient

initially much smaller than the length L of the sample, so that the hydraulic gradient i is much less than unity, where

$$i = \frac{h_1 - h_2}{L} = \frac{\Delta h}{L}$$

The rate of flow of water through the sample, q $(= Q/t)$, is measured when it has reached a steady value. Further values of q are measured as the hydraulic gradient is progressively increased by increasing h_1. A graph is drawn relating the rate of flow, q, to the gradient, i, as in Figure 10.11. From the origin up to the point C the graph is linear, showing that the rate of flow is directly proportional to the hydraulic gradient, i.e. Darcy's law (Equation (10.4)) is valid and the coefficient of permeability k is constant.

At the point represented by C the sand enters into a state of agitation and the rate of flow suddenly increases to the value denoted by B. This is a region of instability, i.e. the 'quick' condition in which the hydraulic gradient is at the critical value i_c.

Liquefaction occurs, the sand and water acting as a dense liquid. A small weight placed initially on the surface of the sand would suddenly sink into the fluid.

Increasing the hydraulic gradient further causes the rate of flow to increase again in proportion to the gradient but at a greater rate than previously (BD in Figure 10.11), indicating a higher permeability than before.

If the hydraulic gradient is then progressively reduced, the rate of flow is represented by the curve DBEO. After passing back through the critical condition denoted by BE, the

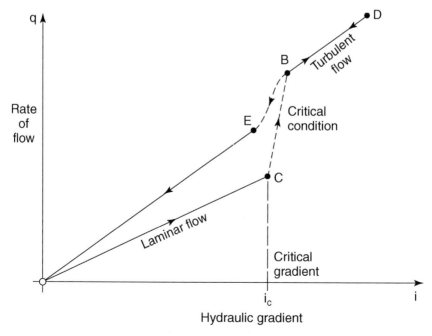

Figure 10.11 Relationship between rate of flow and hydraulic gradient (after Terzaghi and Peck, 1967)

rate of flow again becomes proportional to the gradient but the value of k is greater than the original value. This indicates that in passing through the critical condition an irreversible change in the state of packing of the sand grains occurred, i.e., there was a decrease in density of the sand.

The value of the critical gradient could be increased by applying a longitudinal compressive stress to the sand, provided that the sand is placed at a fairly high relative density. In saturated sand at a low relative density, whether or not an effective stress is applied, liquefaction can result if it is subjected to a sudden shock, even under zero hydraulic gradient. This results in the sudden collapse of the grain structure, after which, in the absence of a hydraulic gradient, the sand settles down at a greater relative density than before.

10.3.9 Erosion of cohesionless soils

The flow of water through a cohesionless soil can initiate movement of soil particles, which may be considered in three degrees of intensity:

- Local motion of a few fine particles, which may give the first indication that more vigorous movement is imminent
- Migration of fine particles through the voids between larger particles in the soil, either to be carried away with the water, or to percolate into the voids in an adjacent mass of soil. This process is known as 'suffusion'
- Internal erosion and removal of particles, known as 'piping', which occurs when the hydraulic gradient exceeds the critical value
- The initiation of piping by increasing the hydraulic gradient to the critical value and beyond is described in the previous section. Some practical aspects of piping are referred to in Section 10.4.1.

Two mechanisms involving suffusion are illustrated in Figure 10.12 (Clough and Davidson, 1977). A fine-grained soil (A) overlying a coarse grained soil (B) is shown in Figure 10.12(a), and water flowing downwards is likely to carry fine particles from soil A into the relatively large voids in soil B. A soil consisting of fine particles contained in a matrix of coarser particles is shown in Figure 10.12(b), and water emerging from this soil may carry fine particles away with it if the velocity of flow is great enough.

Suffusion of fine particles can occur not only from or into materials which have been placed or compacted as fill, but also from one naturally deposited soil into another if they have never been subjected to a significant hydraulic gradient and a movement of water is artificially imposed.

10.3.10 Design of filters

A filter is an intermediate layer of material placed between two layers of soil, or over a soil from which water is emerging, in order to prevent loss or migration of solid particles by suffusion. Filter materials were investigated by Bertram (1940), and empirical rules for the design of filters were stated by Terzaghi and Peck in 1948 (see Terzhagi and Peck, 1967). Subsequent modifications have been made to these rules, which are based on the particle size distribution curves for the materials, and this application of particle size tests was referred to in Volume 1 (third edition), Section 4.4.4. The ideas outlined below merely illustrate some applications of the test data, and are not intended as design recommendations.

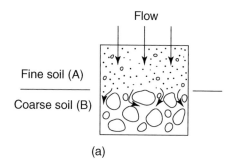

(a)

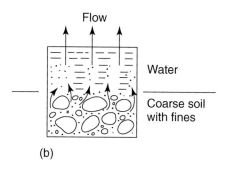

(b)

Figure 10.12 Mechanisms of suffusion: (a) movement of fines into coarse grained soil; (b) piping of fines out of coarse grain structure (after Wittman, 1976)

The symbols used below are similar to those used in Figure 4.1 of Volume 1 (third edition). The soil to be protected is referred to as the 'base' material, and is designated by suffix B; the filter is designated by suffix F. The rules given by Terzaghi and Peck, Art. 12 (1967), are that the 15% size $(D_{15.F})$ of the filter should be at least 4 times larger than the largest 15% size $(D_{15.B})$ of the base, and not more than 4 times larger than the smallest 85% size $(D_{85.B})$ of the base. In addition the grading curves of the filter and base materials should be similar in shape and approximately parallel.

The particle size requirements can be expressed as follows:

$$D_{15.F} > 4(D_{15.B}) \qquad (10.24)$$

$$D_{15.F} \le 4(D_{85.B}) \qquad (10.25)$$

These recommendations are also shown graphically in the form of grading curves in Figure 10.13, in which the grading of the base material may vary within the range indicated. A suitable mean grading for filter material is shown by the broken line curve.

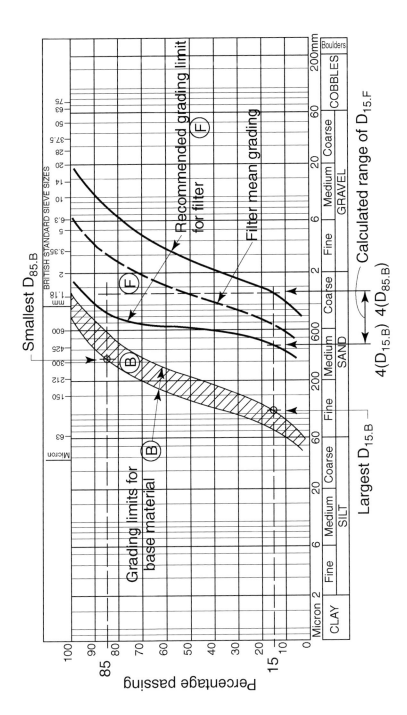

Figure 10.13 Typical particle size requirements for a filter material

An extended set of criteria is given by Lambe and Whitman (1979), based on recommendations of the US Corps of Engineers, Vicksburg. They can be stated briefly as follows:

$$4 < \frac{D_{15.F}}{D_{15.B}} < 20$$

$$(10.26)$$

$$D_{15.F} < 5(D_{85.B})$$

$$(10.27)$$

$$\frac{D_{50.F}}{D_{50.B}} < 25$$

$$(10.28)$$

$$k_F \text{ should be much greater than } k_B \qquad (10.29)$$

It may be necessary, especially for large filters, to use several layers of filter material, each related to the next in the manner referred to above. A composite arrangement of this type is known as a graded filter.

The ratio $D_{15.F}/D_{85.B}$, which the Terzaghi rule states should not exceed 4, is referred to as the 'critical ratio'. Laboratory tests of the type referred to in Section 10.6.6 have indicated that instability starts at a critical ratio of about 8 for well-graded materials, but the ratio decreases as the uniformity coefficient D_{60}/D_{10} of the filter increases. For most practical purposes the conventional ratio of 4 provides an adequate margin of safety.

10.3.11 Erodibility of cohesive soils

Certain clay soils which contain a high sodium content are highly erodible under the action of water flowing through them, and are known as 'dispersive' soils. Erosion takes place because individual clay particles of these soils are capable of going into suspension in practically still water, whereas a flow of water of considerable velocity is required to cause erosion of ordinary clays. Properties of these soils are described by Sherard *et al.* (1976a). Dispersive soils cannot be identified from conventional soil classification tests, but the tests given in Section 10.8 enable them to be recognised by fairly simple means.

10.4 Applications

10.4.1 Relevance of permeability to earthworks

Knowledge of the permeability characteristics of soil is required for many construction projects in which drainage is an important feature. In many applications the use of a flow net analysis (described for instance by Lambe and Whitman (1979, Chapter 18)), together with permeability data, enables the rate of seepage of water through or under a structure to be estimated, and seepage pressures to be calculated. Some important applications are outlined below.

Excavations in water-bearing ground

The permeability of the soil can have a decisive effect on the difficulties encountered in excavations, and on the cost of dealing with them, and is relevant to the following factors:

- Estimation of the quantity of water likely to flow into the excavation, and hence the pumping capacity to be provided
- Whether groundwater lowering is feasible
- Design of sheet-pile walls, and the depth to which they should be extended
- Prevention of 'boiling' or heave of sand strata (or any non-cohesive soil) at the bottom of an excavation below the water table

Earth dams

The permeabilities of the different types of soil, including filter zones, with which an earth dam is constructed need to be known in order to estimate the likely quantity of seepage flow through the dam; to provide adequate drainage capacity for the filters; and to prevent the development of excessive seepage pressures.

The permeabilities of the foundation strata, both horizontally and vertically, are required for the design of a cut-off beneath the dam to minimise seepage losses through the foundations. The significance of the permeability ratio, which can vary within a very wide range (from less than 2 to several hundred), is described by Kenney (1963). The effect of other discontinuities such as fissures is discussed by Thorne (1975).

Seepage pressures

Water percolating through any porous material exerts a pressure, known as the seepage pressure, which can be very high even when the rate of percolation is extremely small. Seepage pressures can be estimated from a flow net analysis, for which the permeability of the material must be known.

Seepage pressures affect the stability of earth structures such as embankments and cuttings, earth and concrete dams, retaining walls; and of sub-surface structures such as basements, pumping stations and dry dock floors. Seepage pressures can develop in concrete and rock as well as in soils.

Piping and erosion

An excessive hydraulic gradient in the ground, for instance near the downstream toe of an earth dam, can cause local instability leading to gradual erosion and the formation of well-defined channels or 'pipes' beneath the structure. This effect is known as 'piping', and the progressive sub-surface erosion of soil can rapidly lead to a disastrous failure. Piping can be avoided by ensuring that the hydraulic gradient is kept well within the critical value, such as by loading the danger zone with a suitably graded filter, or by extending the base width of a dam.

Another cause of piping leading to failure in dams has been the presence of dispersive clays, which can be readily eroded. Some examples are given by Sherard *et al.* (1972). Empirical tests such as the pinhole test (Section 10.8.2) enable these clays to be identified easily so that their use can be restricted or prohibited.

Other applications

- Drainage of highway and airfield bases and sub-bases
- Estimations of the yield of water and the rate of extraction from aquifers (i.e. porous strata from which water is extracted for general supply purposes)
- Design of graded filters. Tests in a permeameter cell can supplement calculations based on empirical design rules

10.4.2 Limitations

Sands

The results of laboratory permeability tests on cohesionless soils such as sands, from which it is not practicable to obtain undisturbed samples, are of limited value for determining the true permeability of these soils in their natural state. There are two main reasons for this

- Without specialised equipment it is very difficult to measure the density, and hence the voids ratio, of granular soils in-situ, especially below the water table, therefore the voids ratio at which to set up samples for test can only be surmised
- Even if the voids ratio is approximately assessed, the features of the soil fabric referred to in Section 10.1.5, item 6, cannot be reproduced when a sample is recompacted in the laboratory

The only satisfactory way of taking these factors into account is to carry out permeability tests in-situ. However, laboratory permeability tests can give an indication of the likely range of permeability values.

Soft soils

In some soft cohesive soils, including silts, it is possible to obtain large diameter undisturbed samples of good quality on which laboratory tests can be performed in large diameter consolidation cells (to be described in Volume 3).

Layered systems

Water-bearing layers of very high permeability within a low permeability deposit, such as layers of sand or gravel in clay, can give rise to unexpected difficulties especially in tunnelling operations (May and Thomson, 1978). Under these circumstances, field observations and measurements are essential, and isolated laboratory tests can be highly misleading.

10.4.3 Relevance to consolidation of clay

The rate at which a clay consolidates under the application of an imposed load is controlled by its permeability. The very low permeabilities of clays are responsible for the prolonged time periods during which settlement takes place in these soils. This aspect is discussed further in Chapter 14. Equation (14.28) of Section 14.3.11 states the relationship between the coefficient of consolidation (from which the rate of settlement can be calculated) and the permeability of the soil, which in some instances needs to be determined from in-situ tests.

10.4.4 Applicability of test procedures

The constant head permeameter is suitable for measuring the permeability of clean sands, and the large version is suitable for gravels, or sands containing gravel. The presence of small quantities of silt can reduce the permeability considerably, and the constant head procedure is applicable to soils with permeabilities not less than about 10^{-4} m/s. This apparatus can be used to observe suffusion, erosion and piping effects in granular soils.

If an attempt were made to measure the permeability of silts and clays in a constant head permeameter, the rate of percolation of water, even under the highest practicable hydraulic gradient, would be extremely small and difficult to measure. The falling head permeameter is used for these soils, especially for clayey soils of low and very low permeability, i.e. less than 10^{-6} m/s. By using a large bore standpipe, as described in Section 10.7.2, acceptable results can be obtained on silts in the medium permeability range ($10^{-4} - 10^{-6}$ m/s). An alternative method for silts, if a tube sample of reasonable length is available, is to test the sample in the tube using the falling head principle (as described in Section 10.7.3). A more satisfactory procedure, in which the stress conditions can be controlled, is to carry out permeability measurements in a triaxial cell, as part of an effective stress compression test or pore pressure dissipation test. This procedure will be described in Volume 3.

The falling head test, whether carried out in a standard permeameter cell or in a sample tube, can be performed either on an undisturbed sample or on recompacted material. Tests on sands in the constant head permeameter can be carried out only on recompacted soil, but it is possible to apply a known constant axial stress to the sample during the test, if necessary.

For soils that are to be compacted for use as fill, laboratory permeability tests can provide valuable data provided that the samples are prepared in an appropriate manner. For example the permeability will depend upon the degree of compaction applied to the sample, and whether the soil is compacted dry or wet of the optimum moisture content. The field condition should be reproduced as closely as practicable.

10.4.5 Typical permeability values

The classification of soils on the basis of permeability is given in Table 10.1, which is derived from a table by Terzaghi and Peck (1967).

The permeability and drainage characteristics of the main soil types, in general terms, are indicated diagrammatically in Figure 10.14, which includes an indication of the type of

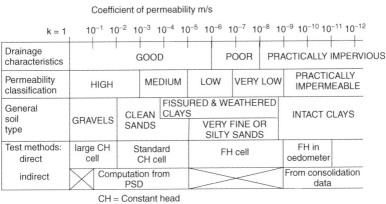

Figure 10.14 Permeability and drainage characteristics of main soil types

Table 10.1 Classification of soil according to permeability

Degree of permeability	Range of coefficient of permeability, k(m/s)
High	Greater than 10^{-3}
Medium	$10^{-3} - 10^{-5}$
Low	$10^{-5} - 10^{-7}$
Very low	$10^{-7} - 10^{-9}$
Practically impermeable	Less than 10^{-9}

test which is most appropriate for each category. These data are shown in a different way, related to effective particle size, in Figure 10.15.

10.5 Permeability by indirect methods

10.5.1 General requirements

Procedures for the evaluation of the coefficient of permeability by the two methods of calculation based on particles size and shape characteristics referred to in Section 10.3.7 are given below, together with worked examples. In both procedures it is first necessary to obtain a truly representative sample, and to carry out a particle size analysis using the wet sieving procedure (Volume 1 (third edition), Section 4.6.4). It is essential to obtain a particle size distribution curve that accurately reflects the proportions of the finest particles present.

The equation for calculating permeability from consolidation test data is derived in Chapter 14 (Section 14.3.11), where an example is given.

10.5.2 Hazen's method

The particle size curve shown in Figure 10.16, obtained from a wet sieving analysis, is used for this example. This method is applicable to fine, uniform sands, and the result may be in error by a factor of 2 either way.

The effective size D_{10}, representing 10% passing, is read off from the grading curve as shown in Figure 10.16, giving $D_{10} = 0.12$ mm.

The coefficient of permeability, k, is calculated from Equation (10.17a), Section 10.3.7

$$k = 0.01 \times (0.12)^2 = 1.44 \times 10^{-4} \text{ m/s}$$

The result is reported to one significant figure only, i.e. $k = 1 \times 10^{-4}$ m/s, stating that the result was derived from the Hazen formula. The grading curve should be included with the report.

10.5.3 Kozeny–Carman method

After completing the wet sieving analysis the size fraction retained on each sieve is kept separated. Each fraction is reduced by riffling (Volume 1 (third edition), Section 1.5.5) to give a representative portion small enough for examination of individual grains. Using a hand lens or

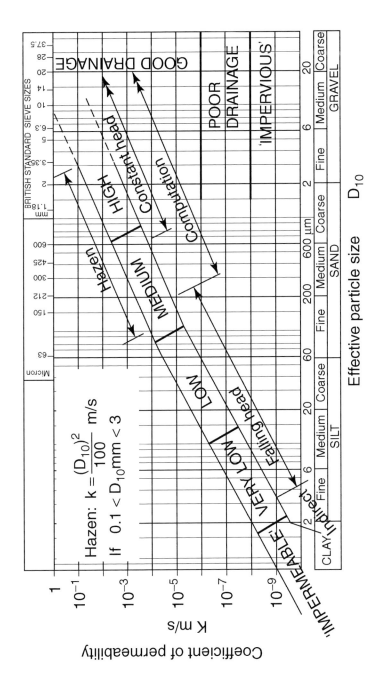

Figure 10.15 Permeability classification related to effective particle size

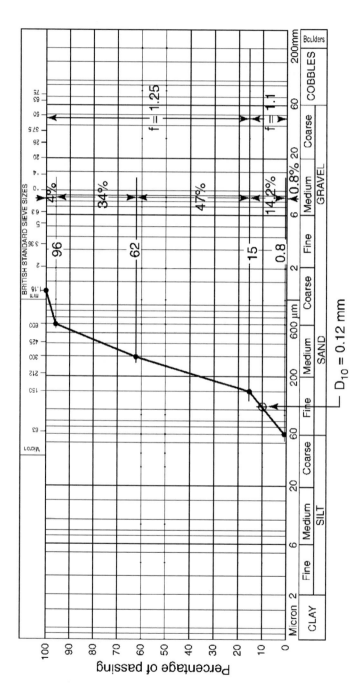

Figure 10.16 Particle size curve used for examples to illustrate indirect determination of permeability

low-power microscope, the angularity of grains is assessed by inspection and classification on the lines indicated in Figure 7.3 (Volume 1, third edition), so that the grains of each size fraction can be assigned to one of three groups designated 'rounded', 'sub-rounded', or 'angular'.

The mean particle density, ρ_s of the grains is determined as described in Volume 1 (third edition), Section 3.6.2. If the dry density, ρ_D, of the soil is known, the voids ratio e can be calculated by applying Equation (3.4) in Volume 1 (third edition), Section 3.3.2, i.e.

$$e = \frac{\rho_s}{\rho_D} - 1$$

The effect of the surface area and angularity of the grains is calculated as described below, using as an example the grading curve given in Figure 10.16. The percentages within each size range tabulated alongside the grading curve are used for the calculation, which is summarised in Figure 10.17.

Angularity characteristics, and the angularity coefficients, f, for this example are as follows:
Grains larger than 0.150 mm: 'sub-rounded', $f = 1.25$
Grains smaller than 0.150 mm: 'rounded', $f = 1.1$
The calculated value of (fS^2) for the whole sample is 1279 mm^{-2}
If the dry density of the sand is 1.85 Mg/m^3, and the mean particle density is 2.65 Mg/m^3, the voids ratio is given by

$$e = \frac{2.65}{1.85} - 1 = 0.432$$

The coefficient of permeability at 20°C, k_{20}, is calculated by using Equation (10.22) as follows:

Particle size range d_1 to d_2	Proportion of total by mass	Specific surface $S = \dfrac{6}{\sqrt{d_1 d_2}}$	Angularity factor f	$\dfrac{p}{100} \times S^2 \times f$
mm	P %	mm^{-1}		mm^{-2}
1. 18 to 0. 60	4	7.1	1.25	2.5
0. 60 to 0. 30	34	14.1	1.25	84.5
0. 30 to 0. 15	47	28.3	1.25	470.5
0. 15 to 0. 063	14.2	61.7	1.1	594.6
0. 063 to 0. 04	0.8	120	1.1	139.4
Total	100.0		Total	1291.5

Value of (fS^2) for the whole sample = 1292 mm^{-2}

Figure 10.17 Calculation of specific surface for Kozeny–Carman method (data based on grading curve in Figure 10.16)

$$k_{20} = \frac{2}{1279} \times \frac{(0.432)^3}{1.432} = 8.80 \times 10^{-5} \text{ m/s}$$

which would be reported as $k_{20} = 0.88 \times 10^{-4}$ m/s.

If particles finer than coarse silt are present, the magnitude of the bottom line of the calculation in Figure 10.17 can be so far out of proportion to the others that the formula becomes unworkable and should not be used.

The result is reported to two significant figures, stating that the Kozeny–Carman method of calculation was used. The grading curve, voids ratio, and specific gravity of particles should also be reported, together with the method by which the dry density was determined.

10.5.4 Calculation from consolidation characteristics

The coefficient of permeability of clays can be calculated from the consolidation parameters m_v and c_v, which are obtained from a standard oedometer consolidation test as described in Chapter 14. The calculation uses Equation (14.29) of Section 14.3.11.

$$k_T = 0.31 \times 10^{-9} \, c_v m_v \text{ m/s}$$

where cv = coefficient of consolidation (m²/year); m_v = coefficient of volume compressibility (m²/MN); and k_T = coefficient of permeability (m/s), at the test temperature of $T°C$, during an increment of load. The result is reported to two significant figures, together with details of the loading stage from which the data were derived.

10.6 Constant head permeability tests

10.6.1 Outline

The constant head procedure is used for the measurement of the permeability of sands and gravels containing little or no silt. In the UK the most common permeability cell (permeameter), which is described in Section 10.6.3, is 75 mm diameter and is intended for sands containing particles up to about 5 mm. A larger cell, 114 mm diameter, can be used for testing sands containing particles up to about 10 mm, i.e. medium gravel size. In the ASTM Standard reference is also made to permeameter cells of 152 mm and 229 mm diameter. A much larger apparatus, designed and built specially to accommodate samples containing gravel particles up to 75 mm, is described in Section 10.6.5. As a general rule the ratio of the cell diameter to the diameter of the largest size of particle in significant quantity should be at least 12.

The constant head permeability cell is intended for testing disturbed granular soils which are recompacted into the cell, either by using a specified compactive effort, or to achieve a certain dry density, i.e. voids ratio.

In the constant head test, water is made to flow through a column of soil under the application of a pressure difference which remains constant, i.e. under a constant head, such that the flow is laminar. The amount of water passing through the soil in a known time is measured, and the permeability of the sample is calculated by using Equation (10.11) in Section 10.3.5.

If the connections to the cell are arranged so that water flows upwards through the sample, the critical hydraulic gradient (see Section 10.3.8) can be determined after measuring the steady state permeability, and the effects of instability (boiling and piping) can be observed. The importance of using air-free water, and measures for preventing air bubbling out of solution during these tests, are discussed in Section 10.6.2.

10.6.2 Water for test

Ideally, the water used for a permeability test should be the groundwater found in-situ. This is seldom practicable, and some treatment of available water is usually desirable. The main objections to the use of untreated tap water are the presence of dissolved air, dissolved solids and possibly bacteria.

Air contained in the water tends to form bubbles as it flows through the narrow voids between soil particles. Bubbles of air in the voids impede the flow of water, thereby giving an erroneously low measurement of permeability. If the air is first removed from the water, this effect can be eliminated. For the same reason the sample itself must be de-aired and fully saturated before starting the test. A vacuum procedure for effecting this is described in Section 10.6.3, stage 8. Natural groundwater generally contains very little dissolved air.

Dissolved salts and organic matter in the water can have an effect on fine-grained soils, especially those containing certain clay minerals, which would alter the permeability characteristics (see Section 10.1.5). It is desirable to use distilled or de-ionised water for all permeability tests, and is essential for soils containing fine silt and clay. However, for tests on clean sands containing no fines, ordinary tap water is usually suitable if it is first made reasonably free of air.

De-aired water can be obtained from tap water, distilled water, de-ionised water or water from site, by three methods:

1. Subjecting to vacuum
2. Using a de-aerator
3. Boiling

Three further methods can be used for partial de-airing as a simple expedient when testing medium to coarse grained sands.

4. Passing through a sand filter
5. Warming
6. Allowing to stand

These methods are described below.

1. Use of vacuum

An arrangement for de-airing water by means of a vacuum line is shown in Figure 10.18. The de-airing vessel must be capable of withstanding full external atmospheric pressure (see Volume 1 (third edition), Section 1.2.5, item 3).

Water is fed into a supply tank which is open to atmosphere. If tap water is to be used, the supply tank can be connected to the mains, and a standard ball-float valve used to maintain a constant level. Water is drawn up from the supply tank to the storage vessel by means of the vacuum which is applied to the vessel. The inlet pipe is fitted with a spray nozzle and baffle plate, so that most of the air is removed from the water while it is in the form of small drops. The vacuum pipe terminates above the level of the end of the supply pipe, and is turned away

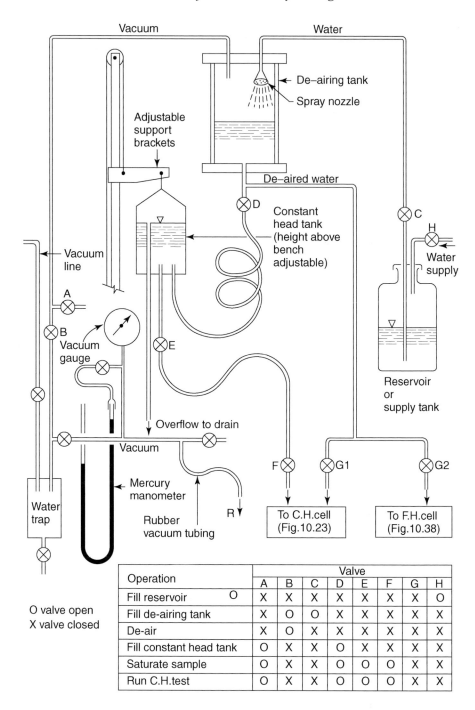

Operation	Valve							
	A	B	C	D	E	F	G	H
Fill reservoir O	X	X	X	X	X	X	X	O
Fill de-airing tank	X	O	O	X	X	X	X	X
De-air	X	O	X	X	X	X	X	X
Fill constant head tank	O	X	X	O	X	X	X	X
Saturate sample	O	X	X	O	O	O	X	X
Run C.H.test	O	X	X	O	O	O	X	X

O valve open
X valve closed

Figure 10.18 Arrangement for de-airing water under vacuum, with summary of valve positions for six operations

from it to minimise the possibility of water being sucked into the vacuum line. It is desirable to fit a water trap on the vacuum line close to the de-airing vessel.

When drawing off water from the storage vessel, the space above the water must be opened to atmosphere. At normal laboratory temperatures the water will re-absorb very little air during the first few hours' exposure, but after that the vacuum should be applied again. Water should be drawn off from the bottom of the vessel, where the concentration of any dissolved air is likely to be lowest.

The arrangement of valves suitable for operating the de-airing unit is shown in Figure 10.18 and the table beneath indicates the valve positions required for six different operations: fill reservoir; fill vessel; de-air; fill constant head tank; saturate sample; run constant head test.

2. De-aerator

An apparatus for rapidly removing air from water, known as the Nold de-aerator (see Figure 10.19(a)) was developed by the Walter Nold Company of Natick, MA, USA in 1971. It consists of a vacuum chamber, fitted with a magnetically motorised impeller. Rotation of the impeller at high speed causes cavitation and rapid release of air bubbles, which are removed by the vacuum line. The vessel holds about 6 litres of water, which it is claimed can be made virtually air-free within about 5 min. The de-aired water is transferred to a supply tank in which it should be sealed from the atmosphere until required for use.

A more recent apparatus, shown in Figure 10.19(b), removes air from the water by continuous spraying in the vacuum chamber, which holds 14 litres. A pump enables de-aired water to be extracted when required, without having to release the vacuum. Water is

(a) (b)

Figure 10.19 (a) Nold de-aerator (photograph courtesy of Geotest Instrument Corp.); (b) automatic de-airing apparatus providing a continuous supply

automatically replenished by means of an internal valve and level control, thereby providing a continuous supply.

3. Boiling

Dissolved air can be removed from water by boiling. The water should be allowed to cool in a vessel in which it can be sealed from the atmosphere, and which is capable of withstanding full atmospheric external pressure. The vessel should be well filled, to exclude as much air as possible.

Boiling reduces the dissolved air to a level which is low enough for most practical purposes, but if it has contact with air, hot water can dissolve air again as it cools. The solubility of air in water increases appreciably with decreasing temperature.

4. Sand filter

During the permeability test the presence of a small amount of dissolved air in the water is of no consequence provided that it does not come out of solution and create bubbles, either in the sample, the filters, or the connecting tubes. Grains of sand, especially fine grains, tend to trap air bubbles, and this effect can be utilised by passing the water through a layer of sand immediately before it enters the permeameter. The grain size of the sand filter should be similar to or finer than that of the sample under test.

After a time the air trapped in the filter reduces the rate of flow, and the filter loses its capacity to absorb more air. Two or more filters can be used in parallel to provide a sufficient rate of flow, or to allow for one to be replaced while the other remains in use. A suggested arrangement is shown in Figure 10.20.

5. Warming

A simple way of minimising the formation of air bubbles in the test sample without the use of vacuum is to maintain the water in the supply vessel at a slightly higher temperature than the

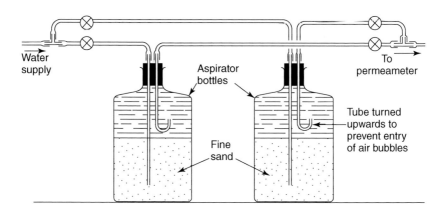

Figure 10.20 Sand filters for removing air bubbles from water

sample. A current of warm air around the supply vessel may be sufficient. Warming the water initially will take some of the air out of the solution, and the slight cooling as the water enters the sample will ensure that no more air is released as bubbles, because the air solubility of cold water is greater than that of warm water. However, if the testing period is prolonged the effect of pre-warming may be counteracted as the sample itself gradually warms up to the supply temperature. The change in temperature must also be taken into account when assessing the results.

A device for providing a continuous flow of de-aired water by using an electronically controlled electric heater was described by Klementev and Novák (1978), but the output (up to 0.5 l/h) would be too small for practical use in soil testing.

6. Allowing to stand

If the water to be used in the test is allowed to stand for an hour or more in a header tank at room temperature, it will lose some of its air. This might be sufficient for tests using relatively high hydraulic gradients with materials of high permeability, when large volumes of water are required.

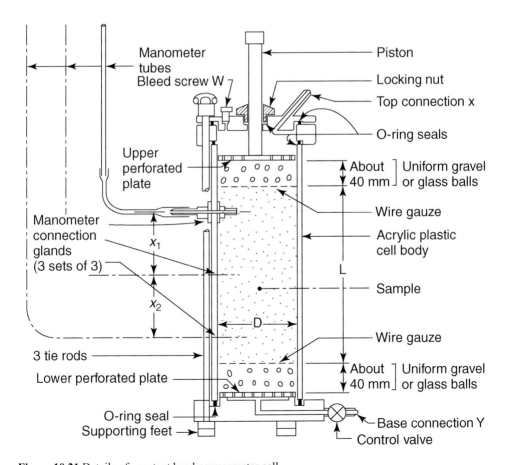

Figure 10.21 Details of constant head permeameter cell

10.6.3 Constant head test in permeameter cell (BS 1377:Part 5:1990:5, and ASTM D 2434)

This test procedure is for the determination of the coefficient of permeability of non-cohesive soils, especially sands, using the constant head permeameter in which the flow of water is laminar.

Figure 10.22 75 mm permeameter cell for constant head test

Apparatus

1.   Permeameter cell, fitted with lockable loading position, perforated top and base plates, flow tube connections, piezometer glands and connections, air bleed valve, control valve, sealing rings. Details of a typical cell are shown in Figure 10.21. The control valve (at Y) should be capable of providing a fine adjustment of the rate of flow of water. An ordinary on/off valve is not suitable.

     The usual type of cell is 75 mm diameter and 260 mm long internally between perforated plates. The larger cell is 114 mm diameter and 460 mm long internally. (The smaller cell is shown in Figure 10.22).

2.   Glass manometer tubes (one for each piezometer point) mounted on a stand with a graduated scale, marked in mm and cm. At least three piezometer points in a vertical line are required, in order to provide a check on the uniformity of the hydraulic gradient.

3.   Rubber tubing (including vacuum tubing) for water flow and manometer connections, fitted with pinch clips or valves.

4.   Uniform fine gravel, or glass balls, for end filter layers. The largest grain size which can pass through a collection of uniform spheres of diameter D is $D/6.47$ (Lund, 1949), but natural sands containing particles down to $D/8$ are usually stable. Glass balls of 3.5 mm diameter could therefore be placed directly in contact with sands containing particles down to 0.4 mm, but in practice a separating gauze is normally used.

5.   Two discs of wire gauze, of the same diameter as the internal cell diameter. The aperture of the gauze should not be greater than the D_{85} size of the sand, i.e. up to 85% of the sand grains may be smaller than the gauze aperture, provided the layer

thickness is 50 mm or more. A 63 µm mesh is therefore adequate for sands and silty material containing not less than 15% fine to medium sand.

6. Two porous stone or sintered bronze discs of the same diameter (required only if the gauze aperture exceeds the D_{85} particle size).

7. Measuring cylinders: 100 ml, 500 ml and 1000 ml.

8. Constant head reservoir, with a means of adjusting the water level from about 0.3 m to about 3 m above bench level (depending upon available headroom) fitted with a supply valve.

9. Discharge reservoir with overflow to maintain a constant water level.

10. Supply of clean water, sufficiently air-free for the soil to be tested (see Section 10.6.2). A suitable de-aired water supply system is shown in Figure 10.18.

11. Small tools: funnel, tamping rod, scoops, etc.

12. Thermometer reading to 0.5°C.

13. Stop-clock (minutes timer).

14. Steel rule graduated to 0.5 mm.

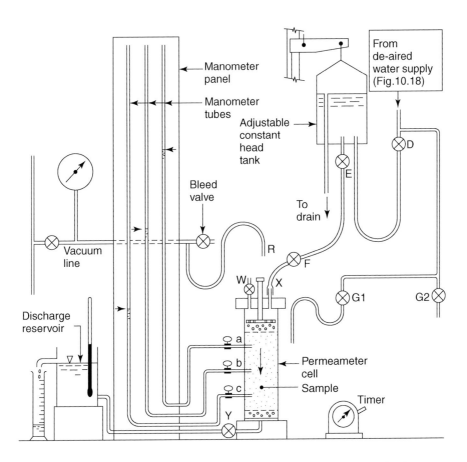

Figure 10.23 General arrangement for constant head permeability test (downward flow)

15. Internal callipers.
16. Suitable balance.

The general arrangement diagram of the test system is shown in Figure 10.23, and the apparatus is shown in Figure 10.24. The control valve is fitted at the outlet (at the base for downward flow, shown as valve Y in Figure 10.23), not on the inlet line. This allows any air bubbles released as a result of the pressure drop across the flow restriction to escape to atmosphere, instead of being taken into the sample. The surface of the discharge reservoir should be slightly above the outlet end of the sample so that the water in the sample is always under a small pressure.

Procedural stages

1. Prepare ancillary apparatus
2. Prepare permeameter cell
3. Select sample
4. Prepare sample
5. Place sample in cell: (a) by compaction or (b) by dry pouring or (c) by pouring through water
6. Assemble cell
7. Connect up cell
8. Saturate and de-air sample
9. Connect up for test
10. Run test
11. Repeat test
12. Dismantle cell
13. Calculate results
14. Report

Test procedure

1. *Preparation of ancillary apparatus*

The constant head device is connected to the permeameter as shown in Figure 10.23. The connecting tube, R, which is to be subjected to vacuum, should be rigid enough to withstand the external atmospheric pressure without collapsing. It is essential to ensure that all joints are airtight.

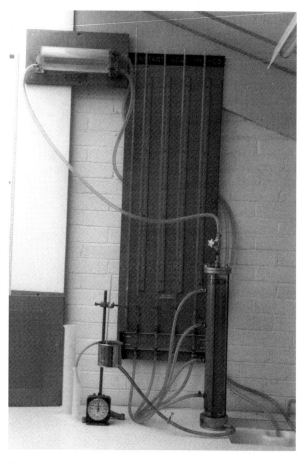

Figure 10.24 Apparatus for constant head permeability test

If a vacuum is applied after filling the system with de-aired water, any leak will be indicated by the formation of bubbles.

At the start of the test the water surface in the constant head reservoir should be at a low level, i.e. at a height above the discharge reservoir which is not greater than about half the sample height, so as to give a hydraulic gradient across the sample of not greater than 0.5. The sudden application of a high gradient can cause disturbance or instability within the sample right from the start. Alternatively, if the sample has a high permeability, the hydraulic gradient can be controlled to a certain extent by restricting the rate of flow with the control valve at the cell outlet (valve Y in Figure 10.23).

2. Preparation of permeameter cell
The main features of the standard cell are shown in Figure 10.21. Remove the top plate assembly from the cell. Measure the following dimensions if they are not already known:

    Mean internal diameter (D mm) (using internal calipers at several positions)
    Distance between centres of each set of manometer connection points along the axis of the cell (L mm)
    Overall internal length of cell (H_1 mm)
The following can be calculated:

Area of cross-section of sample, $A = \pi D^2 / 4$ mm^2

Volume of the permeameter cell, $V = AH_1 / 1000$ cm^3
Approximate mass of soil required, if placed at a density
ρ Mg/m^3, mass $= \rho\, AH_1 / 1000$ g

Check that the manometer connection glands are watertight and that the piezometer tips extending into the cell are not damaged or blocked.

Place the balls or graded gravel filter material in the bottom of the cell to a depth of about 40 mm, level the surface and place a wire gauze disc on top. If the sample contains more than about 80% of particles finer than the mesh of the wire gauze, place a porous stone or sintered bronze disc over the gauze.

3. Selection of sample
The soil sample used for the test should not be dried. Any particles larger than one-twelfth of the internal diameter of the cell should be removed by sieving. The sieve aperture sizes for removing coarse material are as follows:

Diameter of permeameter cell	75 mm	114 mm
Sieve aperture size	6.3 mm	10 mm

The ASTM allows maximum particle sizes up to one-eighth to one-twelfth of the cell diameter, depending on percentages retained on the 2 mm and 9.5 mm sieves.

The material is then reduced by the usual riffling process to produce several batches of samples each about equal to the mass required to fill the permeameter cell, as determined in stage 2. At least two batches should be prepared, for duplicate tests, but it may be necessary to perform several tests covering a range of densities. After thoroughly mixing each batch, take a small portion for the measurement of moisture content (w_a%), and weigh the remainder to the nearest gram. Each batch should be sealed in a bag until required.

4. *Preparation of test sample*

The height of the sample as tested should not be less than twice its diameter. The sample may be placed in the permeameter cell by one of the following three methods:

    (a) Compacting by rodding

    (b) Dry pouring

    (c) Placing under water

Method (a) is used when the test is to be carried out at a dry density in the medium to high relative density range (Volume 1 (third edition), Section 3.4.4), which includes the maximum dry density corresponding to BS 'light' compaction. Low relative densities can be produced by methods (b) and (c). For fine sands and silty sands, method (c) is the only satisfactory means of avoiding entrapped bubbles of air. Methods (a) and (c) are included in BS 1377:Part 5:1990, Clause 5.4.2. Application of vacuum in method (b) is similar in principle to the method described in ASTM D 2434. It might be necessary to apply a partial vacuum during methods (a) and (c) to eliminate persistent air pockets, but this should be done with care.

Dry densities close to a relative density of 1 (i.e. the maximum possible) are difficult to obtain in the permeameter cell. A small vibrator of the type referred to in Section 9.1.2 (see Figure 9.14), fitted with a suitable extension foot, may be a useful aid.

Whichever method is used, a process of trial and error may be necessary in order to obtain a density close to the required value, which is why a series of tests covering a range of densities is preferred to a single test.

The preparation of the soil for each of the above methods of placing is described below.

(a) *For compaction* Compaction by rodding will normally be carried out at a moisture content equal to or close to the optimum value, which can be obtained from the test described in Volume 1 (third edition), Section 6.5.3. The quantity of water to add to the sample to bring it to the required value is calculated as follows: where m_a = mass of riffled sample (g); $w_a\%$ = moisture content of as-received sample; and $w_p\%$ = moisture content required for compaction. The amount of water to be added is

$$= \left(\frac{w_p - w_a}{100 + w_a} \right) m_a \text{ g (or ml)}$$

The water should be thoroughly mixed into the soil, and a little extra water (about 5–10 ml per kg of soil) should be added to allow for evaporation loss while mixing. As a check, take a small representation portion for determination of the actual moisture content (w%) after mixing.

Weigh each batch prepared as above to the nearest 1 g (m_1), and seal in a bag until required. A separate batch, not to be used for the permeability test, should be prepared and used for a trial test to determine the degree of compaction required to obtain the density for the test.

(b) *For dry pouring* If necessary, partially air-dry the soil until individual grains do not stick together when poured slowly. Oven drying should be avoided because this may make it more difficult to remove bubbles of air later. Determine the final moisture content (w%), weigh each batch (m_1 grams), and seal in a bag until required.

(c) *For placing under water* Weigh each batch of soil prepared as in (a) above (m_1 grams), and determine the moisture content (w%). Thoroughly mix the whole sample from each batch with enough de-aired water to submerge it in a large container such as a bucket.

5. *Placing sample in cell*

(a) *Compaction* Compact the soil in the permeameter cell in at least four layers. Place the first layer of soil from the sealed container carefully on the wire gauze at the bottom of the permeameter. Ideally, a small quantity at a time should be lowered into the cell in a small container on a wire or string, which is emptied by tilting. However, segregation of particles is unlikely in 'damp' soil if it is poured in carefully.

Compact the soil with the appropriate number of blows of the tamping rod or hand compactor, evenly distributed. Avoid damaging the tips of the piezometer connections which extend into the cell. The compacted thickness should be about one-quarter of the final height of the sample in the cell (or the appropriate fraction if there are to be more than four layers).

Repeat the process with each of the remaining layers, lightly scarifying the top of each compacted layer before adding the next. Trim the surface of the final layer level. The finished surface should allow for an upper layer of glass balls or gravel filter of about 50 mm thickness.

Weigh the material left over, together with any which may have been spilled (m_2). The mass of soil used in the sample, m, is equal to $(m_1 - m_2)$ grams.

(b) *Dry pouring* Use a funnel fitted with a length of flexible tubing, long enough to reach the bottom of the permeameter cell, for pouring the sample (see Figure 10.25). Maintain a steady rate of pouring and move the tube around in a spiral motion from the periphery towards the centre as the cell is filled, keeping the end of the tube about 15 mm above the sand already placed. Keep the funnel topped up. This procedure should minimise the tendency to segregation.

Continue pouring until the surface of the sand is at the correct level to allow for about 50 mm of the filter material. Level the surface carefully with the minimum of disturbance, and do not jolt the cell or agitate the sample in any way if a low density sample is being prepared.

In general, for a uniform sand a fast rate of pouring with a low drop will give a low density, and a slower rate of pouring with a higher drop will give a greater density (Kolbuszewski, 1948).

Weigh the material left over, together with any which may have been spilled (m_2). The difference $(m_1 - m_2)$ grams gives the mass of soil in the sample.

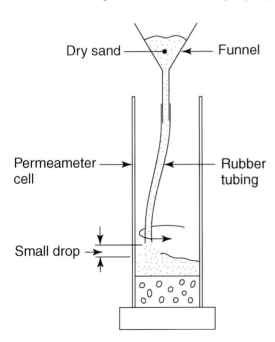

Dry sand —
Funnel
Permeameter cell —
Rubber tubing
Small drop →

Figure 10.25 Pouring dry sand into permeameter cell

(c) *Placing under water* Connect the valve on the base of the permeameter cell to the de-aired water supply and open the valve to allow water to enter the cell to about 15 mm above the wire gauze or porous disc. Avoid trapping air bubbles.

Support a large funnel, fitted with a bung attached to a string or wire, and a length of flexible tubing, over the top of the cell so that the tubing reaches to the surface of the water in the cell (see Figure 10.26). Pour the mixture of soil and water from the container into the funnel.

Remove the bung carefully, releasing the soil and water mixture into the cell. Raise the funnel so that the end of the tubing is just at the water surface. Maintain the water surface at about 15 mm above the surface of the placed soil by admitting more water through the base valve. Continue until the required amount of soil has been deposited in the cell.

Retain any soil not used, dry it in the oven and weigh it (m_3). Calculate the mass, in grams, of dry soil (m_D) in the sample from the equation

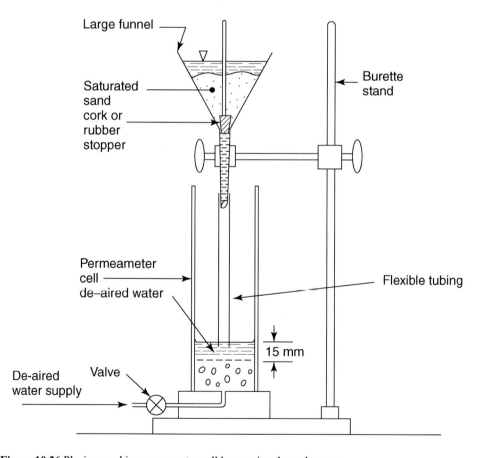

Figure 10.26 Placing sand in permeameter cell by pouring through water

$$m_D = m_1 \times \left(\frac{100}{100 + w} \right) - m_3$$

in which w is the actual measured moisture content of the test sample.

The saturated sample in the cell will be of uniform density in a loose condition. Do not jolt the cell or disturb the soil if the low density is to be maintained for the test.

During either of the pouring procedures (b) and (c), a higher density may be obtained by rodding or vibrating the soil as it is being poured or by placing a layer at a time and rodding each layer.

6. Assembling cell

Place a second porous disc (if necessary) and the second wire gauze disc on top of the soil, followed by about 40 mm thickness of glass balls or gravel filter material (see Figure 10.21). The level of the top surface of the filter should be within the limits required to accommodate the top plate. Avoid disturbance of the sample.

Slacken the piston locking collar on the cell top, pull the piston up as far as it will go, and re-tighten the locking collar. Fit the cell top on the cell and tighten it down into place by progressively tightening the clamping screws. Release the piston locking collar and push the piston down until the perforated plate beds on to the filter material. Hold it down firmly while the locking collar is re-tightened.

Connect valve Y at the base of the cell (see Figure 10.27) to the de-aired water supply (for placing methods (a) and (b)), with valve Y closed. Connect each piezometer gland to

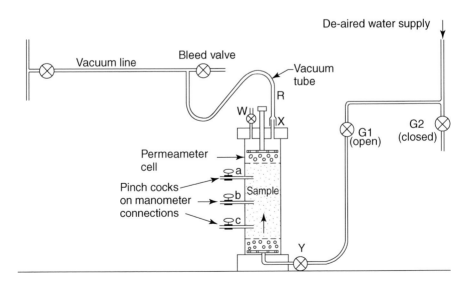

Figure 10.27 Connections to permeameter cell for saturating sample under vacuum

a manometer tube, using flexible tubing, and close them off with pinch cocks (a, b and c in Figure 10.27) adjacent to the cell.

7. Measurements

Measure the distances between the upper and lower wire gauzes or porous discs at three or more positions round the perimeter to the nearest 1 mm, using a steel rule. The average of these measurements gives the mean initial height of the sample, L_1 (mm).

This enables the placement density (ρ) to be calculated from the equation

$$\rho = \frac{m}{AL_1} \times 1000 \ \text{Mg/m}^3$$

where m is the mass of soil in the sample (g) and A is the area of cross-section of the permeameter cell (mm²).

The dry density is calculated from the equation

$$\rho_D = \frac{100\rho}{100 + w} \ \text{Mg/m}^3$$

where $w\%$ is the actual moisture content of the sample. If dry soil is used (method (b)), then $m = m_d$, $w = 0$ and $\rho = \rho_d$.

The measurements and calculations provide a check on whether or not the desired dry density has been achieved.

8. Saturation of sample

Placing procedure 5(a) Open the top outlet X, and the air bleed W, (see Figure 10.27) to atmosphere.

Allow de-aired water to percolate slowly upwards through the sample, by regulating valve Y, until water emerges first from the air bleed W, which is then closed, and then from the top connection at X. Ensure that the water level rises slowly enough not to cause sample disturbance or piping.

Close valve Y.

Placing procedure 5(b) Connect the top outlet of the cell, X, to the vacuum line, fitted with a water trap, using rigid plastic or thick-walled rubber tubing, R (see Figure 10.27). Close the air bleed screw, W, on the cell top. Apply a low vacuum gradually to the top of the permeameter cell by adjustment of the vacuum line and air bleed valves. When this is established open valve Y at the cell base slightly so that water can percolate slowly upwards into the sample. Avoid any sudden movement of water, which may cause disturbance of the sample. When the water level reaches about one-third of the height of the cell, close the inlet valve and leave under partial vacuum for about 10 minutes, or longer if air bubbles are still escaping. Repeat this operation when the cell is two-thirds full; when the surface of the sample is just submerged; and finally when the filter layer is completely submerged. Then gradually apply the highest vacuum attainable, and maintain it until no further bubbles can be observed. Check that no air is entering through connections, including the manometer connections.

Open valve Y slightly until the space above the top filter is completely filled with water, then close it. Turn off the vacuum and disconnect the vacuum tubing, R, from the cell outlet, X. Open valve Y and open the air bleed, W, on the cell top, until de-aired water displaces the last pocket of air from the top of the cell. When water emerges from the bleed valve, close it. Allow water to enter the cell until it emerges from the outlet nozzle, X, then close valve Y.

The sample and cell should by then be completely de-aired.

Placing procedure 5(c) This procedure results in a saturated sample. Continue filling the permeameter cell by very carefully allowing de-aired water to enter from the bottom, as for placing procedure 5(a).

9. *Connections for test*

(i) Immediately after saturation, slacken the piston locking collar and push the piston down to ensure that the perforated plate is still bedded firmly on the filter material, as in step 6 above. Re-tighten the locking collar.

(ii) Measure the length of the sample again, as in step 7 above, and record the new average measurement, L (mm), as the height of the sample as tested.

(iii) Connect the de-aired water supply to the cell top connection at X, and connect valve Y to the base of the discharge reservoir. Ensure that no air is trapped when making connections. Arrange the discharge reservoir so that the overflow feeds into a glass measuring cylinder.

(iv) Set the constant head inlet reservoir so that the water level is a little above the top of the permeameter cell, and open the supply valves E and F (see Figure 10.23). An initial hydraulic gradient of about 0.2 is often sufficient, but a higher value might be more suitable for fine-grained or densely compacted soil samples.

(v) Open the pinchcocks on the manometer tube connections one by one, and allow water to flow into the manometer tubes. Ensure that no air is trapped in the flexible tubing. Squeezing the tubing with the pinchcocks re-closed should expel any air to atmosphere. The water levels in all manometer tubes should come to rest at the level of the inlet reservoir surface.

The arrangement of the apparatus is shown diagrammatically in Figure 10.23, in which the designation of valves is the same as in Figures 10.27 and 10.18. A photograph of a typical assembly is shown in Figure 10.24.

The arrangement shown is for downward flow through the sample. For upward flow the top and bottom connections are reversed so that the de-aired water supply is connected to the base inlet and the regulator valve Y, leading to the discharge reservoir, is connected to the top outlet.

10. *Running the test*

Open the control valve Y (see Figure 10.23) and adjust it to regulate the flow of water through the sample such that the hydraulic gradient is appreciably less than unity. Allow water to flow through the sample until the conditions appear to be steady and the water levels in the manometer tubes become stable. Adjust valve D on the supply line to the constant head device so that there is a continuous small overflow; if this is excessive, the de-aired water will be wasted.

To start a test run, empty the measuring cylinder and start the timer at the instant the measuring cylinder is placed under the outlet overflow. Record the clock time at which the first run is started.

Read the levels of the water in the manometer tubes (h_1, h_2, etc.) and measure the water temperature ($T°C$) in the outlet reservoir. When the level in the cylinder reaches a predetermined mark (such as 50 ml or 200 ml) stop the clock, record the elapsed time to the nearest second, empty the cylinder, and make four to six repeat runs at about 5 min intervals. If the rate of flow is quite fast, such that the measured volume is collected in 30 s or less, use a stopwatch and observe elapsed times to the nearest fifth of a second. An alternative procedure, if the rate of flow is fairly small, is to remove the measuring cylinder after a fixed time, preferably an exact number of minutes, and to read the volume of water received by the cylinder in that time, or to determine the volume from weighings. Four to six such measurements should be made, together with observations of manometer levels and water temperature, as above.

Calculate the rate of flow, q (ml/min), for each reading, and plot a graph of q against the time from the start of the first run. The rate of flow may vary at first, but will usually decrease slightly to reach a constant steady state value, which can be derived from the graph. An example is shown in Figure 10.28(a).

A more positive determination of the steady flow rate can be obtained by plotting the rate of flow, q, at any time t (min), against $1/\sqrt{t}$ (Al-Dhahir and Tan, 1968). The example referred to above is re-plotted in this way in Figure 10.28(b). In this plot the earliest observations lie furthest from the q axis, and steady flow is approached as t becomes very large, i.e. as $1/\sqrt{t}$ approaches zero. The curve is extrapolated back to the q axis, where the intersection clearly defines the long-term steady rate of flow.

If any solid particles are observed in the water emerging from the sample, this should be reported.

Measure the length of the sample after the tests. If the length has decreased due to collapse of the grain structure, the voids ratio relevant to the test should be calculated from the new length.

Possible difficulties If there is a uniform pressure gradient throughout the length of the test sample, this should be indicated by a uniform difference in water level between one manometer tube and the next, assuming that the points to which they are connected are spaced equally apart.

Sometimes the manometers do not indicate a uniform gradient, and manometers at two consecutive levels may give the same, or almost the same reading. This is due to non-uniform compaction of the soil locally providing a relatively free passage for water, or to the formation of a 'pipe' within the soil structure in that region. The latter is more likely to occur when water flows upwards. A 'pipe' adjacent to the wall of the cell can be observed, but not one which forms inside the sample. The permeability can be calculated only on the basis of the hydraulic gradient in the section between apparently reliable piezometers. The only satisfactory way to deal with the discrepancy is to remove the sample and re-compact it.

Sometimes a piezometer gives an unrealistically low reading because the tip in the sample has been blocked by a particle of soil. It may be possible to free the blockage by applying a pressure towards the sample from the manometer tube (perhaps by blowing down it) to momentarily force water back into the sample.

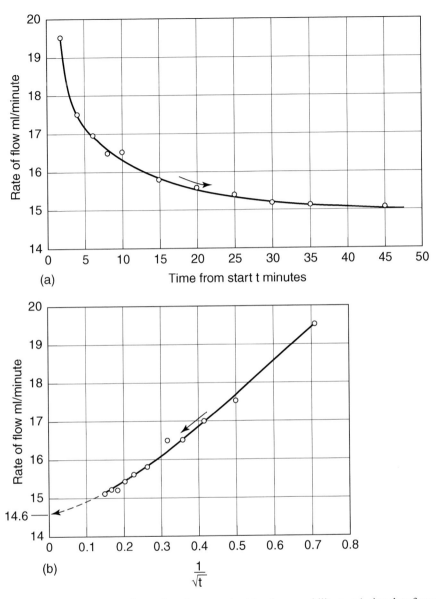

Figure 10.28 Graphical plots of rates flow from constant head permeability test (using data from Figure 10.29): (a) q plotted against time t from start; (b) q plotted against 1/√t

In some soils, such as uniformly graded sand containing a little silt, movement of fine particles through the voids between larger particles may occur even under a hydraulic gradient less than the critical value. This effect can be observed by close examination through the perspex wall of the cell, and is confirmed by the turbidity of the discharge water. The permeability then measured is that of the remaining coarser material only.

Loss of fines can be prevented by placing a suitably graded filter between the sample and the glass balls or gravel filter layer, but after a time this is likely to become partially clogged by a layer of fines at the interface, thereby restricting the flow of water. However, the manometer tubes still indicate the actual hydraulic gradient within the sample corresponding to the observed rate of flow.

11. *Repeat tests*

To run the test at an increased hydraulic gradient, increase the rate of flow by opening the control valve Y further or by raising the level of the constant head tank. Repeat the procedure given in stage 10 when conditions have become steady.

Carry out further repeat tests as appropriate over the required range of hydraulic gradients. The rate of flow should increase approximately uniformly with the overall hydraulic gradient.

If flow through the sample is downwards, the test can be extended to quite high hydraulic gradients without loss of stability. If flow is upwards, instability of the soil may be observed when the hydraulic gradient reaches the critical value (i.e. $i = i_c$). This may be accompanied by the formation of 'pipes' within the sample, or by heave, and Darcy's law is no longer valid (see Section 10.3.8). If this effect is to be investigated, several runs can first be carried out at hydraulic gradients less than ic to obtain the coefficient of permeability under Darcy's law conditions, before the sample is allowed to suffer disturbance.

If it is required to measure the coefficient of permeability over a range of voids ratios in order to obtain a relationship between voids ratio and permeability, the whole procedure should be repeated from step 5 using separate batches of the prepared soil compacted to different densities.

If the sample was placed at a low or medium relative density, it might be possible to achieve a higher density without removing it by tapping the cell with a wood mallet. The top plate should be brought back firmly into contact with the top of the sample, and the new length of the sample is measured, enabling the new density and void ratio to be calculated. A repeat test can then be carried out at the new void ratio.

12. *Dismantling cell*

When several consistent sets of readings have been obtained, close the inlet valve F (see Figure 10.23) at the top of the cell and wind down the constant head tank to a low level. Disconnect the inlet pipe and open the valve Y at the base and the air bleed screw W, allowing water to drain out of the cell at the lower end (see Figure 10.27).

Remove the cell top, take out the filter layer and wash away any adhering fines. Empty the sample out of the cell, and take out and wash the lower filter layer. Clean out the cell, making sure that no fine material is left in the piezometer glands and connecting tubing.

13. *Calculations*

If a quantity of water Q ml flows through a sample in a time of t min, the mean rate of flow q is equal to Q/t ml/min or $Q/60t$ ml/s.

If the manometer readings at three or more gland points along the vertical axis of the sample indicate that the hydraulic gradient is reasonably uniform, calculate the hydraulic gradient between the outermost gland points form the equation

$$i = \frac{h}{y}$$

where h is the difference in height between the two outermost manometer levels (mm) ($h = h_a - h_c$) in Figure 10.5) and y is the distance between the corresponding gland points (mm) ($y = x_1 + x_2$ in Figure 10.21, x_1 and x_2 usually being equal).

If the area of cross-section of the sample is equal to A mm^2, the permeability kT (m/s) of the sample at $T°C$ is calculated from Equation (10.5)

$$k_T = \frac{Q}{60\ Ait}$$

The permeability at 20°C, if the test was not run at that temperature, is calculated from Equation (10.9) using the graph in Figure 10.4.

Calculate the density, ρ, and dry density, ρ_D (Mg/m^3) of the sample as tested, using the equations in step (7) but with the length L (from step (9)(ii)) in place of L_1.

Calculate the void ratio, e, of the test sample from the equation

$$e = \frac{\rho_s}{\rho_D} - 1$$

where ρs is the particle density (Mg/m^3).

If the coefficient of permeability was determined at several different densities, plot the derived values of k, to a logarithmic scale, against corresponding values of dry density or void ratio, e, to a linear scale.

14. Reporting results

The mean permeability of the test sample is reported to two significant figures, in the form $k_T = 2.3 \times 10^{-4}$ m/s. The temperature of the test, $T°C$, is also reported, together with the range of hydraulic gradients applied.

The condition of the sample should form part of the test result, which should include:

- Particle size distribution curve
- Proportion of oversize material removed from original sample before test
- Method of placing sample
- Dry density
- Void ratio (with a statement whether the particle density was measured or assumed)
- Observations relating to migration of particles, or any form of instability

The following should also be reported:

- Dimensions of the permeameter
- Method of placing, compacting and de-airing the test sample
- Whether or not de-aired water was used
- If appropriate, state that the test was carried out in accordance with Clause 5 of BS 1377:Part 5:1990
- If the sample was tested at various voids ratios, the permeability corresponding to each voids ratio should be tabulated

A typical set of laboratory data and tests results is given in Figure 10.29.

Constant Head Permeability Test

Location .... *Bromsbury* .... Sample No .... *P2-8* ....

Operator .... *G.G.B.* .... Date .... *6.2.79* ....

Soil description .... *Fine to medium light brown sand* ....

Method of preparation .... *Placed dry in 3 layers, lightly tamped* ....

Sample diameter ....*75*.... mm   area A ...*4418*... mm^2 dry mass ....*1471*....... g

       length ....*164*.... mm   volume ...*724.6*... cm^3 dry density ....*2.03*.... Mg/m^3

S.G. assumed ....*2.65*.......   voids ratio = $\dfrac{\rho_S}{\rho_D} - 1 =$ ....*0.305*.......

Heights above datum: inlet ....*535*.... mm   manometer   a ....*493*.... mm

               outlet ....*360*.... mm   b   *452* mm

Temperature ....*19.5*.......... °C   c ....*406*....... mm

                                     Head difference ....a to c ....*87*.... mm

                                       Distance between a to c ....*100*.... mm

Flow downwards   Hydraulic gradient $i =$ ..$\dfrac{87}{100} = 0.87$..

Readings

Time from start min.	Time interval t min.	Measured flow Q ml	Rate of flow q ml/min	$\dfrac{1}{\sqrt{t}}$	Remarks
2	2	39	19.5	0.707	
4	2	35	17.5	0.5	
6	2	34	17	0.408	
8	2	33	16.5	0.354	
10	2	33	16.5	0.316	
15	5	79	15.8	0.258	
20	5	78	15.6	0.224	
25	5	77	15.4	0.2	
30	5	76	15.2	0.183	
35	5	76	15.2	0.169	*Steady state rate of flow*
45	10	151	15.1	0.149	*(from graph) q = 14.6 ml/min*

Permeability k = $\dfrac{q}{Ai \times 60}$ = $\dfrac{14.6}{4418 \times 0.87 \times 60}$ = 6.33×10^{-5} m/s

Temperature correction *negligible*

Dry density ....*2.03*.......... Mg/m^3

Voids ratio ....*0.305*..............

Permeability (20°C) *6.3 × 10^{-5}* m/s

Figure 10.29 Typical data from constant head permeability test

10.6.4 Test under constant axial stress

The plunger fitted to the top of the permeameter cell can be used to apply a constant axial stress to the sample during the test, at a known value. A loading yoke carrying a weight hanger is hung from the top end of the plunger, and weights are added to the hanger to give the desired axial stress. The arrangement is shown in Figure 10.30. The apparatus described in ASTM D 2434 uses a spring to provide an axial force of 22–45 N.

If total mass of hanger and weights = m kg

Area of cross-section of sample = A mm²

$$\text{Axial stress} = \frac{mg}{A} \times 1000 = 9807 \times \frac{m}{A} \ \text{kN}/\text{m}^2$$

The piston should initially be near the upper end of its range of movement, so that it is free to move downwards if collapse of the grain structure leading to a decrease in volume of the sample occurs.

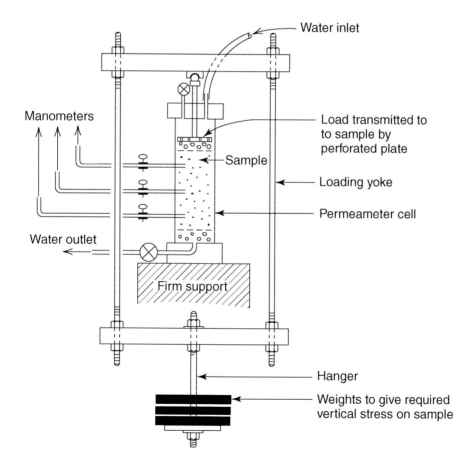

Figure 10.30 Constant head permeameter cell fitted with loading yoke for test under known axial stress

Figure 10.31 Permeameter cell 16 in diameter for tests on gravels

10.6.5 Permeameter cell for gravels

For measuring the permeability of gravel soils in the laboratory, a cell much larger than the 114 mm diameter cell referred to in Section 10.6.1 is required. A suitable type which was used on a large dam construction project for testing various types of gravel fill and filter zone materials is shown in Figure 10.31. This permeameter cell was 16 in (406 mm) diameter, and could take a sample up to 34 in (964 mm) long containing particles up to 75 mm. The sample could be compacted or vibrated into the cell, which was designed to pivot about its mid-height on the support frame to facilitate emptying and cleaning.

A header tank of some 900 litres was specially installed to supply the permeameter, together with a constant head device consisting of a 200 mm diameter pipe fitted with several overflow levels, so that the inlet head could be varied. Water was led to and from the permeameter by means of 75 mm diameter flexible kink-free rubber pipes.

Piezometers consisted of ordinary glass tubes, 27 in all, mounted on a panel, and connected to the permeameter by 2 mm bore nylon tubing. Each piezometer tube passed through a gland in the wall of the permeameter and extended to the centre-line of the cell, surrounded by 20 mm of fine sand. The end 50 mm of the nylon tube was cut in half along the centre-line and wrapped with a 63 μm wire mesh, as shown in Figure 10.32.

The principle of the test was the same as for the standard permeability test on sands. The main applications of this permeameter, in addition to permeability measurements, were to examine the stability of some of the gravel fill materials, with a view to assessing the possible loss of fines due to flow of water, i.e. suffusion, and to assess visually the behaviour of particles at the interface between two materials such as gravel fill and a filter layer, as described in Section 10.6.6. The large scale of the apparatus provided a fairly realistic representation of the likely behaviour of these materials in-situ.

10.6.6 Tests on filter materials

The constant head permeability apparatus can be used to obtain a visual indication of the behaviour of particles in the vicinity of the surface of contact between a filter material and the soil which it is intended to protect against erosion, referred to as the 'base material'. The principle applies equally to the surface between one filter layer and another in a multiple filter system. From these tests an assessment of the stability of the interface between the two materials can be made, for comparison with a theoretical assessment based on the particle size distribution of each material (see Section 10.3.10).

The procedure outlined below is based on carefully planned experimental work described by Lund (1949). It is not intended as a standard laboratory procedure, but is presented as a good approach for a laboratory investigation.

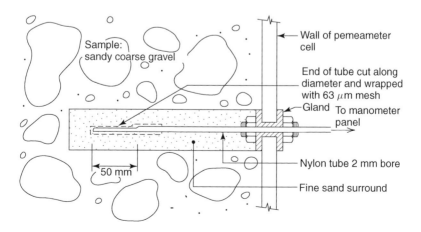

Figure 10.32 Details of piezometer tips in large permeameter cell

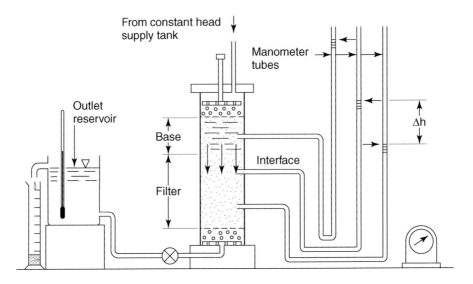

Figure 10.33 Arrangement of permeameter for tests on filter materials

The constant head apparatus is arranged as shown in Figure 10.33. The two materials are placed in the permeameter cell to form a composite sample, about two-thirds of which comprises the filter material. The interface surface should be about halfway between the middle and upper manometer tapping point, and the latter should be at about the centre of the base material. Each material is placed and compacted to its appropriate dry density, using a suitable method (Section 10.6.3, stage 5). The cell is fitted together as described in Section 10.6.3, stages 6 and 7, with downward flow as shown in Figure 10.33.

The composite sample is saturated by allowing water to percolate upwards (stage 8), and during this process (and throughout the test) the interface surface is observed and any migration of particles is described. After saturation, the cell wall at the interface is tapped gently, and any further movement of particles is observed. The cell is connected up as in stage 9.

A permeability test with downward flow of water is then performed, as described in Section 10.6.3, stages 1) and 11. Initially a low hydraulic gradient is applied, and is increased in increments up to the required value for the test, which may be several times the likely maximum field value. At each stage any observed migration of particles is recorded. The manometer readings are recorded when a steady state has been reached, so that the permeability of the filter can be calculated (see Section 10.6.3, stage 13). At the completion of each stage the cell is again tapped, and any further movement of particles is observed and recorded. If there is any visible movement the permeability is measured again.

In this type of test, visual descriptions of particle movement at each stage are at least as important as measurements of permeability values, and both should be faithfully reported. The suitability of the filter material for use in contact with the base material may be summarised in general terms as follows:

- Stable: no appreciable change in permeability over the range of hydraulic gradients applied. No visible migration of base material into filter on being tapped. Base material penetrates no more than 5 or 10 mm, provided that permeability remains unaltered
- Unstable: reduction of permeability after tapping, or with increasing hydraulic gradient. Visible loss of base material into filter
- Completely unstable: most or all of base material washed into filter

10.6.7 Horizontal permeameter

A horizontal permeameter for determining the permeability of road sub-base layers containing particles up to 30 mm size was developed by the Department of Transport (Jones and Jones, 1989). The apparatus and its use are described in detail in Department of Transport Advice Note HA 41/90 (1990).

Essentially the permeameter consists of a rectangular galvanised steel box, fitted with connections to allow water to flow under controlled heads, into which the material can be compacted at the required density and moisture content in four or five layers. The resulting test specimen is 1 m long and 300 × 300 mm in cross-section (see Figure 10.34). The specimen is covered with an impermeable neoprene foam sheet, and the pressure of bars fitted to the cover prevents flow of water around the specimen. A vacuum connection enables air to be removed prior to saturating the specimen with de-aired water. An apparatus of this type is shown in Figure 10.35.

Hydraulic gradients can be altered by changing the levels of the inlet and outlet weirs. For small adjustments the permeameter may be raised at one end; the flow remains parallel to the axis of the specimen.

The principle on which calculations are based is the same as that shown in Figure 10.2.

10.7 Falling head permeability tests

10.7.1 Outline

For measuring the permeability of soils of intermediate and low permeability (less than 10^{-4} m/s), i.e. silts and clays, the falling head procedure is used. In the falling head test a relatively short sample is connected to a standpipe, which provides both the head of water and the means of measuring the quantity of water flowing through the sample. Several standpipes of different diameters are normally available from which can be selected the diameter most suitable for the type of material being tested. The test procedure is described in Section 10.7.2. Calculations are a little more difficult than those for the falling head test; the equation used is derived in Section 10.3.6.

In permeability tests on clays, much higher hydraulic gradients than are normally used with sands can be applied, and are often necessary to induce any measurable flow. The cohesion of clays provides resistance to failure by piping at gradients of up to several hundred, even under quite low confining or surcharge pressures (Zaslavsky and Kassiff, 1965). Dispersive clays, however, are very susceptible to erosion at much lower gradient (see Section 10.3.11).

The falling head principle can be applied to an undisturbed sample in a sampling tube (see Section 10.7.3) and to a sample in an oedometer consolidation cell (see Section 10.7.4).

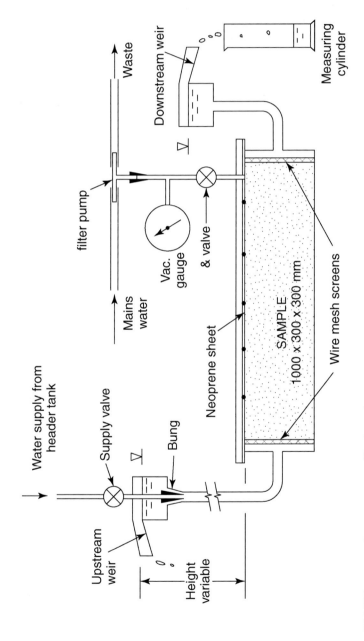

Figure 10.34 Diagrammatic arrangement of horizontal permeameter

Figure 10.35 Horizontal permeability apparatus

10.7.2 Test in falling head cell

This test is neither included in BS 1377:1990, nor in the ASTM Standards. The procedure described below follows generally accepted practice.

Apparatus

1. Permeameter cell, comprising:
    Cell body, with cutting edge (core cutter), 100 mm diameter and 130 mm long
        Perforated base plate with straining rods and wing nuts
        Top clamping plate
        Connecting tube and fittings
    Details of the cell are given in Figure 10.36, and the cell and components are shown in Figure 10.37. A longer cell body can be used if straining rods of the appropriate length are fitted. Smaller diameter sample tubes can also be accommodated.
2. Standpipe panel fitted with glass standpipe tubes of different diameters, each with a valve at its base and connecting tubing. A typical range of tubes comprises diameters of 1.5, 3 and 4.5 mm, which are suitable for low permeability soils such as silty clays. Larger diameter tubes (e.g. 10, 15, 20 mm) are more suitable for soils of intermediate permeability such as silts. For fine silty sands a standpipe diameter equal to that of the sample may be appropriate.
    The diameter of a standpipe tube for clay soils should not be less than about 1.5 mm, otherwise the effect of the capillary rise of water may be appreciable.

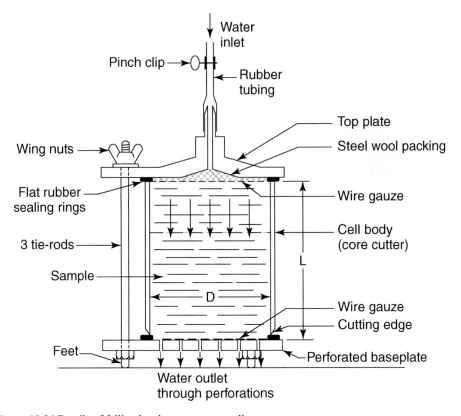

Figure 10.36 Details of falling head permeameter cell

The diameter of the standpipe determines the duration of the test, which should not be so short that timing becomes inaccurate, nor inconveniently long. Permeabilities in the range from 10^{-4}–10^{-10} m/s can be measured by this method within a period ranging from a few minutes to the working day, if a suitable standpipe diameter is selected.

    3. Source of de-aired and distilled or de-ionised water (see Section 10.6.2).

    4. Vacuum line and gauge or mercury manometer, and water trap.

    5. Steel wool.

    6. Small tools: funnel, trimming knife, spatula, etc.

    7. Thermometer.

    8. Stop-clock (minutes timer) or stopwatch.

    9. Immersion tank, with overflow.

The arrangement of the whole test assembly is shown in Figure 10.38 and the principle of the test is illustrated in Figure 10.8.

Procedural stages

    1. Assemble apparatus

    2. Calibrate manometer tubes

Figure 10.37 Permeameter cell 100 mm diameter and components for falling head test

3. Prepare cell
4. Prepare sample
5. Assemble cell
6. Connect cell
7. Saturate sample
8. Fill manometer system
9. Run test
10. Calculate permeability
11. Report result

Test procedure

1. *Assembly of apparatus*

   The apparatus is set up as shown in Figure 10.38, and assembly of the cell is described below. The volume of water passing through a sample of low permeability is quite small, and a continuous supply of de-aired water is not necessary, but the reservoir supplying the de-airing tank should be filled with distilled or de-ionised water. Otherwise, the arrangement is as shown in Figure 10.18. The flexible vacuum tubing at J, K, M (Figure 10.38) connected to the glass tee-piece must be rigid enough to withstand collapse when subjected to vacuum.

2. *Calibration of manometer tubes*

   If the areas of cross-section of the three manometer tubes are not known, they should be etermined as follows for each tube:

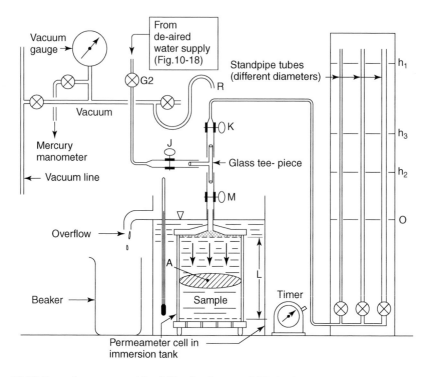

Figure 10.38 General arrangement for falling head permeability test

Fill the tube with water up to a mark near the top of the scale, and measure its height above bench level (l_1) to the nearest mm.

Run off water from the tube into a weighed beaker, until the level in the tube has fallen by about 500 mm or more.

Measure the height of the new water level (l_2) above bench level to the nearest mm.

Weigh the beaker containing water from the tube (weighings should be to the nearest 0.01 g).

If mw = mass of water (g); l_1 = initial level in tube (mm); l_2 = final level in tube (mm); and a = area of cross-section of tube (mm²).

$$\text{Volume of water run off} = m_w \text{ cm}^3$$
$$\text{Measured volume} = (l_1 - l_2)a \text{ mm}^3$$
$$= \frac{(l_1 - l_2)}{1000}a \text{ cm}^3$$
$$a = \frac{1000 m_w}{l_1 - l_2} \text{ mm}^2$$

Repeat the measurements two or three times for each tube, and average the results.

Reference marks on the manometer scale which facilitate the performance of the test are made as follows. The bench working surface forms a convenient datum level.

Measure the height y_0 (mm) of the overflow from the soaking tank above datum. Make a mark about 50 mm below the top of the manometer tubes on the scale, and measure its height y_1 (mm) above datum. Make another mark near the bottom of the tubes, but at least 200 mm above the overflow level of the soaking tank, y_2 mm above datum. Calculate the following:

$$h_1 = y_1 - y_0$$
$$h_2 = y_2 - y_0$$
$$h_3 = \sqrt{(h_1 \cdot h_2)}$$

The intermediate mark referred to in Section 10.3.6 is made at a height of $y_3 = (h_3 + y_0)$ mm above datum level (see Figure 10.8, Section 10.3.6). These marks are denoted by h_1, h_3 and h_2 in Figure 10.38.

3. *Preparation of cell*

Dismantle the cell. Check that the sealing rings are in good condition and lightly coat them with silicone grease. Ensure that the ends of the cell body are true and free from distortion, and undamaged, and that watertight joints are made when the sealing rings and end plates are clamped in position (see Figure 10.36).

See that the cell body is clean and dry, and weigh it to the nearest 0.1 g (m_1). Measure the mean internal diameter (D) and length (L) to the nearest 0.5 mm.

4. *Preparation of sample*

The core cutter type of cell body is designed for taking an undisturbed sample of cohesive soil in-situ. It can also be used to take a test sample from a block sample, or from a conventional tube or piston sample. It is essential in all instances to ensure that the sample is a tight fit in the body, and that there are no cavities around the perimeter through which water could pass. Gaps or cavities should be well packed with the fine 'matrix' portion of the soil, or with plasticine.

The sample is prepared as described in Section 9.3.3, and the ends are trimmed flush with the ends of the tube. The sample may be prepared in the usual manner with its axis vertical, for measurement of vertical permeability; or with its axis horizontal (or parallel to bedding) for measurement of horizontal permeability, or permeability parallel to bedding.

Weigh the sample in the cell to the nearest 0.1 g (m_2). Use some of the soil trimmings for determining the moisture content of the sample.

5. *Assemble cell*

Fit a wire gauze disc to each end of the sample. Place the cell body, cutting edge downwards, on the baseplate with the sealing ring in place. Fill the space in the cell top with wire wool, so that it is compressed fairly tightly when the top is screwed down on to the cell. Ensure that the sealing ring is in place so that a watertight joint is made. Tighten down the wing nuts on the straining rods progressively and evenly.

Place the assembled cell in the immersion tank, and fill with de-aired, distilled or de-ionised water up to the overflow level. Tilt the cell to release any entrapped air from underneath the cell top. If the cell has a flush base without projecting feet, it should be stood on flat spacer pieces to allow free access of water.

6. *Connecting cell*

   Connect the top inlet of the permeameter cell to the glass tee-piece with a short length of thick rubber or rigid plastic tubing fitted with a screw clip, M (see Figure 10.38). To the other branches of the tee-piece are fitted short lengths of similar tubing, each with a screw clip (J and K). Apply a smear of grease at the joints, and use connecting clips if necessary, to ensure that the joints are airtight.

7. *Saturating sample*

   With screw clips M and K open (see Figure 10.38), allow water to flow upwards through the sample under the small external head in the soaking tank and by capillary action. If the water level in the tank falls by an amount greater than that due to evaporation loss it is a positive indication that the sample is taking up water. It may be necessary to allow this process to continue overnight, or for 24 hours or longer, for a low permeability soil.

   Connect the vacuum tube R to branch K of the glass tee-piece, and close screw clip J. Continue the saturation process by applying a low suction (about 50 mm mercury) to the top of the sample by adjusting the vacuum line and air bleed valves. Maintain the suction, and increase it slightly if necessary, until water is drawn up into the glass tee-piece above M. This indicates that the sample is saturated. If air bubbles are present, maintain the suction until the system is air-free.

8. *Filling manometer system*

   Connect the de-aired water supply, with valve A (see Figure 10.18) open and valve G2 (see Figure 10.38) fractionally open, to the tee-piece at J, making sure that no air is entrapped. Open screw clip J carefully to allow water from the supply to fill the glass tee-piece and the tube at K while there is a small suction there. Close screw clip K, shut off the vacuum and disconnect the vacuum tube R. Open screw clip K slightly so as to connect to the manometer tubes panel, first making sure that the tubes at the point of connection are filled with water so that no air becomes entrapped.

   Select the manometer tube to be used for the test and open the valve at its base. Allow water to fill the tube to a level a few centimetres above the h_1 mark by opening valves G2, J and K. If any air bubbles are observed in the manometer or connecting tubes, they can be removed by applying a low suction to the top end of the standpipe.

   Close valve G2 and screw clip J, and fully open screw clip K. Top up the water in the soaking tank to bring it level with the overflow outlet.

9. *Running permeability test*

   Open screw clip M to allow water to flow down through the sample, and observe the water level in the standpipe. As soon as it reaches the level h_1, start the timer clock. Observe and record the time when the level reaches h_3, and again when it reaches h_2, then stop the clock. Close screw clip M.

   The standpipe can be re-filled for a repeat run by opening valves G2 and J. Three or four test runs should be done consecutively.

   Record the temperature of the water in the soaking tank ($T°C$).

10. Calculations

During each run the time taken for the standpipe water level to fall from h_1 to h_3 should be the same as from h_3 to h_2 (see Section 10.3.6), to within about 10%. If the difference is much more than this, the test run should be repeated. Calculate the average time for each set of test runs (*t* minutes).

The permeability *kT* of the sample is calculated from Equation (10.15) derived in Section 10.3.6.

$$k_T = 3.84 \frac{aL}{At} \log_{10}\left(\frac{h_1}{h_2}\right) \times 10^{-5} \ \text{m/s}$$

where *a*, *L* and *A* are as determined in stages 2 and 3, and (h_1/h_2) will be represented in turn by (h_1/h_3) and (h_3/h_2).

11. Report result

The result is reported as the permeability of the sample at the temperature of the test, to two significant figures, in the form

$$k_T = 2.3 \times 10^{-6} \ \text{m/s}$$

If the test temperature was not 20°C, the result can be expressed as the permeability at 20°C by multiplying by a factor obtained from the temperature conversion graph (see Figure 10.4), as explained in Section 10.3.4.

The density and moisture content of the sample are also reported, together with the voids ratio if the specific gravity of particles is known. A full description of the sample should also be given, indicating the presence of any fabric features such as laminations and fissures. The type of sample from which it was obtained, the way the test sample was prepared and its orientation should be reported.

Typical laboratory data and test results are given in Figure 10.39.

10.7.3 Test in sample tube

A falling head permeability test can be carried out on an undisturbed sample in a sample tube, such as a U-100 tube, if suitably adapted end caps are available. These can be made from standard end caps, the upper one being provided with a watertight seal and an inlet tube which can be connected to rubber or plastic tubing attached to the glass tee-piece in the same way as in Figure 10.38. The lower end cap is perforated and stands on three or four thin packing pieces, such as small coins, immersed in water in the soaking tank. The arrangement is shown in Figure 10.40.

As in the permeameter cell, a wire gauze is placed in contact with each trimmed end of the sample, and are held in place by packing the ends of the tube with steel wool. The length *L* of the sample is determined by measuring the distance to each face of the sample from each end of the tube.

The remainder of the apparatus, and the procedure for carrying out the test, are as described in Section 10.7.2.

Falling Head Permeability Test

Location .... *WEDNESFORD* .... Sample No .... *3/10* ....

Operator .... *T.Y.P.* .... Date .... *14-5-81* ....

Soil description ....*Firm grey clayey silt*...

Method of preparation ....*Hand trimmed into mould (some disturbance)*.......................

Sample diameter D ....*100.7*.........mm   area A ....*7964*...........mm^2

length L ....*127*...........mm   volume V ....*1011*........... cm^3

Mass of sample + mould ....*3098*...........g   particle density ρ_S ....*2.67*.....

mould ....*1033*...........g   measured /assumed

sample ....*2065*...........g   Bulk density ρ .....*2.04*.....Mg/m^3

Moisture content....................*12.7*...........%   Dry density ρ_D .....*1.81*......Mg/m^3

Voids ratio $= \dfrac{\rho_S}{\rho_D} - 1 = $ ....*0.475*..............   Test temperature ......*20.5*.....°C

Standpipe diameter.....*1.50*.....mm   area a ......*1.77*.....mm^2

Reference point	Height above datum y mm	Height above outlet h mm	Test		Height ratios
			min-sec	t min	
(1)	1100	$h_1 = 915$	(1–3) 4–25	4.42	$\dfrac{915}{750} = 1.22$
(3)	935	$h_3 = 750$	(3–2) 4–0.5	4.08	$\dfrac{750}{615} = 1.22$
(2)	800	$h_2 = 615$			
(0)	185	0			$\log_{10}(1.22) = 0.08636$

Permeability $k = 3.84 \times \dfrac{aL}{At} \log_{10}\left(\dfrac{h_1}{h_2}\right) \times 10^{-5}$ m/s

$$= 3.84 \times \frac{1.77 \times 127}{7964 \times t} \times 0.08636 \times 10^{-5} \text{ m/s}$$

$$= \frac{9.36}{t} \times 10^{-8} \text{ m/s}$$

Test run (1–3) $k = \dfrac{9.36}{4.42} \times 10^{-8} = 2.12 \times 10^{-8}$ m/s

Test run (3–2) $k = \dfrac{9.36}{4.08} \times 10^{-8} = 2.29 \times 10^{-8}$ m/s

Temperature correction negligible | Permeability (20°C) $= 2.2 \times 10^{-8}$ m/s |

Figure 10.39 Typical data from falling head permeability test

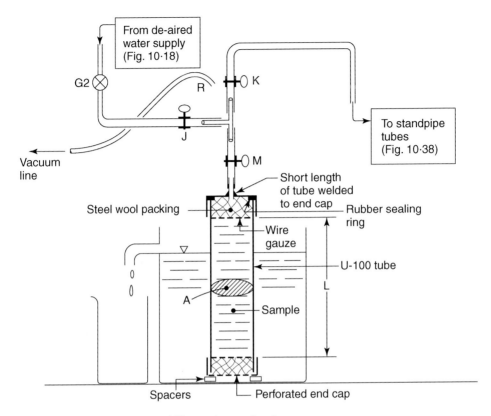

Figure 10.40 Falling head permeability test in sample tube

10.7.4 Test in oedometer consolidation cell

The arrangement for carrying out a falling head permeability test in an oedometer consolidation cell, referred to in Section 14.6.6, is shown in Figure 10.41. It is essential that the 'O' ring seals are completely watertight.

The base outlet of the oedometer cell is connected to a glass standpipe or burette by a length of flexible tubing fitted with a screw clip P (see Figure 10.41). The tube is filled with de-aired water, without trapping any air, and is supported alongside the oedometer cell by a burette stand. The bore of the tube should be appropriate to the rate of flow expected through the type of sample under test. The oedometer cell is filled with water to the level of the overflow, or to the top of the cell body if no overflow is fitted.

At any stage during the consolidation test when 100% primary consolidation has been achieved, the permeability may be measured by opening valve P. The fall of water level in the standpipe is observed and timed as for the falling head test (see Section 10.7.2), using three marked levels h_1, h_2 and h_3 as described in Section 10.7.2, stage 2.

The length of sample during the test is the consolidated height of sample at the end of the load increment, i.e. $(H_0 - \Delta H)$. Otherwise the calculations and presentation of results are as

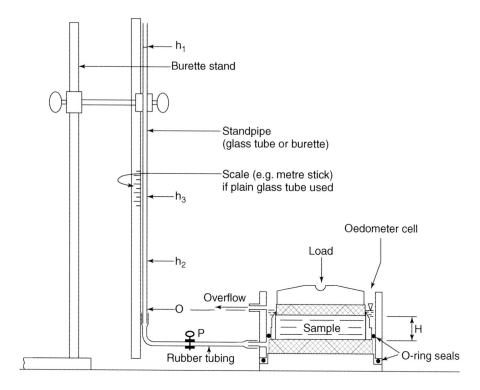

Figure 10.41 Falling head permeability test in oedometer consolidation cell

described in Section 10.7.2. It should be reported that the test was carried out in the oedometer consolidation cell, and the vertical stress should be stated.

10.7.5 Test on recompacted sample

A falling head permeability test can be carried out on a recompacted sample in a BS compaction mould by using end fittings similar to those of the falling head test apparatus. These consist of a perforated base and a top cap with an inlet tube, which are secured to the mould and extension collar by tie-rods and wing nuts of the correct length. The apparatus is shown in Figure 10.42, and the general arrangement is shown diagrammatically in Figure 10.43.

The soil is brought to the required moisture content and is compacted into the mould either to provide the desired density or by using the

Figure 10.42 Apparatus for falling head permeability test in BS compaction mould

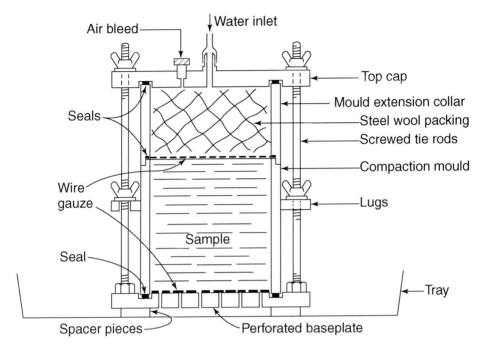

Figure 10.43 Arrangement for falling head permeability test in compaction mould

appropriate degree of compaction. The sample is trimmed, and weighed with the mould (see Chapter 6, Volume 1 (third edition)). The extension collar is fitted.

A wire gauze is inserted between the lower end of the sample and the perforated base. Another wire gauze separates the sample from steel wool which fills the extension collar. The arrangement and test procedure are as described in Section 10.7.2. It is essential first to displace the air from the sample, which will generally be only partially saturated.

A series of tests is normally carried out so as to cover a range of densities. The particle density of the soil grains should be determined, so that the permeability measurements can be related to voids ratio or porosity.

10.8 Erodibility tests

10.8.1 Scope of Tests

The three tests described in Sections 10.8.2–10.8.4 are those which were used in investigations by Sherard *et al.* (1976a), referred to in Section 10.3.11, for the identification of dispersive clays, i.e. clays susceptible to being eroded. They can be carried out easily in a soil laboratory, only the first requiring special apparatus which is quite simple to make. These tests are described in Clause 6 of BS 1377:1990. An additional test is outlined in Section 10.8.6.

Chemical tests for the measurement of the effective amount of sodium cations present in the pore water of the clay are referred to in Section 10.8.5. It is the presence of sodium and the relationship of the concentration of sodium cations to other metallic cations which is the prime factor responsible for a clay being dispersive. The extraction of the pore water from the clay is briefly described, but the tests themselves require the services of a specialist chemical testing laboratory.

10.8.2 Pinhole test (BS 1377:Part 5:1990:6.2 and ASTM D 4647)

Principle

Details of this test were originally given by Sherard *et al.* (1976b), and the procedure is based on extensive trials and observational experience. Distilled water is caused to flow through a 1 mm diameter hole formed in a specimen of recompacted clay. The water emerging from a dispersive clay carries a suspension of colloidal particles, whereas water from an erosion-resistant clay is crystal clear. The test is based on visual assessment of the presence of turbidity in the emerging water, and on measurement of the rates of flow.

Distilled water is used as a basis of comparison, but it has been found that the effect of using natural ground or river water is generally somewhat less severe.

Apparatus

1. Pinhole test apparatus, as shown diagrammatically in Figure 10.44(a) comprises
   (a) Cylindrical plastic body about 33 mm internal diameter and 100 mm long
   (b) End plates with O-ring seals to make watertight fit to body, fitted with water inlet and outlet connections and a standpipe connection
   (c) Wire mesh discs, 1.18 mm aperture, 33 mm diameter (three required)
   (d) Plastic or metal nipple, in the form of a truncated cone 13 mm long with a 1.5 mm hole (see Figure 10.44(b))

The component parts of the pinhole apparatus are shown in Figure 10.45, and the apparatus set up for a test is shown in Figure 10.46

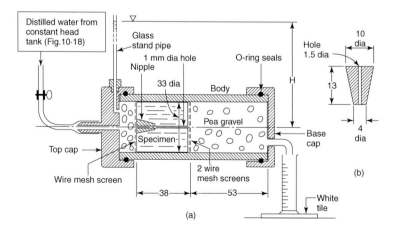

Figure 10.44 Apparatus for pinhole test: (a) general arrangement; (b) details of nipple (dimensions in mm) (after Sherard et al., 1976b)

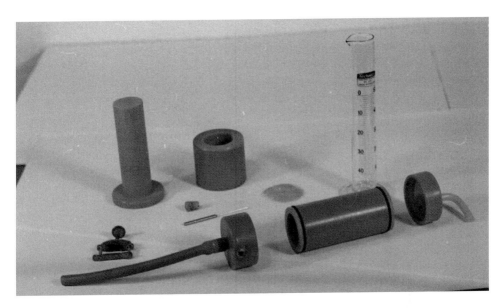

Figure 10.45 Component parts of apparatus for pinhole test

2. Hypodermic needle 1.00 mm outside diameter, about 100 mm long
3. Pea gravel, about 5 mm size
4. Supply of distilled water from an adjustable constant head supply tank. The head required ranges from 50 mm to 1.02 m
5. BS sieve, 2 mm aperture
6. Apparatus for moisture content determination
7. Seconds timer or stop-clock
8. Graduated glass measuring cylinders: 10 ml, 25 ml, 50 ml (at least two of each)
9. White ceramic tile
10. Glass standpipe and rubber tube connection to pinhole apparatus, about 1200 mm long with scale graduated in millimetres
11. Burette stand for supporting the pinhole test apparatus, standpipe and graduated scale
12. (Optional) Harvard miniature compaction tamper (see Volume 1 (third edition), Section 6.5.10 and Figure 6.24), fitted with a 15 lb (6.8 kg) spring (not included in BS 1377)

Test procedure
Tests should be carried out on samples of soil which have been preserved at their natural moisture content and not allowed to dry out. About 150 g is required for the test sample.

1. Determine the natural moisture content and Atterberg limits of the soil, using a separate portion identical to that which is to be tested.
2. Remove any particles retained on a 2 mm sieve.
3. Bring the soil to about the plastic limit, either by mixing in distilled water, or by gradual drying. Check by using the thread-rolling procedure described in Section 2.6.8 of Volume 1 (third edition). Determine the resulting moisture content by using a representative small portion of the sample.

Figure 10.46 Pinhole test in progress

4.   Fit the base to the body of the pinhole apparatus, making sure that the seal is clean and correctly positioned to give a watertight joint.

5.   Support the body of the apparatus with its axis vertical. Place pea gravel to a depth of 53 mm in the cylinder, level the surface, and place two wire mesh discs on top.

6.   Compact the test sample into the cylinder to a depth of 38 mm, in 5 layers, so as to give a dry density about 95% of the maximum dry density achieved by BS 'light' compaction (Section 6.5.3 of Volume 1 (third edition)). (If the Harvard tamper is used, about 16 strokes per layer should be suitable.) Level the surface of the top of the sample.

7.   Using finger pressure, push the plastic nipple into the top of the compacted soil at the centre until the upper face is flush with the sample surface.

8.   Punch a 1 mm diameter hole through the sample by inserting the hypodermic needle through the hole in the nipple (see Figure 10.44(b)).

9.   Place a wire mesh disc over the sample, followed by pea gravel to the top of the cylinder.

10. Fit the top cap, ensuring that the seal is clean and correctly positioned to give a waterproof joint. Support the cylinder with its axis horizontal.

11. Set the constant head distilled water supply to give a head H (see Figure 10.44(a)) of 50 mm, measured from the centre-line of the apparatus. Connect the inlet on the pinhole apparatus to the supply, and the standpipe connection to the glass standpipe supported by a burette stand. Place a glass measuring cylinder on a white tile or sheet of white paper under the outlet pipe.

12. Open the inlet valve to allow water to fill the apparatus, and then to flow through the system for a few minutes to obtain a steady flow. Observe and record the colour of the water collected in the measuring cylinder; if perfectly clear, record that fact. If there is no flow, disconnect the distilled water supply, remove the top end cap and wire disc, re-form the 1 mm hole, then resume from step 9.

13. Measure the rate of flow by observing the time required to fill the 10 ml measuring cylinder several times within a period of 5–10 min.

Observe and record the clarity and colour of the water by looking through the side of the measuring cylinder against a sheet of white paper, and vertically through the column of water in the cylinder. If individual particles are observed this should be recorded, together with an indication of their intensity. If the water is not substantially clear and the rate of flow has increased to more than 1 ml/s, stop the

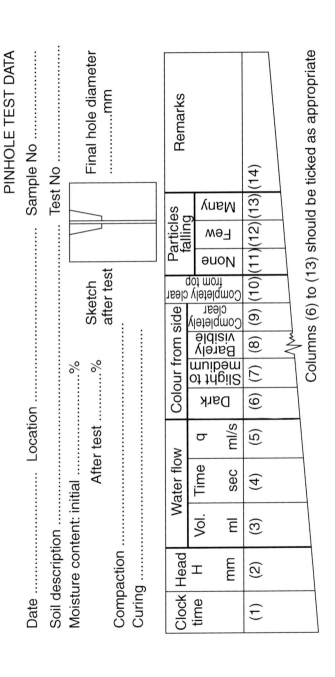

Figure 10.47 Headings for recording observations during pinhole test (after Sherard et al., 1976b)

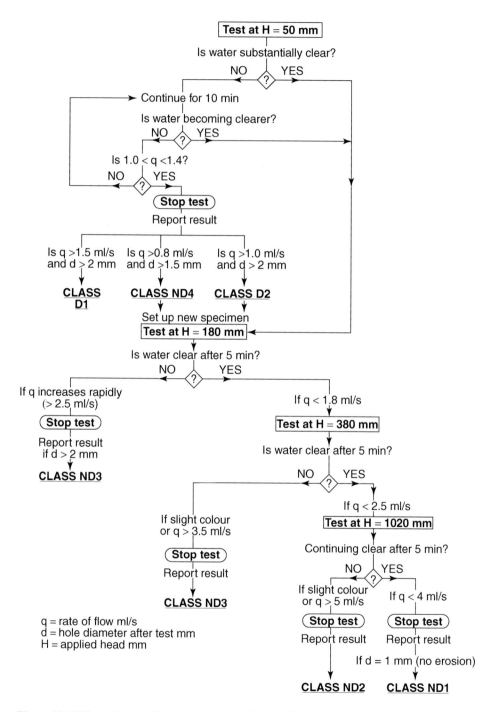

Figure 10.48 Flow diagram illustrating sequence for pinhole test (Courtesy of British Standards Institution)

test and proceed to step 15. If the water is clear, record that fact. Suitable headings under which data may be recorded are indicated in Figure 10.47.

14. Raise the inlet head level to 180 mm, then to 380 mm and to 1020 mm. Repeat the measurements and observations described in step 13 at each level, changing to the 25 ml or 50 ml measuring cylinder as the flow rate increases. At each stage refer to the flow diagram, Figure 10.48, for guidance whether to continue or stop the test. The limiting rate of flow imposed by the apparatus for each inlet head is approximately as follows:

Inlet head (mm)	Limiting rate of flow (ml/s)
50	1.2–1.3
180	about 2.7
380	about 3.7
1020	5 or more

15. When the flow tests have been completed, disconnect the distilled water supply and dismantle the apparatus.
16. Remove the specimen intact from the mould, using an extruder if necessary. Break or cut the specimen open and examine the hole. Measure its approximate diameter by comparison with the hypodermic needle, or to the nearest 0.5 mm using a steel rule. Sketch the configuration of the hole if it is not of uniform diameter along its length.

Analysis of data

The results of the test are evaluated from the following criteria:
- Appearance of the collected water
- Rate of flow of water
- Final diameter of the hole in the specimen

Table 10.2 Classification of soils from pinhole test data (taken from Table 2 of BS 1377:Part 5:1990)

Dispersive classification	Head (mm)	Test time for given head (min.)	Final flow rate through specimen (ml/s)	Cloudiness of flow at end of test		Hole size after test (mm)
				from side	from top	
D1	50	5	1.0–1.4	Dark	Very dark	≥2.0
D2	50	10	1.0–1.4	Moderately dark	Dark	>1.5
ND4	50	10	0.8–1.0	Slightly dark	Moderately dark	≤1.5
ND3	180	5	1.4–2.7	Barely visible	Slightly dark	≥1.5
	380	5	1.8–3.2			
ND2	1020	5	> 3.0	Clear	Barely visible	<1.5
ND1	1020	5	≤ 3.0	Perfectly clear	Perfectly clear	1.0

Soils are classified according to the results of the test, into 'Dispersive soils' (categories D1 and D2), and 'non-dispersive soils' (categories ND1–ND4), as shown in Table 10.2. The principal differentiation between dispersive and non-dispersive soils is obtained from the result under 50 mm head of water. Detailed criteria for evaluating results are given in Table 10.2, which forms the basis for the method of reporting results.

Reporting results

The test report should include the following:
- Method used, with reference to Clause 6.2 of BS 1377:Part 5:1990 (or to ASTM S 4647)
- Identification and description of soil, and whether any coarse particles were removed for the test
- Soil properties: liquid limit, plastic limit, moisture content, dry density to which compacted
- Rates of flow during test, and appearance of collected water during each hydrostatic head applied
- Diameter and configuration of the hole after test
- Classification of soil according to the categories referred to above

10.8.3 Crumb test (BS 1377:Part 5:1990:6.3)

Purpose

This test was originally described by Rallings (1966), and was modified by Sherard *et al.* (1976a). It provides a very simple means of identifying dispersive clay soils without requiring special equipment. A test for a similar purpose was described by Emerson (1967).

Apparatus and materials

Glass beaker 100 ml
Sodium hydroxide solution (c(NaOH) = 0.001 mol/l)
Dissolve 0.04 grams of anhydrous sodium hydroxide in distilled water to make 1 l of solution.

Procedure

Take a few 'crumbs', each of about 6–10 mm diameter, representative of the sample at the natural moisture content. Drop them into the beaker containing the sodium hydroxide solution. Observe the reaction after allowing to stand for 5–10 min.

For many soils, use of distilled water is as good an indicator as the sodium hydroxide solution, in that the soil is probably dispersive if the test indicates dispersion. However, many dispersive clays do not show a reaction in distilled water, but do react in the solution.

Results

Report the observations in accordance with the following guide to interpretation:
- Grade 1: no reaction. Crumbs may slake and run out on bottom of beaker in a shallow heap, but no sign of cloudiness caused by colloids in suspension

Figure 10.49 Crumb test for identification of dispersive soils: (a) non-dispersive; (b) dispersive

- Grade 2: slight reaction. Bare hint of cloudiness in water at surface of crumb. (If cloud is easily visible, use Grade 3)
- Grade 3: moderate reaction. Easily recognisable cloud of colloids in suspension, usually spreading out in thin streaks on bottom of beaker
- Grade 4: strong reaction. Colloidal cloud covers nearly the whole of bottom of beaker, usually as thin skin. In extreme cases, all water in beaker becomes cloudy.

Grades 1 and 2 are reported as giving a 'non-dispersive' reaction, and Grades 3 and 4 a 'dispersive' reaction. Crumbs of a non-dispersive soil are shown in Figure 10.49(a), and of a dispersive soil in Figure 10.49(b).

With some soils, partial drying may influence the results obtained.

10.8.4 Dispersion test (BS 1377:Part 5:1990:6.4 and ASTM D 4221)

Principle

This procedure was referred to by Sherard *et al.* as the SCS dispersion test. It was developed by Volk (1937) and has been widely used by the US Soil Conservation Service. It is sometimes referred to as the double hydrometer test, because it is based on the degree of dispersion of clay particles achieved during the pre-treatment stage of a hydrometer sedimentation test.

Apparatus

As for the hydrometer sedimentation test (Volume 1 (third edition), Sections 4.8.1 and 4.8.3). Two sedimentation cylinders are required for each sample to be tested.

Procedure

Take two identical representative portions, referred to as specimens A and B, from the sample to be tested. On specimen B, carry out the standard hydrometer sedimentation test as

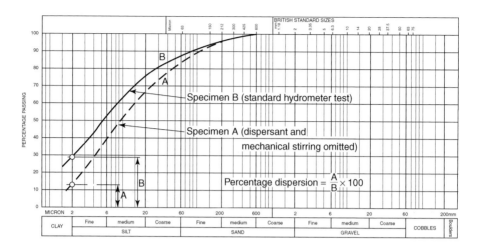

Figure 10.50 Typical particle size curves from SCS dispersion test (double hydrometer test)

described in Volume 1 (third edition), Sections 4.8.1 and 4.8.3. Draw the particle size curve on a standard sheet (Curve B, Figure 10.50).

Carry out a similar test on specimen A, but omitting the mechanical stirring (see Section 4.8.1, stage 6), and use distilled water instead of a dispersant solution. The additional 1000 ml cylinder placed in the constant temperature bath (see Volume 1 (third edition), Section 4.8.3, stage 1), should contain distilled water only.

Draw the resulting particle size curve on the same sheet as before (see Curve A, Figure 10.50). Read off the percentage of clay size particles, i.e. the percentages finer than 0.002 mm, on each curve. The 'percentage dispersion' is defined as the ratio of these percentages, multiplied by 100. From Figure 10.50, percent dispersion = (A/B) × 100%.

10.8.5 Chemical tests

Scope

Two types of chemical test, one on the pore water and one on the clay itself, were described by Sherard *et al.* (1972), for the determination of the relative amount of sodium cations present in the clay. They concluded that for most practical purposes, it was necessary to analyse the pore water only. The pore water can be extracted in the soil laboratory as outlined below, and is sent to a specialist chemical testing laboratory for analysis. The test on the clay itself is not included here.

Preparation of water extract

Mix the soil at natural moisture content with sufficient distilled water to bring it to the liquid limit. This can be verified by using the cone penetrometer described in Volume 1 (third edition), Section 2.6.4, to obtain a penetration of about 20 mm.

Allow to stand for several hours, or overnight, to enable equilibrium to be established in the soil/water system. Using vacuum filtration with a Büchner funnel, extract about 10–25 ml of pore water from the saturated soil paste. This water is known as the 'saturation extract'.

Analysis of water extract

The saturation extract is analysed in the chemical laboratory to determine the quantities of the four main metallic cations in solution, i.e. calcium (Ca), magnesium (Mg), sodium (Na) and potassium (K), in milli-equivalents per litre (meq/l). The total dissolved salts (TDS) is taken to be equal to the sum of the quantities of these four cations, and the 'percentage sodium' is the ratio of the quantity of sodium to the TDS. These can be expressed by the following equations:

$$TDS = Ca + Mg + Na + K$$
$$\text{Percentage sodium} = \frac{Na}{TDS} \times 100\%$$

The sodium absorption ratio (SAR) is calculated from

$$SAR = \frac{Na}{\sqrt{\left(\dfrac{Ca + Mg}{2}\right)}}$$

Results

The SAR is a measure of the amount of sodium in the pore water relative to other cations and is the main factor in determining whether or not a clay is dispersive. Generally, if the SAR exceeds 1 the clay will be dispersive, but the SAR criterion increases with increasing TDS. Details of the analysis of results are given in Sherard *et al.* (1972).

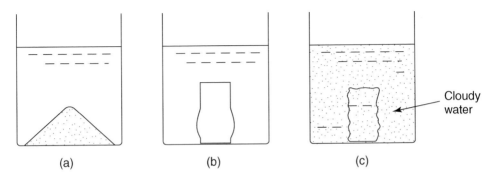

Figure 10.51 Typical behaviour of samples in cylinder dispersion test: (a) non-dispersive cohesionless (type N); (b) non-dispersive cohesive (type C); (c) dispersive (type D)

10.8.6 Cylinder dispersion test

Purpose

This test was developed at City University, London, as an extension to the crumb test described in Section 10.8.3. The objective was to examine soil behaviour in the 'fully softened' state when submerged in water, i.e. under zero effective stress (Atkinson *et al.*, 1990). The test is carried out on a cylindrical specimen of remoulded soil which has been consolidated from a slurry.

Specimen preparation

In the method described by Atkinson *et al.* (1990), the soil is first dried, ground to a powder, then mixed with de-aired water to form a slurry at a moisture content of about twice the liquid limit. Distilled water, or water of the appropriate chemistry, is used. However, the author Suggests that the soil should not be dried, because of the changes in certain properties which can occur as a result of drying (see Volume 1 (third edition), Section 2.6.3).

The slurry is poured into a suitable perspex cylinder (e.g. 38 mm diameter), taking care to prevent inclusion of air bubbles. The cylinder is fitted at either end with a close-fitting but free-running piston with porous stone filters and holes for drainage. The assembly is placed in a water bath, and axial load can be applied by means of dead weights hanging from a yoke. (The principle is similar to that shown in Figure 14.49.) The sample is consolidated in stages, starting with the weight of the top piston only, so that the ends of the sample become stiff enough to prevent material from being squeezed out under the next stage of loading. The sample should be consolidated enough to enable it to be handled, and the load maintained long enough to ensure that consolidation is substantially complete. The final height of the sample should be between one and two times its diameter. The pistons are removed and the specimen is carefully extruded from the cylinder.

Test procedure

Place the prepared sample into a beaker containing the appropriate water, or place it in an empty beaker and add water carefully. Avoid excessive turbulence by either method.

Observe the behaviour of the sample until pore pressure equilibrium is reached. This period depends on the coefficient of consolidation (see Chapter 14, Section 14.3.5) and on the size of sample; up to one week is usually enough for a 38 mm diameter sample.

Results

Three basic types of behaviour are usually observed, as shown diagrammatically in Figure 10.51(a)–(c) as types N, C, D, respectively. Type N is characteristic of non-dispersive cohesionless soils (true cohesion equal to zero). The angle of the cone to which the sample eventually slumps approaches the 'fully softened' friction angle. Type C relates to non-dispersive cohesive soils (true cohesion positive), and might show some plastic deformation by bulging near the bottom. In both N and C types the water remains clear. Type D behaviour, in which the water becomes cloudy and opaque, indicates dispersive soils (true cohesion negative) in which inter-particle repulsive forces exceed the pore water suction. Some soils might show characteristics of two types, especially N and D for a soil containing dispersive clay together with non-cohesive grains. The result should be reported according to characteristic type letter (or types, if relevant;

e.g. N/D). The size of sample, and chemistry of the water both in the slurry and in the beaker, should be reported. The reference in which the procedure is described should also be given.

References

Al-Dhahir, Z. A. and Tan, S. B. (1968) A note on one-dimensional constant-head permeability tests. *Géotechnique*, Vol. 18, No. 4

ASTM D 2434-68 R06, Standard test method for permeability of granular soils (constant head)

ASTM D 4221-99 R05, Dispersion characteristics of clay soil by double hydrometer. American Society for Testing and Materials, Philadelphia, PA, USA

ASTM D 4647-06 E01, Identification and classification of dispersive clay soils by the pinhole test. American Society for Testing and Materials, Philadelphia, PA, USA

Atkinson, J. H., Charles, J. A. and Mhach, H. K. (1990) Examination of erosion resistance of clays in embankment dams. *Q. J. Eng. Geol.*, Vol. 23, No. 2, pp. 103–108

Bertram, G. E. (1940) An experimental investigation of protective filters. Harvard University, Graduate School of Engineering, Soil Mechanics Series 7.

Carman, P. S. (1939) Permeability of saturated sands, soils and clays. *J. Agric. Sci.*, Vol. XXIX, No. 11

Clough, G. W. and Davidson, R. R. (1977) The effects of construction on geotechnical performance. Speciality Session III. *Proc. 9th Int. Conf. on Soil Mech. and Found. Eng.*, Tokyo, July 1977

Darcy, H. (1856). *Les fontaines publique de la ville de Dijon*. Dalmont, Paris

Department of Transport (1990) A permeameter for road drainage layers, Departmental Advice Note HA 41/90. Department of Transport, London

Emerson, W. W. (1967) A classification of soil aggregates based on their coherence in water. *Aust. J. Sci.*, Vol. 5,

Hazen, A. (1982) Some physical properties of sands and gravels with special reference to their use in filtration. 24th Annual Report, Massachusetts State Board of Health, Boston, MA, USA.

Jones, H. A. and Jones, R. H. (1989) Horizontal permeability of compacted aggregates. Chapter 11 of *Unbound Aggregates in Roads* (eds. R. H. Jones and A. R. Dawson). Butterworths, London

Kaye, G. W. C. and Laby, T. H. (1973) *Tables of Physical and Chemical Constants* (14th edition). Longman, London

Kenney, T. C. (1963) Permeability ratio of repeatedly layered soils. *Géotechnique*, Vol. 13, No. 4

Klementev, I. and Novák, J. (1978) Continuously water de-airing device. *Géotechnique*, Vol. 28, No. 3

Kolbuszewski, J. (1948) An experimental study of the maximum and minimum porosities of sands. *Proc. 2nd Int. Conf. on Soil Mech. and Found. Eng.*, Vol. 1, p. 158

Kozeny, J. (1927) Über kapillare Leitung des Wassers in Boden. *Ber. Wien Akad.*, 136a-271

Lambe, T. W. and Whitman, R. V. (1979) *Soil Mechanics, S.I. Version*. Wiley, New York

Lane, K. S. and Washburn, D. E. (1946) Capillarity tests by capillarimeter and soil filled tubes. *Proc. Highw. Res. Board*, Vol. 26, pp. 460–473

Loudon, A. G. (1952) The computation of permeability from simple soil tests. *Géotechnique*, Vol. 3, No. 4

Lund, Agnete (1949) An experimental study of graded filters. MSc. thesis, Imperial College, London

May, R. W. and Thomson, S. (1978) The geology and geotechnical properties of till and related deposits in the Edmonton, Alberta, area. *Can. Geotechl J.*, Vol. 15, No. 3

Rallings, R. A. (1966) An investigation into the causes of failure of farm dams in the Brigalow belt of Central Queensland. Water Research Foundation of Australia, Bulletin No. 10, Appendix 4, October 1966

Sherard, J. L., Dunnigan, L. P. Decker, R. S. (1976a) Identification and nature of dispersive soils. *J. Geol. Eng. Div., ASCE*, Paper 12052, April 1976

Sherard, J. L., Dunnigan, L. P., Decker, R. S. and Steele, E. F. (1976b) Pinhole test for identifying dispersive soils. *J. Geol. Eng. Div., ASCE*, Paper 11846, January 1976

Sherard, J. L., Ryker, N. L. and Decker, R. S. (1972) Piping in earth dams of dispersive clay. *Proc. ASCE Speciality Conf.: The performance of earth and earth-supported structures*, Vol. 1, pp. 602–611

Taylor, D. W. (1948) *Fundamentals of Soil Mechanics*. Chapman & Hall, London

Terzaghi, K. and Peck, R. B. (1967) *Soil Mechanics in Engineering Practice*. Wiley, New York

Terzaghi, K., Peck, R. B. and Mesri, G. (1996) *Soil Mechanics in Engineering Practice* (third edition). Wiley, New York

Thorne, C. P. (1975) In-situ properties of fissured clays. *Symposium on in-situ testing for design parameters*, Melbourne, Australia, November 1975

Volk, G. M. (1937) Method of determination of the degree of dispersion of the clay fraction of soils. *Proc. Soil Sci. Soc. Am.*, Vol. 2, pp. 561–567

Wittman, L. (1976) Stabilität Hydrodynamisch Beanspruchter Böden. Institut für Bodenmechanik und Felsmechanik, Abteilung Erddammbau und Grundbau, Universität Karlsruhe, Germany.

Zaslavsky, D. and Kassif, G. (1965) Theoretical formulation of piping mechanism in cohesive soils. *Géotechnique*, Vol. 15, No. 3

Chapter 11

California bearing ratio

11.1 Introduction

11.1.1 Purpose and scope

The California Bearing Ratio test, or CBR test as it is usually called, is an empirical test which was first developed in California, USA, to estimate the bearing value of highway sub-base and subgrade, hence its name. The test follows a standardised procedure, which is described in this chapter. There is little difference between British and American standards for the test. However, there are numerous ways of preparing samples for the test, and in this respect American practice differs in detail from British practice. The methods most often used are described.

11.1.2 Principle

The test is performed by pushing a standard plunger into the soil at a fixed rate of penetration, and measuring the force required to maintain that rate. Using the resulting load–penetration relationship, the CBR can be derived for the soil in the condition in which it was tested.

It is important to appreciate that this test, being of an empirical nature, is valid only for the application for which it was developed, i.e. the design of highway base thicknesses. Some of the terms frequently used in this connection are illustrated in Figure 11.1, and are defined in Section 11.2 (see also BS 6100-4:2008, Section 3.2).

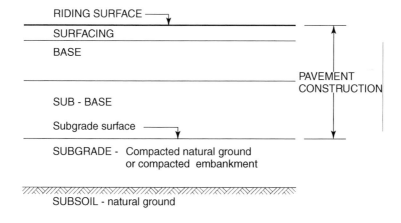

Figure 11.1 Some terms used in pavement construction

11.1.3 Historical development

The test was developed during the 1930s at the laboratory of the Materials Research Department of the California Division of Highways, USA, and was reported by Porter (1938). Previously the assessment of the quality of materials for use in highway bases and sub-bases had depended upon indirect methods such as identification of the soil, and analysis of the fine fraction only. With the increasing volume and weight of traffic during the early 1930s it became apparent that these procedures were no longer adequate. Porter was able to show that a favourable correlation existed between his test data and the observed performance of roads under the action of traffic, and the value of his procedure was soon recognised.

The CBR test was recommended to the American Society for Testing and Materials as a standard test by Stanton (1944), and is now designated ASTM D 1883. The procedure was further developed for use on airfields construction by Porter (1949). Davis (1949) showed that the procedure was applicable to the design of roads and airfield pavements in Britain, and a standard procedure for use in Britain was given by the Transport Research Laboratory (1952). The test first appeared in British Standards in 1953 when it was termed the 'cylinder penetration test', and was used for stabilised soils (BS 1924:1953). It was called the 'California bearing ratio test' in Test 15 of BS 1377:1967, and is now specified in Clause 7 of BS 1337: Part 4:1990 as the CBR test. The test principle is still recognised and used in most parts of the world as an important criterion in pavement design.

11.1.4 Types of sample for test

The test to be described is the standard laboratory test carried out on a sample of soil in a special container known as a CBR mould. The sample may be either undisturbed or remoulded, and can be prepared in one of several ways. Remoulded samples may be compressed into the mould under a static load, or dynamically compacted into it, at the required moisture content, either to achieve a specified density or by using a standard compactive effort. Undisturbed samples may be taken on site in a CBR mould, either from natural ground or from recompacted soil such as in an embankment or a road sub-base. Specimens may be tested in the mould either as prepared (or as received), or after soaking in water for several days.

A similar type of test may be carried out on the soil in-situ, by using a test rig mounted on a vehicle or trailer. The principle is the same as the laboratory test, but the details are beyond the scope of this book. The results of in-situ tests are not directly comparable with laboratory test results, and the laboratory test in the CBR mould is recognised as the standard test.

The various types of sample on which CBR tests are carried out are summarised diagrammatically in Figure 11.2.

The BS and ASTM tests require different types of sample mould, and the two types should not be confused. Mould dimensions, rammer dimensions and rammer masses all differ, and spacer discs are used for different purposes in the two procedures (see Section 11.6.3, item 1).

11.2 Definitions

California bearing ratio (CBR), or *bearing ratio* The ratio of the force required to penetrate a circular piston of 1935 mm^2 (3 in^2)* cross-section into soil in a special container at a rate

---

*(*The numbers given in inch units are the original standard Imperial values which gained general recognition. The numbers in millimetre units, though not exact conversions, are those used in current British Standards.*)*

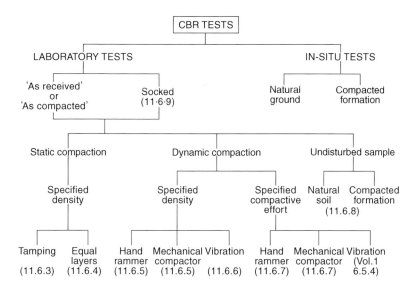

Figure 11.2 Types of sample for CBR tests, according to methods of preparation (section references given in brackets)

of 1 mm/min (0.05 in/min), to that required for similar penetration into a standard sample of compacted crushed rock. The ratio is determined at penetrations of 2.5 mm and 5.0 mm (0.1 in and 0.2 in) and the higher value is used.

$$\text{CBR} \frac{\text{measured force}}{\text{'standard' force}} \times 100\%$$

Penetration resistance Force or pressure required to maintain a constant rate of penetration of a probe, such as a CBR piston, into the soil.

Subgrade Natural soil or embankment construction prepared and compacted to support a pavement (see Figure 11.1).

Subgrade surface Surface of the earth or embankment prepared to support a pavement.

Subsoil Soil below the subgrade or fill.

Sub-base Layer of selected material of specified thickness in a pavement system between subgrade and base course, or between subgrade and pavement construction.

Base course (*base*) Layer of high-grade selected material of specified thickness constructed on the subgrade or sub-base to spread the load from the pavement and provide drainage.

Pavement Constructed layer of durable material of specified thickness, usually of concrete, asphalt or bituminous materials, designed to carry wheeled vehicles. This term covers roads, and airfield runways and taxiways.

Rigid pavement Pavement constructed of concrete, whether reinforced or not.

Flexible pavement Pavement constructed by using asphaltic or bituminous materials as a binder between pieces and particles of crushed stone or rock, or unbound granular materials with a bituminous surface.

Surfacing Topmost layer of the pavement construction, providing a durable surface and smooth riding qualities.

Iso-CBR lines Contours of constant CBR value obtained from a series of tests plotted on a set of dry density–moisture content relationship graphs for several degrees of compaction (see Figure 11.7, Section 11.3.4).

11.3 Principles and theory

11.3.1 Basis of test

The CBR test is a constant rate of penetration shear test in which a standard plunger is pushed into the soil at a constant rate and the force required to maintain that rate is measured at suitable intervals. The load–penetration relationship is drawn as a graph from which the loads corresponding to standard penetrations are read off and expressed as ratios (percent) of standard loads. The accepted percentage is known as the CBR value of the soil in the condition at which it was tested. The CBR value can be regarded as an indirect measure of the shear strength of the soil, but it cannot be related directly to shear strength parameters. An assumed mechanism of failure of the soil beneath the plunger (Black, 1961) is indicated in Figure 11.3.

The CBR is derived from an *ad hoc* test and is not based on theoretical concepts. The only calculation necessary is to express the measured force for a certain penetration as a percentage of the 'standard' force for the same penetration.

$$CBR = \frac{\text{measured force}}{\text{standard force}} \times 100\% \tag{11.1}$$

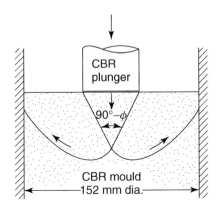

Figure 11.3 Assumed mechanism of failure of soil beneath CBR plunger (after Black, 1961)

The standard forces corresponding to penetrations in the range 2–12 mm, as given in the BS, are shown in Table 11.1. The forces shown in heavy type, corresponding to penetrations of 2.5 mm and 5 mm, are those used in the standard calculations of CBR value. These are rounded equivalents to the original criteria for contact pressures under a plunger of 3 in^2 cross-section, of 1000lb/in^2 at 0.1 in penetration and 1500 1b/in^2 at 0.2 in penetration, respectively. These standard forces were based on tests on samples of compacted crushed rock, and by definition relate to a CBR of 100%. The corresponding load–penetration relationship is shown in Figure 11.4 and by the thick curve in Figure 11.5. Curves corresponding to several other

Table 11.1 Standard force–penetration relationship for CBR tests

Penetration		Force		Pressure
(in)	*(mm)*	*(kN)*	*(lbf)*	*(lb/in²)*
	2	11.5		
(0.1)	**2.5**	**13.2**	(3000)	(1000)
	4	17.6		
(0.2)	**5**	**20.0**	(4500)	(1500)
	6	22.2		
	8	26.3		

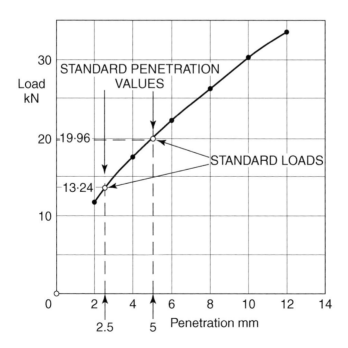

Figure 11.4 Standard load–penetration curve for CBR of 100% (standard loads shown are before rounding)

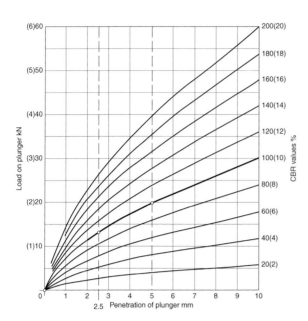

Figure 11.5 Curves of constant CBR value. Intermediate values may be obtained by interpolation (figures in brackets refer to CBR values below 20%)

CBR values, in the range 20–200%, are shown in Figure 11.5. Curves of intermediate values of CBR may be obtained by interpolation. CBR values in excess of 100% are possible, for instance on crushed slag, hoggin or stabilised soil.

The CBR value is essentially a means of expressing the data obtained from the load–penetration curve as a single numerical quantity.

11.3.2 Compaction criteria

A CBR test is normally carried out on a sample which reproduces as closely as possible the conditions likely to occur in the field. If the in-situ density and moisture content are known, a test specimen can be prepared to fulfil these conditions.

However, specifications for road embankments and sub-bases are often worded in terms of a permitted moisture content range and a maximum air voids content. The moisture content range is usually specified in relation to the optimum moisture content for one of the standard laboratory methods of compaction. The maximum permitted air voids may be typically 5% of the total volume.

The dry density, ρ_D, of a soil at a moisture content of $w\%$ and with an air content of $V_a\%$ (expressed as a percentage of total volume) is given by

$$\rho_D = \frac{1 - \dfrac{V_a}{100}}{\dfrac{1}{\rho_s} + \dfrac{w}{100}} \tag{11.2}$$

The derivation of this equation is given in Volume 1 (third edition), Section 6.3.2, Equation (6.6).

The corresponding wet density, ρ, is obtained from

$$\rho = \frac{100+w}{100}\rho_D \qquad (11.3)$$

which is a re-arrangement of Equation (3.12) in Section 3.3.2 of Volume 1 (third edition). Combining Equations (11.2) and (11.3), the wet density from which the mass of soil required to just fill the CBR mould after compaction or compression, as explained in Section 11.6.2, can be calculated by using the following equation:

$$\rho = \frac{\left(1-\dfrac{V_a}{100}\right)\left(1+\dfrac{w}{100}\right)}{\dfrac{1}{\rho_s}+\dfrac{w}{100}} \qquad (11.4)$$

11.3.3 Relationship to density and moisture content

The CBR value for a given soil depends upon its dry density and moisture content. It is convenient to relate CBR values for a recompacted soil to the moisture–density curve derived from one of the standard laboratory compaction tests (see Volume 1 (third edition), Section 6.5).

For a given degree of compaction, the CBR value decreases with increasing moisture content and this decrease becomes more rapid above the optimum value. The rate of decrease is particularly sharp for granular soils (Davis, 1949).

A general relationship between CBR and moisture content is shown diagrammatically in Figure 11.6 where CBR values are plotted on a logarithmic scale corresponding to points on the BS 'light' compaction curve plotted in the usual way. The two peaks shown on curve C often occur with clays compacted dry of optimum, especially for low levels of compaction. Similar relationships could be obtained for other degrees of compaction.

When the CBR corresponding to a certain density and moisture content is to be determined it is good practice to obtain CBR values over a range of conditions and to interpolate for the relevant field conditions, as described in the next section.

11.3.4 Derivation of iso-CBR lines

The procedure recommended by the Transport Research Laboratory (TRL) for relating CBR values to variations in density and moisture content for a particular soil is outlined below (TRL, Road Note 31, 1977).

Compaction tests are carried out on the soil in 152 mm diameter CBR moulds, usually at three different levels of compaction. These correspond to BS 'light' compaction (using a 2.5 kg rammer); BS 'heavy' compaction (using a 4.5 kg rammer); and some intermediate degree of compaction. Procedures for compacting the soil using 'light' and 'heavy' compaction are similar to those for the compaction tests described in Volume 1 (third edition), Sections 6.5.3 and 6.5.4, respectively, except that the number of blows applied to each layer must be increased from 27 to 62 to allow for the larger size of sample in the CBR mould. A suitable 'intermediate' compaction, recommended in the Note to Clause 7.2.4.4.1 of BS 1377:Part 4:1990, is obtained by applying 30 blows per layer using a 4.5 kg rammer. Compaction details for all three conditions are summarised in Table 11.2.

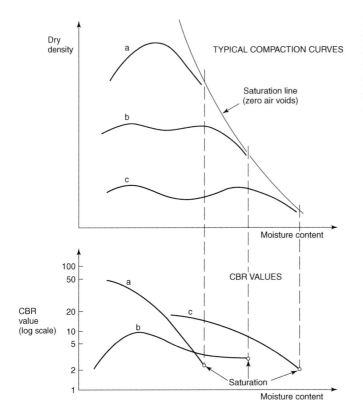

Figure 11.6 CBR values related to moisture content and compaction curves for typical soils: (a) well-graded silty sand with clay; (b) uniform fine sand; (c) heavy clay

Table 11.2 Compaction in CBR mould equivalent to BS compaction mould

Type of compaction (clause in BS 1377:Part 4:1990)	Rammer		No. of layers	Blows per layer
	mass (kg)	drop (mm)		
BS 'light' (3.4)	2.5	300	3	62
BS 'heavy' (3.6)	4.5	450	5	62
'Intermediate' (Note to 7.2.4.4.1)	4.5	450	5	30
Vibrating hammer (3.7)	30–40*	(vibration)	3	(60 s)
ASTM 'standard'	5.5 lb	12 in	3	56
'Modified AASHO'	10.0 lb	18 in	5	56

* Downward force (kgf) to be applied

For each level of compaction at least five specimens are compacted, so as to cover a range of moisture contents suitable for deriving the moisture/density curves. After compaction each specimen is tested to obtain its CBR value, as described in Section 11.7.2. Silty and clayey soils compacted at moisture contents at or above optimum should be protected against moisture loss and left for at least 24 hours before testing, to enable excess pore pressure to dissipate.

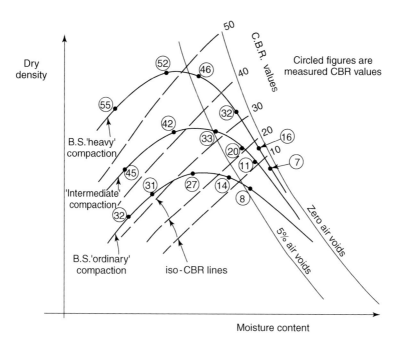

Figure 11.7 Construction of iso-CBR lines from tests on compacted samples

Each CBR value is recorded against the corresponding point on the relevant compaction curve, giving a 'map' of CBR values. By interpolation, it is possible to sketch in 'contour lines' of constant CBR values, for instance 10%, 20% and so on. These contours are known as 'iso-CBR lines' and a typical set of them is shown in Figure 11.7.

11.4 Applications

11.4.1 Purpose of CBR test

The CBR test is the most widely used of a number of empirical penetration type tests. It is perhaps the most adaptable of these tests, as it can be carried out on most types of soil ranging from heavy clay to material of medium gravel size.

The CBR test was originally devised to provide a rational method of design for flexible pavements (such as macadam or asphalt), but it can also be applied to the design of rigid (concrete) pavements and granular base courses. Test data are applicable to the design of airfield runways and taxiways as well as roads. The CBR value enables a suitable thickness of sub-base construction to be determined to withstand the anticipated traffic conditions (vehicles or aircraft), in terms of axle loadings and traffic frequency, over the design life span of the pavement. Principles and methods of design are set out in Highways Agency Interim Advice Note IAN 73/06 (Highways Agency, 2009), which supersedes previous TRL publications.

It is generally considered that the design charts and the forecasting of traffic intensity are more likely sources of error in design than is the CBR test itself.

11.4.2 Merits of CBR test

The test can be applied to a wide variety of soil types and to the design of various kinds of pavement construction, as indicated above. It can be used on the subgrade, as well as on sub-base and base course materials, and the results obtained enable maximum utilisation to be made of low-cost materials where better quality material is not available.

The test is relatively quick and simple to operate, and gives an immediate result. It can be carried out on undisturbed or recompacted materials, and can be performed in the field, in a small site laboratory, or in a main laboratory.

11.4.3 Iso-CBR lines

From a graphical plot of iso-CBR lines of the type shown in Figure 11.7, it is possible to estimate CBR values for the soil over a wide range of densities and moisture contents. Densities measured or anticipated in the field can be related to the laboratory compaction curves and the subsequent estimation of subgrade strengths for the same soil may require only a few check tests.

This type of plot is also of value in estimating the loss of strength due to flood conditions. The zero air voids line enables the maximum possible moisture content for any given density to be read off. It has been found in fact that most compact soils do not saturate much beyond the 5% air voids level.

A complete study of the effects of compaction on CBR value, especially for an unfamiliar soil, requires a comparison of the results of tests on laboratory-compacted specimens with those on undisturbed samples of compacted soil taken from site.

11.4.4 Limitations

The CBR test was devised for the particular purpose referred to above. For it to be valid the standard procedure must be strictly adhered to, otherwise results cannot be compared with those obtained elsewhere. The results are applicable only to pavement design, for which the procedure was devised.

The CBR value is a dimensionless number, and is not related to fundamental soil properties governing shear strength or compressibility. That is why the test was included in Part 4 of BS 1377, and not under strength tests. Attempts have been made (some quite successfully, for instance Black (1962)) to relate CBR values to other parameters for a particular soil, but no satisfactory relationship has been obtained for general application. For instance, the CBR test should not be used to estimate the bearing capacity of ground for foundations. The test result should be regarded as an index property, the application of which is restricted to pavement construction.

11.4.5 Typical CBR values

Some typical values of CBR for compacted British soils, when equilibrium moisture conditions have been reached, are indicated in Table 11.3. Higher equilibrium CBR values are attained with better conditions during construction, and with a lower water table (i.e. more than 300 mm below formation level). The CBR value of compacted soil is very sensitive to variations in moisture content and dry density, as explained in Section 11.3.3. These and other factors affecting measured CBR values are discussed in TRL Report 1132 (Powell *et al.*, 1984).

Table 11.3 Typical range of CBR values for compacted british soils (based on Table 5.1 of Highways Agency Interim Advice Note 73/06 (Highways Agency, 2009))

Type of soil	Plasticity	Range of CBR values
Clay	CH	1.5–2.5
	CI	1.5–3.5
Silty clay	CL	2.5–6
Sandy clay	PI = 20	2.5–8
	PI = 10	2.5–8 or more
Silt		1–2
Sand:		
poorly graded		20
well graded		40
Sandy gravel: well graded		60

11.5 Practical aspects of the test

The main features which influence the results of CBR tests are discussed below. These relate mainly to laboratory tests.

11.5.1 Effect of method of preparation

Specimens prepared in the laboratory, whether by static compression or dynamic compaction, will not necessarily give the same CBR values as those obtained in the field, or those obtained from undisturbed samples taken after compaction to the same density and at the same moisture content on site. There are several reasons for this.

1. The distribution of densities within the soil layer is not the same on site as in a laboratory-compacted specimen.
2. Moisture changes can occur quite rapidly in exposed formations on site.
3. The edge restraint of the mould imposes boundary conditions which are absent in the field, an effect which varies with the type of soil. In heavy, unsaturated clays the mould usually has little effect. In cohesive soils close to saturation, lower values are usually obtained in the laboratory test than in-situ. In granular soils the additional friction due to the confining action of the mould gives much higher values in the laboratory than those measured in-situ (Croney, 1977).
4. There are also differences between the effects of static, dynamic and vibration procedures for the preparation of laboratory test specimens. Comparisons of CBR results may not be valid if made between one method of preparation and another.
5. Recompaction in the laboratory destroys the fabric structure of natural soil.
6. The strength of remoulded cohesive soils is often less than that of the undisturbed soil and due to thixotropic effects the remoulded strength increases with time (Skempton and Northey, 1952).

The CBR test as performed in the laboratory is the standard procedure used for design purposes, but the method of sample preparation must follow standardised practice if the result is to be of value and should be related to the method of placing used in-situ (Transport Research Laboratory, 1977). The in-situ test is not normally used as a basis for design, but can be used as a rapid means of assessing the uniformity of a prepared formation, whether it is of compacted material or of natural soil.

11.5.2 Surcharge weights

Surcharge weights, in the form of annular steel discs, are usually placed on the top surface of the prepared specimen before testing. The surcharge simulates the effect of the thickness of road construction overlying the layer being tested. Each 2 kg disc is equivalent to about 70 mm thickness of superimposed construction (or each 5 lb disc is equivalent to about 3 in thickness). The exact amount of surcharge is not critical and an incorrect assessment will not affect the results very much. The effect of surcharge is greater for granular soils than for cohesive soils, but granular soils generally provide satisfactory subgrades and pavement bases so this difference is not critical.

If the specimen is to be soaked before testing (see Section 11.6.9), the surcharge rings should be placed on the sample immediately before immersion, so that their presence can control the amount of swelling.

11.5.3 Effect of soaking

The soaking of test specimens was introduced into American practice as a precaution to allow for moisture content increase in the soil due to flooding or elevation of the water table. However, soaking has been shown to give rise to conditions that are too severe in many cases, resulting in unnecessarily conservative designs of pavement thicknesses. Soaking is not normally used for CBR tests in Britain, but may be considered in arid or semi-arid climates, where soaking can result in collapse of the soil structure, as well as swelling. Soaked CBR's are frequently performed on soils that have been stabilised with lime and/or cement.

Soaking tends to produce an uneven distribution of moisture content within the specimen. Friction with the side of the mould results in non-uniform swelling, and the top 10 mm or so of soil tends to soften more than would occur in-situ. The addition of surcharge weights possibly restrains swelling to a certain extent, but with clay soils which develop high swelling pressures the surcharge would have to be quite large to provide appreciable restraint.

Provision of adequate drainage at the sides of the road formation helps to prevent inundation of the soil immediately underlying the pavement construction. The soil base then remains close to an equilibrium moisture content (Croney, 1977, Sections 6.97–6.100) on which would be based the specification for the moisture range of the test specimens.

11.5.4 Limit of penetration

The CBR test should not be taken beyond a penetration of 7.5 mm (0.3 in). If the correction to an initially concave-upward type of curve (see Section 11.7.3, stage 10b) requires a penetration greater than 7.5 mm (0.3 in) to obtain a corrected 5 mm (0.2 in) penetration, the load corresponding to 7.5 mm (0.3 in) penetration should be used instead. A curve which is continuously concave upwards, i.e. steepening throughout the test, must be interpreted with caution (see Section 11.7.3, stage 10c).

When both ends of a specimen are to be tested, the penetration applied to the first end should not be any greater than is necessary to give an unambiguous result. It is advantageous in this case to plot a force–penetration graph as the test proceeds, using a chart marked with standard curves as shown in Figure 11.5. When it is clear that the CBR has passed its maximum value, the test can be terminated. For instance, if the CBR value at 2.5 mm (0.1 in) penetration was seen to be 6%, and at 3.5 mm (0.14 in) was noticeably less, the test could be stopped and the result reported as:

CBR at 2.5 mm (0.1 in) penetration: 6%
CBR at 5 mm (0.2 in) penetration: less than 6%.

11.6 Specimen preparation

11.6.1 Methods

Remoulded specimens for CBR tests may be prepared from a disturbed sample either by static compression or by dynamic compaction or by vibration. Alternatively a CBR test may be carried out on an undisturbed sample (either of natural soil, or of material compacted in-situ) taken in a CBR mould.

The preparation of material for making up all types of remoulded specimens is described in Section 11.6.2. For a dry density or maximum air voids specification, the precise mass of soil required can be calculated. The methods of preparing test specimens are listed below and described in Sections 11.6.3–11.6.8 (see also Figure 11.2).

Static compression
1. Placing in one layer, with hand tamping, followed by static loading to obtain a required dry density (see Section 11.6.3).
2. Placing and static compression in three layers, to obtain a given dry density (see Section 11.6.4).

Dynamic compaction
3. Compaction by hand rammer to a given dry density (see Section 11.6.5).
4. Compaction by hand rammer using a specified compactive effort (see Section 11.6.7).

Compaction by vibration
5. Compaction by vibrating hammer to achieve a given dry density (see Section 11.6.6).
6. Compaction by vibrating hammer using a specified compactive effort (see Section 11.6.7).

Undisturbed
7. Taking an undisturbed sample from site for test, as outlined in Section 11.6.8. For procedures 1–3 and 5, the dry density of the soil at which the test is to be carried out must be known. The dry density (ρ_D) rather than wet (bulk) density (ρ) is used as the criterion because it is the dry density which directly affects the strength, and the condition of the soil can then be related to the compaction curve.

The procedure for soaking CBR test specimens is given in Section 11.6.9. Soaking is less common in the UK than in some other countries.

11.6.2 Preparation of material

Material for making remoulded CBR test specimens must be correctly prepared so that the standard conditions for the test are fulfilled (see Section 11.4.4).

The test is carried out on soil containing particles no larger than 20 mm. The moisture content for the test must be known and is usually specified as being representative of the state of the material which will apply in the field. Alternatively it may be desirable to carry out a series of tests over a given range of moisture contents, in which case several batches of soil are prepared.

The preparation procedures detailed below are generally as specified in BS 1377:Part 4: 1990:Clause 7.2. Initial preparation to obtain the test sample should follow the procedure given in Volume 1 (third edition), Section 1.5.

Apparatus
1. Riffle box, large enough for subdividing the original sample (see Volume 1 (third edition), Section 1.2.5, item 8)
2. Heavy-duty balance, capacity 30 kg reading to 1 g or 5 g (see Volume 1 (third edition), Section 1.2.3)
3. BS sieves, 300 mm diameter, 20 mm and 5 mm aperture, and receiver
4. Large metal tray (e.g. 760 × 760 mm and 63 mm deep)
5. Rubber pestle and mortar
6. Drying oven, 105–110°C, and equipment for moisture content determination (as described in Volume 1 (third edition), Section 2.5.2)
7. Measuring cylinders, 500 ml and 250 ml
8. Small tools such as scoop, trowel, spatula

Procedural stages
1. Remove oversize particles
2. Adjust moisture content
3. Obtain required mass of soil by riffling (only for the following conditions):
   • specification by dry density, or
   • specification by maximum air voids, or
   • specification by compactive effort

Procedure
1. Limitation of particle size
The bulk sample from which test samples are to be prepared is weighed to the nearest 5 g. If the material contains particles larger than 20 mm, these are first removed by passing the soil through the 20 mm sieve. If the soil is cohesive and too sticky to handle it may be advantageous to partially air dry it first.

The material retained on the 20 mm sieve is washed and dried if necessary, and weighed. If its mass does not exceed 25% of the mass of the original sample, no correction is necessary for its removal. If the mass retained is greater than this, the test as specified in BS 1377 is not applicable. However, it could be appropriate in some instances to perform the test on a sample in which the mass in excess of 25% is replaced by a similar mass of particles in the range 5–20 mm, obtained from a separate batch of similar soil.

(2) Moisture content adjustment
First check the moisture content of the sample as received by taking representative portions for the standard oven-drying procedure (see Volume 1 (third edition), Section 2.5.2). If the

moisture content is greater than that required for the test, it should be reduced by carefully controlled partial air drying back to the desired moisture content. Mix frequently to prevent local over-drying, and do not allow the soil to dry more than necessary. The soil should not be dried and then re-wetted. A rough check on the extent to which a wet soil needs to be dried can be obtained as follows:

Let mass of wet soil = W g

Measured moisture content = w_0%

Required moisture content = w_1% (less than w_0)

The soil is allowed to air dry and is weighed at intervals. The moisture content will be w_1% when the mass of wet soil has been reduced to W_1 g, where

$$w_1 = w \frac{(100 + w_1)}{100 + w_0} \text{ grams} \tag{11.5}$$

If the soil is too dry, it should be broken down so that there are no aggregations of particles larger than 5 mm, followed by the addition of distilled water. The amount of water to be added can be estimated as follows:

- Mass of soil and moisture content are denoted by W g and w_0% as before
- Required moisture content (greater than w_0) = w_2%

The amount of water (m_w) to be added in order to increase the moisture content from w_0% to w_2% is given by the following equation:

$$m_w = W \frac{(w_2 - w_0)}{(100 + w_0)} \text{ g (or ml)} \tag{11.6}$$

Add a little extra water (say 0.5–1% of W) to allow for evaporation loss. It is essential to mix the water thoroughly with the soil, which should then be sealed and stored for 24 hours before compacting.

After reducing or increasing the amount of moisture in the soil the actual moisture content should be verified by measurement, and a further adjustment made if necessary.

When several separate batches are required, these should be obtained from the original sample by the riffling or quartering procedure described in Volume 1 (third edition), Section 1.5.5. The amount of soil required for the preparation of each test specimen depends upon whether a given density is to be obtained, or a known compactive effort is to be applied.

3. *Preparation of required mass of soil*

Use whichever of the following procedures is appropriate:

(a) Specification by dry density (see Sections 11.6.3–11.6.6).

The mass of wet soil required to just fill the BS CBR mould (volume nominally 2305 cm³) to provide a dry density ρ_D Mg/m³ at a moisture content of w% is denoted by m_1 g and is calculated from the equation

$$m_1 = 23.05(100 + w)\rho_D \text{ g} \tag{11.7}$$

This mass is weighed out for the test after first riffling the original sample, if necessary, to approximately this quantity. Care should be taken to ensure that the proportion of any coarse material present is the same as that in the original sample.

If the measured volume of the mould differs from the nominal volume, and is denoted by V_m cm³, Equation (11.7) is replaced by the equation

$$m_1 = \frac{V_m}{100}(100 + w)\rho_D \qquad (11.8)$$

The ASTM mould with the spacer disc inserted (see Section 11.6.3) has a nominal volume of 2124 cm³ (0.075 ft³).

(b) Specification by air voids content.

The maximum permitted air voids content is denoted by V_a% for a moisture content of w%, and the particle density of the soil is denoted by ρ_s. The corresponding bulk density ρ Mg/m³ is calculated from Equation (11.4) in Section 11.3.2.

The mass of wet soil required to just fill the CBR mould of the nominal volume is equal to 2305ρ g. The corresponding dry density is given by Equation (11.2) in Section 11.3.2.

(c) Specification by compactive effort (see Section 11.6.7).

The exact mass of soil required to fill the mould is not known, so about 6 kg of soil should be prepared for each specimen as described above. The number of specimens to be tested will depend upon the range of moisture contents to be covered.

11.6.3 Static compression with tamping (BS 1377:Part 4:1990:7.2.3.2)

Apparatus

1. Cylindrical metal mould, (CBR mould, BS type), internal dimensions 152 mm diameter and 127 mm high, with detachable baseplate, top plate, and extension collar 50 mm deep. Details of two types of mould were given in Volume 1 (third edition), Figures 6.20 and 6.21. The first type has screw-thread fittings and is shown in Figure 11.8, and the second has lugs secured by wing nuts.

Figure 11.8 CBR mould, fittings and tools for BS tests

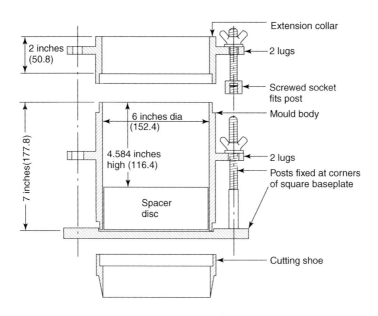

Figure 11.9 Arrangement of CBR mould and fittings (ASTM type) (dimensions in brackets are mm)

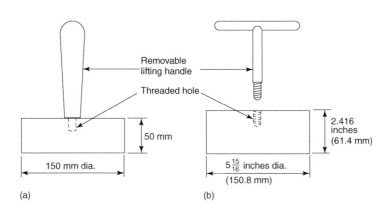

Figure 11.10 Spacer discs and lifters for CBR moulds: (a) BS type; (b) ASTM type

The US type of CBR mould (ASTM D 1883) is 6 in (152 mm) diameter and 7 in (177.8 mm) high, with a separate extension collar (see Figure 11.9.). Before compacting soil into this type of mould a spacer disc 5.94 in (150.8 mm) diameter and 2.416 in (61.4 mm) deep (see Figure 11.10 (b)) is placed at the bottom so that a specimen 4.584 in (116.4 mm) high is obtained when trimmed flush with the top of the mould, the same height as in the ASTM ('Proctor') compaction mould

2. Steel spacer plug 150 mm diameter and 50 mm thick, to which a removable lifting handle can be fitted as shown in Figure 11.10 (a)
3. Steel tamping rod about 20 mm diameter and 400 mm long
4. Filter papers (e.g. Whatman No. 1), 150 mm diameter
5. Compression machine for applying a static load for compressing soil into the mould. The maximum compressive force required is 300 kN. The machine platens should cover a circle not less than 150 mm diameter and be capable of a separation of at least 300 mm
6. Steel straight edge 300 mm long
7. Balance, preferably with a flat pan, 30 kg capacity and accurate to 5 g
8. Two cee-spanners to fit around the body and extension collar of the screw-thread type of mould, and baseplate retaining tool (see Figure 11.8)
9. If CBR tests are carried out frequently in a screwed mould, a circular metal plate fitted with suitable projecting pegs, bolted to the bench, makes for easier and safer tightening and releasing of moulds, caps and bases. Details of an attachment designed by and used at the TRL are shown in Figure 11.11

Procedural stages

1. Prepare apparatus
2. Fill mould
3. Compress soil
4. Cure sample

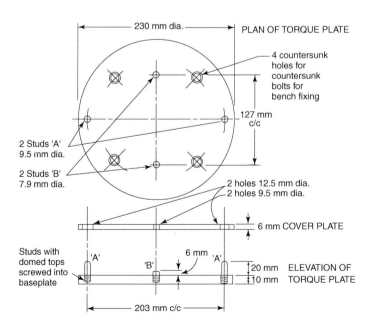

Figure 11.11 Bench-mounted torque plate for gripping a screwed CBR mould base. Cover plate is fitted when holding top cap (diagram courtesy of Transport Research Laboratory, Crowthorne, UK)

Figure 11.12 Assembling CBR mould and baseplate

Procedure

1. Preparation of apparatus

The components parts of the mould should be clean and dry. Screw threads and mating surfaces should be clean and undamaged, and very lightly oiled.

Assemble the mould and baseplate securely (see Figure 11.12) and place a filter paper to cover the base. Weigh to the nearest 1 g (m_2). Measure the internal volume V_m. The dimensions stated at the start of Section 11.6.3 under 'Apparatus', item 1, may change slightly due to wear.

Measure the depth of the extension collar, and the thickness of the spacer plug, to 0.1 mm. Fit the collar securely to the top of the mould.

When fitting together the screw-thread type of mould, avoid cross-threading and hand-tighten securely without leaving any threads exposed, but do not over-tighten.

2. Filling the mould

Pour the prepared and weighed batch of soil slowly into the mould, while at the same time tamping it with the steel rod (see Figure 11.13 (a)). Avoid segregation of particle sizes. Ensure that the largest particles are uniformly distributed within the mould. When all the soil has been added, the levelled surface should be about 5–10 mm above the top of the mould body (see Figure 11.13 (b)). If the level is significantly different from this, remove the soil, break it up again and repeat the process using a suitably modified tamping procedure.

Place a filter paper on the top of the soil, followed by the 50 mm thick spacer plug. Remove the lifting handle after placing the plug.

3. Compression

Place the mould in the compression machine and apply a load to compress the soil until the top of the plug is flush with the top of the collar (see Figure 11.13 (c)). Hold the load constant for at least 30 s before releasing. If any rebound occurs, apply the load again for a longer period. A sample ready for loading is shown in Figure 11.14.

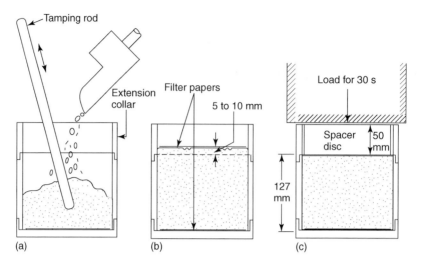

Figure 11.13 Placing and static compression of soil in CBR mould (BS test): (a) place soil in mould and tamp with rod; (b) level surface of soil 5–10 mm above mould body; (c) place spacer disc and apply load for at least 30 s

Figure 11.14 CBR sample in compression machine ready for application of load

4. *Curing*

When the sample has been fully compressed, take the mould out of the compression machine and remove the plug, collar and top filter paper. Weigh the specimen with mould and baseplate to the nearest 5 g (m_3). If the sample is not to be tested immediately, fit the top plate to the mould and store in a cool place to prevent loss of moisture. If a humidified storage cabinet or room is not available, seal the end plates with petroleum jelly, wax or tape.

For clay soils, or other soils that are near full saturation (air voids less than 5%), the sample should be allowed to stand for at least 24 hours before testing to allow equalisation and dissipation of any excess pore pressures set up by the compression process.

11.6.4 Static compression in layers (BS 1377:Part4:1990:7.2.3.3)

Apparatus
As listed in Section 11.6.3, with the addition of two further spacer plugs (item 2), making three in all.

Procedural stages
1. Prepare apparatus
2. Sub-divide sample
3. Fill mould
4. Cure sample

Procedure

1. Preparation of apparatus
    As in Section 11.6.3, stage 1.

2. Sub-division of sample
    Divide the prepared batch of soil into three portions equal to within 50 g. Place each into a sealed bag or container until it is required, to prevent moisture loss.

3. Filling mould
    Place one portion of the batch into the CBR mould, level the surface and place all three spacer plugs on top. Compress the soil in the compression machine until the thickness of soil, after removal of the load, is about one-third of the depth of the mould (42 mm). The surface of the topmost plug is then about 15 mm above the top of the extension collar (see Figure 11.15(a)).
    Remove the spacer plugs, add another portion of soil to form a second layer and repeat the above process, using two plugs. The soil thickness should then be 85 mm and the surface of the upper plug about 8 mm above the extension collar (see Figure 11.15 (b)).

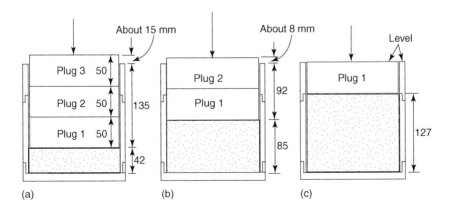

Figure 11.15 Static compression of soil in three layers in BS CBR mould: (a) after compression of first layer; (b) after compression of second layer; (c) after final compression

Repeat for the third layer, using only one plug, which should be pushed down until the top surface remains level with the top of the collar, as in Section 11.6.3, stage 3) (see Figure 11.15 (c)).

4. Curing

As in Section 11.6.3, stage 4.

11.6.5 Compaction by rammer to specified density (BS 1377:Part 4:1990:7.2.4.2)

Compaction to a required density by using a hand rammer or mechanical compactor is one alternative to static compression when a compression machine is not available. The British Standard method is described below. The ASTM method is slightly different and is outlined separately.

Apparatus

Items 1, 4, 6–8 listed in Section 11.6.3, with the addition of the following.

10. Compaction rammer, 2.5 kg with 300 mm drop, as used for the BS 'light' compaction test (see Volume 1 (third edition), Figure 6.9).

11. Compaction rammer, 4.5 kg with 450 mm drop, as used for the BS 'heavy' compaction test (see Volume 1 (third edition), Figure 6.17).

12. Alternatively, an automatic compactor of the type referred to in Volume 1 (third edition), Section 6.5.8 and shown in Figure 6.19, could be used.

Procedural stages

1. Prepare apparatus
2. Sub-divide sample
3. Make preliminary trials
4. Compact into mould
5. Trim sample
6. Cure sample

Procedure

1. Preparation of apparatus

As in Section 11.6.3, stage 1.

2. Sub-division of sample

Divide the prepared batch of soil into five portions equal to within 50 g. Place each into a sealed bag or container until required for use, to prevent moisture loss.

3. Preliminary trials

The degree of compaction required to just fill the mould (stage 4) may have to be determined by trial. It depends upon the type of rammer used, as well as the number of blows applied to each layer.

A separate batch of the required mass of soil should be used for trial tests. For each trial follow the procedure given in stage 4 below, and record the type of rammer, number of layers, number of blows per layer, and final thickness of compacted soil. The required degree of compaction can then be assessed by interpolation.

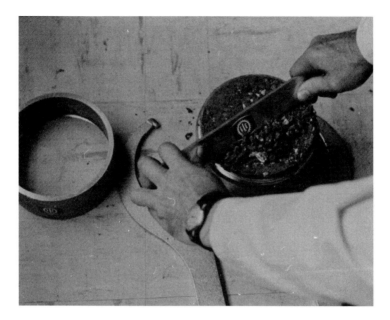

Figure 11.16 Trimming soil surface in CBR mould

4. Compaction into mould
Place the first portion of soil into the mould and compact it with the appropriate hand rammer, applying blows evenly over the surface, until the layer occupies one-fifth of the height of the mould (about 25 mm). Repeat the process using the other four portions, so that the compacted level of the fifth layer of soil is level with or only just above the top of the mould. Lightly scarify the top surface of each compacted layer before placing the next.

5 Trimming sample
Carefully remove the extension collar. Using the scraper, trim the soil flush with the top edge of the mould (see Figure 11.16) and check with the straight edge. Weigh the sample with mould and baseplate to the nearest 5 g (m_3).

6. Curing
As in Section 11.6.3, stage 4.

Use of ASTM mould
Before compacting into the ASTM type of CBR mould (shown in Figure 11.17), the spacer disc of 2.416 in (61.4 mm) thickness (see Figure 11.10(b)) must be placed inside the mould. After compaction and trimming off the top surface, the baseplate and spacer disc are removed. The mould is inverted so that the trimmed face of the sample can be placed on a perforated baseplate, separated from it by a sheet of filter paper. Finally the mould is clamped to the baseplate. The procedure is illustrated in Figure 11.18.

Soaking, which is required by the ASTM procedure, is then carried out as described in Section 11.6.9, and the penetration test is similar to that given in Section 11.7.

Figure 11.17 CBR mould, fittings and accessories (ASTM type)

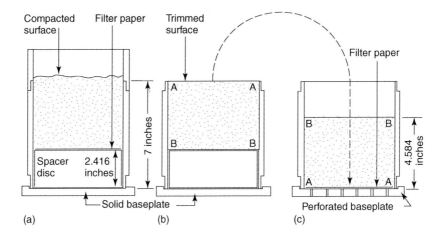

Figure 11.18 Compaction into ASTM CBR mould: (a) compact soil into mould on spacer disc; (b) remove extension collar, trim level, disconnect baseplate; (c) invert mould on to perforated baseplate and clamp together

11.6.6 Compaction by vibration (BS 1377:Part 4:1990:7.2.4.3)

Compaction to a required density under vibration is suitable for granular soils and is another alternative to static compression where a compression machine is not available.

Apparatus

Items 1, 4, 6–8 listed in Section 11.6.3, with the addition of the following:
   13. Electric vibrating hammer, as used for the BS vibrating hammer compaction test (see Volume 1 (third edition), Section 6.5.9, shown in Figure 6.22).
   14. Steel tamper with circular foot 145 mm diameter (see Volume 1 (third edition), Figure 6.23).
   15. Support frame for vibrating hammer, as shown in Figure 6.22 of Volume 1 (third edition). This item is not essential, but it enables a more uniform pressure to be exerted on the sample and is less tiring for the operator than supporting the vibrating hammer by hand.

Procedural stages
1. Prepare apparatus
2. Sub-divide sample
3. Compact into mould
4. Trim sample
5. Cure sample

Procedure

1. Preparation of apparatus
As in Section 11.6.3, stage 1.

Ensure that the vibrating hammer is in good order, and is set up in accordance with the manufacturer's instructions. It should be properly connected to the mains electricity supply, via an earth-leakage circuit breaker, by a cable in sound condition. If the support frame is used the sliding parts should move freely without jerking or sticking.

The tamper must fit properly into the hammer adaptor, and the foot should fit inside the CBR mould with the necessary 3.5 mm clearance all round.

2. Sub-division of sample
Divide the prepared batch of soil into three portions equal to within 50 g. Place each in a sealed bag or container until required.

3. Compaction into mould
Place the first portion of the soil into the mould, and compact with the vibrating hammer until the soil occupies one-third of the depth of the mould. Take care to keep the hammer vertical. The surface of the soil will then be about 135 mm below the top of the collar (see Figure 11.19(a)).

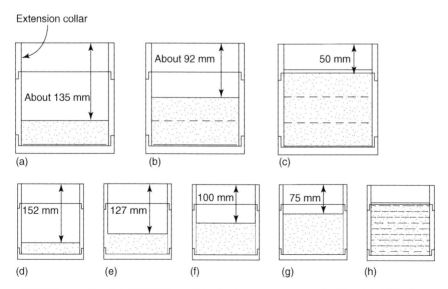

Figure 11.19 Compaction into BS CBR mould using hand rammer or vibration: (a)–(c) in three equal layers; (d)–(h) in five equal layers

Place the second portion of soil in the mould and compact as before so that its surface is about 92 mm below the collar (see Figure 11.19 (b)). Place the third portion and compact again so that the final surface is just above the top of the mould body, i.e. about 45 mm below the top of the collar (see Figure 11.19 (c)).

4. *Trimming sample*
Remove the extension collar and trim the soil level and flush with the top of the mould. Weigh the sample with the mould and baseplate to the nearest 5 g (m_3).

5 *Curing*
As in Section 11.6.3, stage 4.

11.6.7 Compaction using specified effort (BS 1377:Part 4:1990:7.2.4.4 and 7.2.4.5)

This procedure is used when a given amount of compactive effort is to be applied to the soil, equivalent to that used in one of the standard compaction tests. It is the one normally followed when a relationship is to be obtained between dry density and CBR value; or when studying the effect of moisture content on CBR value for a given compactive effort.

The degree of compaction applied is usually equivalent to either the BS 'light' compaction using a 2.5 kg rammer, or the BS 'heavy' compaction using a 4.5 kg rammer (see Volume 1 (third edition), Sections 6.5.3 and 6.5.4, respectively). An intermediate degree of compaction may also be used, such as that suggested in Section 11.3.4 (see Table 11.2).

A vibrating hammer may be used for compaction of granular soils. The 'standard' vibration procedure is that described in Volume 1 (third edition), Section 6.5.9 (see Table 11.2). Two additional intensities of vibration which could be used are, for example, compaction in three layers for 30 s per layer; and in five layers for 2 min per layer. Dynamic and vibration compaction procedures should not be mixed in a test series (see Section 11.5.1, item 4).

Apparatus
As for the compaction procedures described in Section 11.6.5 or 11.6.6. Use of an automatic mechanical compactor is preferable to hand ramming.

Procedural stages
    1.   Prepare apparatus
    2.   Compact into mould ('light' compaction), or
    3.   Compact into mould ('heavy' compaction), or
    4.   Compact into mould by vibration
    5.   Trim sample
    6.   Cure sample

Procedure

1. *Preparation of apparatus*
As in Section 11.6.3, stage 1, or 11.6.6, stage 1.

2. *Compaction into mould ('light' compaction)*
Place a quantity of soil in the mould so that after 62 blows of the 2.5 kg rammer the surface is one-third of the distance up the mould or a little higher (135–130 mm below the top of

the extension collar, as in Figure 11.19 (a)). The blows should be uniformly applied over the area, the first few being applied as shown in Volume 1 (third edition), Figure 6.18. Lightly scarify the surface of the compacted soil. Add the same amount of soil as for the first layer, and compact the second layer in the same way so that the soil surface is about 92 mm below the collar (see Figure 11.19 (b)). Repeat the process again for the third layer. The final level of the soil surface should be just above the top of the mould by not more than 6 mm (see Figure 11.19 (c)). It may be necessary first to carry out trial compactions on a separate batch of soil in order to estimate the correct quantity for each layer.

In the ASTM mould, 56 blows are applied to each layer.

3. Compaction into mould ('heavy' compaction)
The procedure is similar to stage 2 except that the soil is placed in five equal layers instead of three, and the 4.5 kg rammer is used, applying 62 blows for each layer (Figures 11.19 (d)–(h)) (56 blows per layer in the ASTM mould).

If an intermediate degree of compaction is required, it is suggested in the BS that this should be standardised by using the 4.5 kg rammer and applying 30 blows to each of five layers.

4 Compaction into mould (vibration)
The procedure is similar to stage 2 if the soil is vibrated in three layers, or stage 3 if in five layers, except that the vibrating hammer is used. In the 'standard' procedure the vibrator is applied to each layer for 60 s, as described in Volume 1 (third edition), Section 6.5.9, stage (4). Shorter or longer periods of vibration may be applied to obtain different degrees of compaction, as suggested above.

5. Trimming sample
Remove the extension collar and trim the soil level and flush with the top of the mould, checking with the straight-edge. Surface cavities may be filled with fine material lightly pressed or tamped into place using a spatula. Weigh the sample with mould and baseplate to the nearest 5 g (m_3).

6. Curing
As in Section 11.6.3, stage 4.

11.6.8 Undisturbed sample

An undisturbed sample can be obtained in-situ from cohesive soil not containing stones, by fitting a steel cutting shoe to one end of the CBR mould body and forcing it steadily into the ground. Hammering should be avoided, unless the full penetration required can be achieved by a single sharp blow. In that case a lump of wood placed on the mould will protect the metal against damage. If may be necessary to roughly trim the sample in-situ first, leaving the last 2 or 3 mm to be sliced off by the cutting edge as the mould is pushed downwards. The cutting shoe is removed after taking the sample from the ground, and the ends of the sample are cut and trimmed flush with the mould, checking with a straight-edge. Any cavities between sample and mould should be filled with well-compacted fine soil or with molten paraffin wax, just before testing, so that the sample is fully supported all round.

The sample is weighed in the mould to the nearest 5 g before fitting the baseplate, the mould body having been weighed separately. Excess soil which has been trimmed from the ends, provided that it has been protected against moisture loss while in transit, can be used to determine the in-situ moisture content.

If the sample is not to be tested immediately, the top plate is fitted to the mould or the top surface is sealed with polythene sheet to prevent loss of moisture.

11.6.9 Soaking (BS 1377:Part 4:1990:7.3)

Soaking of CBR test samples is not normally employed in British practice, but it may be appropriate for some overseas countries. The soaking procedure given in the BS, which is similar to that given in the ASTM (designation D 1883) is described below.

Apparatus

The following items are required in addition to the CBR mould fittings (item 1 of Section 11.6.3).

16. Perforated baseplate for CBR mould
17. Perforated swell plate with adjustable stem (see Figure 11.20)
18. Tripod mounting for dial gauge indicator (see Figure 11.20)
19. Dial gauge, 25 mm travel, reading to 0.01 mm (see Figure 11.20)
20. Soaking tank, large enough to accommodate the CBR mould with baseplate. A suitable size is about 610 × 610 × 380 mm deep, preferably with an open-mesh platform (see Figure 11.21)
21. Annular surcharge discs, external diameter 145–150 mm, internal diameter 52–54 mm, mass 2 kg. Up to three may be needed. Alternatively split (half-round) segments may be used (see Section 11.5.2 for guidance regarding the number of discs to use)
22. Petroleum jelly (Vaseline)
23. Stop-clock or timer

Figure 11.20 Accessories for measurement of swelling in soaking test

Figure 11.21 Soaking tank

Procedural stages

1. Set up
2. Immerse sample
3. Take readings
4. Plot readings
5. Remove from tank
6. Prepare for test

Procedure

1. *Setting up*

After preparing the sample either by static compression or by dynamic compaction, and weighing, remove the baseplate and replace it by the perforated base. Fit the collar to the other end, packing screw threads and mating surfaces with petroleum jelly to obtain a watertight joint. Gently lower the mould into the empty tank.

Plate a filter paper on top of the specimen, followed by the perforated swell plate with stem. Place the required number of surcharge discs on top of the perforated plate (see Section 11.5.2).

For the ASTM test a surcharge of not less than 4.54 kg is required, and if no surcharge is specified this mass should be applied.

Mount the dial gauge support on top of the extension collar, fit the dial gauge and adjust the level of the stem on the perforated plate so that the gauge reads zero or some convenient value. The assembly is shown diagrammatically in Figure 11.22, and a typical arrangement is shown in Figure 11.23.

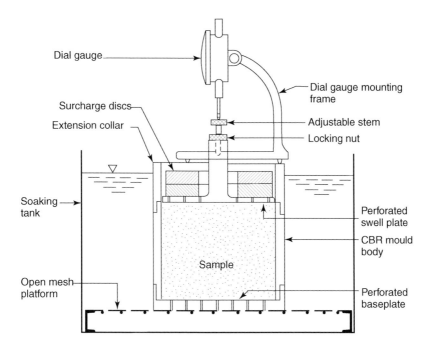

Figure 11.22 Arrangement for soaking and swelling test

Figure 11.23 Swelling test apparatus

2. *Immersion*
Add water to the soaking tank to bring the level to just below the top of the mould collar, without allowing water to splash over on to the top of the sample. Start the timer as soon as the water just covers the baseplate.

In the ASTM procedure the mould is completely immersed so that water has free access to the top and bottom of the sample, and sealing the joint with petroleum jelly is omitted. This procedure has also been used at TRL (Daniel, 1961). The BS method enables air to be displaced by the rising water and probably leads to more uniform saturation.

3. *Taking readings*
Record the reading of the dial gauge at suitable time intervals after immersion, depending upon the rate at which movement takes place. Also record the time when the water is seen to reach the top surface of the sample, but this may not necessarily indicate the end of the swelling stage.

If water does not appear at the top surface after three days' immersion, pour water on to the top surface so that it remains covered and leave to soak. The normal soaking period is four days, but a longer period may be necessary to allow swelling to reach completion.

4. *Plotting readings*
Draw a graph of swelling (dial gauge movement) against elapsed time, or square-root time, since immersion. Swelling is substantially complete when the curve has flattened out, as shown on the typical swelling curve in Figure 11.24.

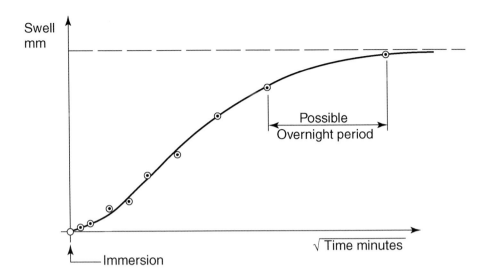

Figure 11.24 Typical swelling curve from CBR soaking test

5. *Removing from tank*

When the soaking is complete take off the dial gauge support, and remove the mould and specimen from the tank. Place the mould on a surface where it can be allowed to drain for about 15 min. Alternatively, if the soaking tank incorporates a perforated tray (as in Figure 11.21), the mould can be left in the tank to drain after the water has been drained or siphoned out.

6. *Preparation for test*

After surplus water has drained away, remove the surcharge discs, perforated plate and extension collar. Take off the perforated baseplate and re-fit the solid baseplate. Weigh the sample with mould and baseplate to the nearest 5 g (m_4).

The ASTM states that it might be necessary to tilt the sample to remove surface water. It is better to siphon off excess water, to avoid the possibility of disturbing the top of the sample.

11.7 Penetration test

11.7.1 General

The procedure given below is based on Clause 7.4 of BS 1377:Part 4:1990. It is covered in ASTM Test Designation D 1883, as the test for the bearing ratio of laboratory-compacted soils.

11.7.2 CBR test procedure (BS 1377:Part 4:1990:7.4, and ASTM D 1883)

The CBR test itself is the same by whatever means the sample has been prepared, and whether it has been soaked or not. The test is often carried out on both ends of the sample, but the BS procedure refers to a repeat test on the base as being optional. It is sometimes convenient to test the upper end without soaking, and then to invert the sample and test the other end after soaking.

Apparatus

1. Load frame (compression testing machine) with a means of applying the test force at a nominal rate of penetration of 1 mm/min. Unloaded, the machine approach speed should be 1.2 mm/min to within ± 0.2 mm/min. The machine should be capable of applying a force of at least 45 kN at this rate of penetration. Hand operation is possible but less convenient.
2. Load measuring device, such as a calibrated load ring or load cell (see Section 8.2.1, subsection on 'Conventional and electronic measuring instruments'). Rings of three capacities are required to cover the range of values normally obtained, as shown in Table 11.4.
3. Cylindrical metal plunger, approximately 250 mm long, the lower end of hardened steel with a cross-sectional area of 1935 mm^2 (3 in^2). This corresponds to a diameter of 49.64 mm, which is rounded to 49.65 ± 0.10 mm in the BS.
4. Dial gauge or displacement transducer (see Section 8.2.1, subsection on 'Conventional and electronic measuring instruments') with a range of 25 mm, reading to 0.01 mm, for measuring the penetration of the plunger into the specimen. The gauge is fitted with an extension stem and a chisel-edge anvil.
5. Mounting bracket for attaching dial gauge to plunger.
6. Surcharge discs, as appropriate (see Section 11.6.9 (21)).

The general arrangement of the apparatus is given in Figure 11.25, and the complete assembly is shown in Figure 11.26.

Procedural stages

1. Set up in load frame
2. Seat plunger
3. Set machine
4. Run test
5. Remove from machine
6. Repeat test on base (if required)
7. Remove sample from mould
8. Measure moisture content
9. Calculate and plot data
10. Correct curve, if necessary
11. Calculate CBR value } see section 11.7.3
12. Calculate density
13. Report result

Table 11.4 Load rings for CBR tests

Expected CBR value	Load ring	
	range (kN)	readability (N)
up to 8%	0–2	2
8–40%	0–10	10
Above 40%	0–50	50

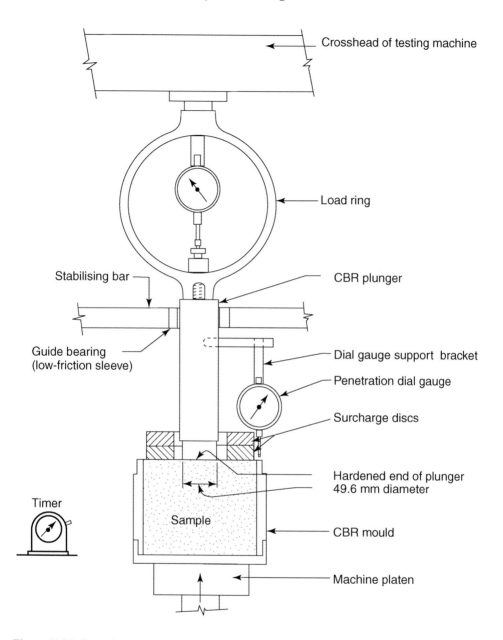

Figure 11.25 General arrangement for CBR penetration test

Procedure

1. *Setting up in load frame*

The following description relates to the use of a load frame of the type shown in Figure 11.26. Procedural details may differ for other types of load frame, but the principles are the same.

Figure 11.26 Load frame set up for CBR test

Place the mould with baseplate containing the sample centrally on the platen of the testing machine, with the platen wound down to near its lowest position. If annular surcharge discs are used they must be placed in position on the exposed face of the sample before fitting the plunger, but semi-circular rings can be added afterwards. If the sample has been soaked, the surcharge should be equal to that used during soaking. The ASTM states that if no surcharge is specified for the test, a mass of 4.54 kg should be applied.

Fit the load ring to the cross-head of the load frame, and attach the plunger to the load ring. To maintain alignment of the plunger it should be supported in the guide bearing of a stabiliser beam fitted to the tension rods of the machine, as shown in Figure 11.25 and Figure 11.26.

Wind up the machine platen so that the end of the plunger is almost in contact with the top of the sample. If the cross-head of the machine had to be raised in order to insert the surcharge discs it should be lowered again, otherwise there may not be enough platen travel left to perform the test. Ensure that the plunger is aligned vertically, and that there is a uniform clearance between it and the surcharge discs. Check that the connections between plunger, load ring and cross-head are tight.

Mount the penetration dial gauge on the bracket attached to the plunger. The stem of the gauge rests on the upper edge of the extension collar or mould, with the chisel-edge anvil across the thickness of the wall (see Figure 11.27 (a)). Alternatively, the chisel edge could bear on the shoulder of the mould (see Figure 11.27 (b). Ensure that the connections are tight, and that the dial gauge stem has at least 10 mm free travel.

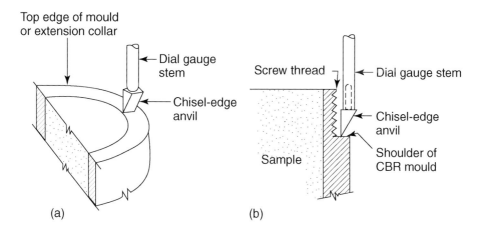

Figure 11.27 Use of chisel-edge (offset) anvil on penetration dial gauge: (a) bearing on upper edge of mould or collar; (b) bearing on shoulder of mould

2. *Seating plunger*

The plunger must be seated on top of the specimen under a 'seating force', the value of which depends on the expected CBR value, as follows:

        CBR value up to 5%: apply 10 N
        CBR value from 5% to 30%: apply 50 N
        CBR value above 30%: apply 250 N

    Wind up the machine platen slowly by hand until the load ring indicates this reading. Then re-set the load dial gauge to zero, because the seating load is not taken into account in the test. Adjust the penetration dial gauge to read zero, or some convenient datum reading.

3. *Setting machine*

Select the machine speed to give a platen speed of 1 mm/min, or the nearest available speed. Change from hand control to motor drive. Make adjustments to the dial gauges to bring them back to the zero or datum readings if they have moved. The complete apparatus ready for testing is shown in Figure 11.26.

4. *Running the test*

Switch on the motor, and as the penetration proceeds record the load ring dial reading at every 0.25 mm interval of the penetration dial gauge. If the gauge makes one revolution for every millimetre, a load reading is taken every quarter revolution (25 divisions). This type of gauge is convenient if the load is being applied by hand, because the correct rate of penetration is ensured if the penetration pointer is made to keep pace with the seconds pointer of a clock. When the penetration reading reaches 7.5 mm, stop the machine (see Section 11.5.4).

    In the ASTM procedure, readings are taken at penetration intervals of 0.025 in (0.64 mm) up to 0.2 in penetration, then at penetrations of 0.3, 0.4, 0.5 in.

    Part of a typical set of readings from a CBR test is shown in Figure 11.28.

CBR Test data

BS 1377: 1975 Test 16	Location _Highbury_	Sample No _3/24_
	Operator ___R.B.S___	Date ___30.4.81___

Soil description _____ *Dark grey silty clay*

SPECIMEN PREPARATION   Recompacted   Dynamic

Recompaction ___*B S Heavy*___	Container No ___B - 3___
No. of layers ___5___ Rammer _4.5_ kg	Diameter _152.6_ mm Volume
Blows per layer ___62___ Drop ___450___ mm	Height ___127___ mm _2323_ cm³

Test on TOP / surface   As compacted   Surcharge rings _2_ No

Load _4_ Kg

PENETRATION TEST   Load ring No ___R512___ Capacity ___10___ KN

Penetration mm	Load dial divs.	Ring factor N/div	Load KN
0	6	8. 56	0.05
0.25	30		0.26
0.5	66		0.56
0.75	115		0.98
1.0	181		1.55
1.25	219		1.87
1.5	259	8.50	2.20
1.75	292		2.48
2.0	324		2.75
2.25	354		3.01
2.5	380		3.23
6.25	634	8.43	5.34
6.5	646		5.45
6.75	655		5.52
7.0	663		5.59
7.25	670		5.65
7.5	676		5.70

DENSITY

Mass of wet soil + mould	_10.075_ kg
Mould	_5.307_ kg
Wet soil	_4.718_ kg
Bulk density	_2.05_ Mg/m³
Moisture content	_21.4_ %
Dry density	_1.69_ Mg/m³

Figure 11.28 Typical data from CBR test

5. Removal from machine

Wind down the machine platen, either by hand or by putting the machine into reverse, and if necessary raise the cross-head, so that the surcharge weights can be removed and the mould taken off the machine.

If a second test on the base of the sample is not required, continue at stage 7. If a repeat test is to be carried out on the base, follow stage 6 first.

6. Repeat test on base

Fill up the depression made by the plunger with similar soil, and remove any projecting material. True up the surface and check with the straight-edge.

Take off the baseplate from the lower end of the mould and fit it securely to the top end. Invert the mould, and trim the exposed surface if necessary.

If the sample is to be soaked before the repeat test, carry out the procedure given in Section 11.6.9, followed by stages 1–5 above. If not, proceed from stage 1 above.

7. Removal from mould

The easiest way of removing the tested sample from the mould is by using a purpose-built extruder of the type shown in Chapter 9, Figure 9.6 (b). After removal of the sample the mould is cleaned and made ready for re-use.

8. Measurement of moisture content

For a cohesive soil containing no gravel-size particles, take a sample of soil about 350 g from immediately below the penetrated surface for the determination of moisture content, before removing the sample from the mould. If tests were carried out at both ends the moisture content sample taken from the end first tested should not include material used for filling the depression.

For a cohesionless soil, or a cohesive soil containing gravel-size particles, extrude the whole sample first, break it in half and weigh, dry and weigh the upper and lower halves separately. If the sample was soaked between the first and second tests, the moisture content measured in this way will not relate to the unsoaked CBR result obtained on the end first tested. In this case the moisture content measured when the soil was placed in the mould is more relevant. If the change in mass due to soaking was measured, this provides a cross-check on the initial moisture content.

11.7.3 Plotting, calculating and reporting

9. Load–penetration curve

Readings of the load ring dial gauge are converted to force units by multiplying by the load ring factor.

If R = load dial reading (divisions)

F = load ring factor (newtons per division, N/div.)

P = force applied to specimen (kN)

then

$$P = \frac{FR}{1000} \text{ kN} \tag{11.9}$$

Values of the force P (kN) are plotted against the corresponding plunger penetration (mm) and a smooth curve is drawn through the points. Three examples, curves A, B and C are shown in Figure 11.29; curve A is derived from the data given in Figure 11.28.

It is not essential to calculate the applied force for every reading. A graph can be plotted of load ring dial readings against penetration, and the force need be calculated only at penetrations of 2.5 mm and 5 mm (corrected if necessary as described below).

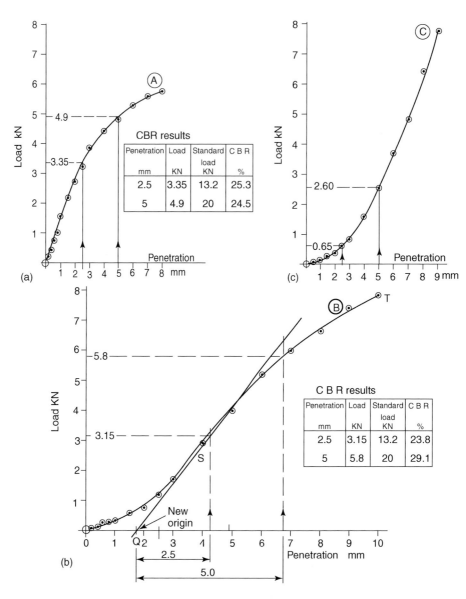

Figure 11.29 Three types of load-penetration curve from CBR tests: (a) no correction required; (b) correction required as indicated; (c) correction as (b) may not be valid

10. *Correction to curve*

    (a) A load–penetration curve similar to curve A in Figure 11.29, i.e. convex upwards, requires no correction and can be used as it stands for the calculations given in stage 11.

    (b) A curve similar to curve B in Figure 11.29, in which the initial part is concave upwards and is followed by curvature in the other direction, indicates that the surface layer may be disturbed or less well compacted than the underlying material. This type of curve should be corrected as follows.

Draw a tangent to the curve where the slope is greatest (the point of inflexion, S). Produce the tangent to cut the penetration axis at the point Q. This point is taken as the origin of the corrected curve represented by QST and a new penetration scale starting with zero at Q is added.

(c) If the graph continues to curve upwards as curve C in Figure 11.29, the probable behaviour of the soil under the plunger should be considered and referred to the engineer. This type of curve is sometimes obtained with granular soils compacted to a low or medium density, and indicates that further compaction and consequent increase of strength may be taking place under the plunger as it penetrates into the soil. If the engineer considers that the steepening curve is a result of continuing compaction, the CBR values at 2.5 mm and 5 mm penetration should be read off without correction (Daniel, 1980), otherwise the correction referred to under (b) above may lead to a grossly exaggerated CBR value.

11. Calculation of CBR

The CBR value is calculated from the loads corresponding to plunger penetrations of 2.5 mm and 5 mm as illustrated by the following examples. For curve A (see Figure 11.29), ordinates are drawn at these penetrations from the original penetration scale on the horizontal axis, and are shown by heavy full lines. Their intersections with the curve enable the corresponding loads to be read off (3.35 kN and 4.9 kN in this example). From Table 11.1 (see Section 11.3.1), the standard loads corresponding to 2.5 mm and 5 mm penetration are 13.2 kN and 20.0 kN, and these are used for the calculation of CBR as follows:

At 2.5 mm penetration

$$CBR = \frac{3.35}{13.2} \times 100\% = 25.4\%$$

At 5 mm penetration

$$CBR = \frac{4.9}{20.0} \times 100\% = 24.5\%$$

The greater of the two values is the accepted result, which is rounded off and reported as 25%.

For curve B, ordinates are drawn at penetrations of 2.5 mm and 5 mm from the corrected penetration scale, with origin at Q, and are shown as heavy dashed lines. Their intersection with the curve give loads of 3.15 kN and 5.8 kN, and the corresponding CBR values are calculated as follows:

At 2.5 mm penetration

$$CBR = \frac{3.15}{13.2} \times 100\% = 23.9\%$$

At 5 mm penetration

$$CBR = \frac{5.8}{20.0} \times 100\% = 29.0\%$$

The accepted CBR is reported as 29%.

The CBR value can be estimated as the test proceeds, without the need for calculations, if the load–penetration curve is plotted on a chart similar to that shown in Figure 11.5. The curve marked CBR = 100% is obtained from the data given in the BS for the standard curve which is shown in Table 11.1 and Figure 11.4. Curves for other CBR values are obtained by multiplying these loads by the CBR value (20%, 40%, etc.) The CBR value from a test curve is estimated by interpolation between the standard curves, as shown in Figure 11.30 for curve A. Low values of CBR are difficult to assess directly unless a similar set of curves is drawn to a larger scale to cover the low range.

If a correction to the test curve is necessary (see Figure 11.30, curve B) the points at which CBR values are interpolated can be obtained by following the procedure as indicated, which is based on that shown in Figure 11.29 (b).

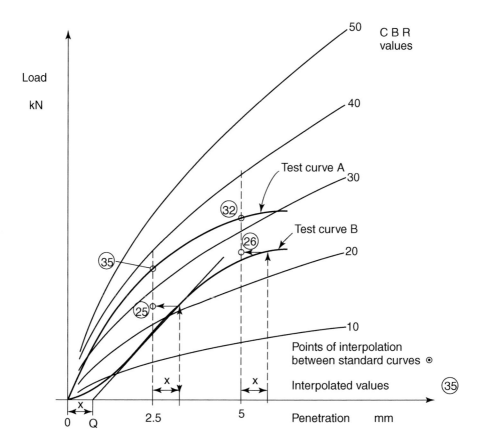

Figure 11.30 Estimation of CBR values by plotting on chart with printed standard curves. Test curve type A: Interpolate between CBR values corresponding to 2.5 mm and 5 mm penetration. Test curve type B: obtain Q as in Figure 11.29(b). Offset by x = OQ and obtain interpolation points from test curve as shown

12. *Density calculations*

The bulk density, ρ, of the test sample is calculated from the dimensions of the CBR mould and the weighed masses. The dimensions given in the BS give a mould volume of 2305 cm³, and in the ASTM 2214 cm³, but if the measured dimensions are different the actual volume should be calculated. Symbols and their units used below are the same as those referred to earlier, and are summarised as follows:

> ρ, bulk density of test sample (Mg/m³)
> ρ_D, dry density of test sample (Mg/m³)
> w, moisture content of test sample (%)
> m_1, mass of soil in mould (g)
> m_2, mass of mould and baseplate (g)
> m_3, mass of sample, mould and baseplate after compaction or compression (g)
> m_4, mass of same after soaking (g)
> For a known mass of soil (m_1 g):

$$\rho = \frac{m_1}{2305} \ \mathrm{Mg/m^3} \quad \text{or} \quad \frac{m_1}{2214} \ \mathrm{Mg/m^3}$$

For a given degree of compaction:

$$\rho = \frac{m_3 - m_2}{2305} \ \mathrm{Mg/m^3} \quad \text{or} \frac{m_3 - m_2}{2214} \ \mathrm{Mg/m^3}$$

For a soaked specimen:

$$\rho = \frac{m_4 - m_2}{2305} \ \mathrm{Mg/m^3} \quad \text{or} \frac{m_4 - m_2}{2214} \ \mathrm{Mg/m^3}$$

The dry density, ρ_D, is calculated from ρ and the moisture content $w\%$ by the equation

$$\rho_D = \frac{100}{100 + w}\rho$$

13. *Reporting results*

The CBR is reported to two significant figures. Results of tests on each end of the sample are reported separately, but if they are within 10% of the mean value the average result may be reported.

The load–penetration curve, showing corrections if appropriate, should be presented as part of the test results. Curves for tests on the two ends of the sample can be shown on the same graph sheet. If a test was discontinued for the reason given in Section 11.5.4, this should be explained. The test report should also include the following:

- Method of test, with reference to BS 1377: Part 4:1990, Clause 7 (or to ASTM D 1883)
- Initial density, moisture content and dry density of test sample
- Method of sample preparation
- Proportion by mass of any oversize particles removed

- Whether or not the sample was soaked, and if so the period of soaking, the amount of swell and the swell–time curve
- Surcharge applied, for soaking and for the test.
- Moisture contents after test, taken either below the plunger or from the two halves of the sample

A suitable form for reporting results is shown in Figure 11.31.

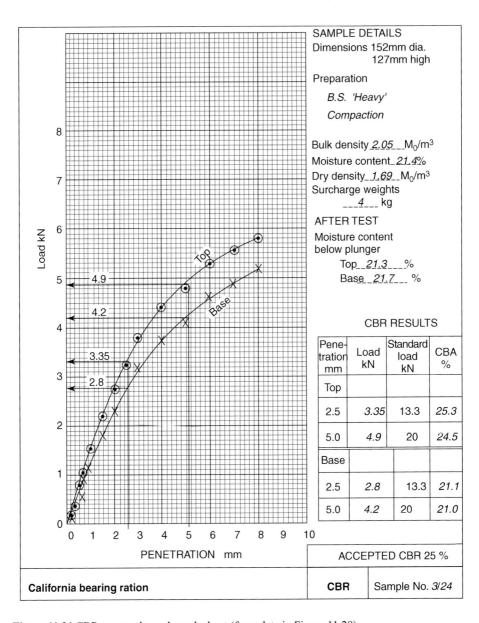

Figure 11.31 CBR test results and graph sheet (from data in Figure 11.28)

References

ASTM D 1883-07 E02, Standard test method for CBR (California Bearing Ratio) of laboratory-compacted soils. American Society for Testing and Materials, Philadelphia, PA, USA

BS 6100-4:2008 Building and civil engineering vocabulary: Part 4:Transport. British Standards Institution, London

Black, W. P. M. (1961). Calculation of laboratory and in-situ values of California bearing ratio from bearing capacity date. *Géotechnique*, Vol. 11, No. 1, pp. 14–21

Black, W. P. M. (1962). A method of estimating the California bearing ratio of cohesive soils from plasticity data. *Géotechnique*, Vol. 12, No. 4.

Croney, D. (1977). *The Design and Performance of Road Pavements*. HMSO, London.

Daniel, M. N. (1961). An investigation of the effects of soaking and compacting procedure on the results of California bearing ratio tests on two tropical soils and stabilised soils. Research Note No. RN/4088/MND, November 1961. Transport Research Laboratory, Crowthorne, UK

Daniel, M. N. (1980). Private communication

Davis, E. H. (1949). The California Bearing Ratio method for the design of flexible roads and runways. *Géotechnique*, Vol. 1, No. 4, pp. 249–263

Highways Agency (2009). Design Guidance for Road Pavement Foundations (Draft HD25). Interim Advice Note 73/06 Revision 1 (2009). Highways Agency, London

Porter, O. J. (1938). The preparation of subgrades. *Proc. Highw. Res. Bd.*, Vol. 18, No. 2, pp. 324–331

Porter, O. J. (1949). Development of CBR flexible pavement design for airfields. Development of the original method for highway design. *Proc. ASCE*, Vol. 75, pp. 11–17

Powell, W. D., Potter, J. F., Mayhew, H. C. and Nunn, M. E. (1984). *The Structural Design of Bituminous Roads*. TRL Report 1132. Transport Research Laboratory, Crowthorne, UK

Skempton, A W. and Northey, R. D. (1952). Sensitivity of clays. *Geotechnique*, Vol. 3, No. 1, pp. 40–51

Stanton, T. E. (1944). Suggested method of test for the California bearing ratio procedures for testing soils. ASTM, Philadelphia, PA, USA

Transport Research Laboratory (1952). *Soil Mechanics for Road Engineers*, Chapters 19 and 20. HMSO, London

Transport Research Laboratory (1977). Road Note 31, *A Guide to the Structural Design of Bitumen-Surfaced Roads in Tropical and Sub-tropical Countries*. HMSO, London

Chapter 12

Direct shear tests

12.1 Introduction

12.1.1 Scope

This chapter deals with the measurement of the shear strength of soils in the laboratory by two direct methods, both of which involve the sliding of one portion of soil on another. The first is the shearbox test, in which the relative movement of two halves of a square block of soil takes place along a horizontal surface. The second is the vane test, in which a relative rotational movement takes place between a cylindrical volume of soil and the surrounding material. (The measurement of shear strength by means of compression tests is covered in Chapter 13.)

Some theoretical background knowledge is necessary for a proper understanding of the test principles. Basic concepts such as force, stress, strain, are explained at the outset in order to clarify the proper usage of these terms. The theory of shear strength in soils, as exemplified by the shearbox test, is presented, leading to the Coulomb equation relating shear strength to normal stress. Shear strength properties of dense and loose sands, and of saturated clays, are described.

The principle of effective stress, which will be covered in Volume 3, is introduced in relation to drained shearbox tests, including the measurement of residual drained shear strength which is relevant to OC clays. Measurement of pore water pressure is not necessary because in these tests the effective stresses are equal to the total stresses.

An indirect method of measuring shear strength of soil, the fall cone test, described by Hansbo (1957), is included at the end of this chapter.

12.1.2 Purpose

Every building or structure that is founded in or on the earth imposes loads on the soil which supports the foundations. The stresses set up in the soil cause deformations of the soil, which can occur in three ways.

- By elastic deformation of the soil particles
- By the change in volume of the soil resulting from the expulsion of fluid (water and/ or gas) from the voids between the solid particles
- By the slippage of soil particles, one on another, which may lead to the sliding of one body of soil relative to the surrounding mass

The first of these is negligible for most soils at the usual levels of stress which occur in practice. The second is known as consolidation, and is dealt with in Chapter 14. The third is the process known as shear failure and occurs when shear stresses set up in the soil mass exceed the maximum shear resistance which the soil can offer, i.e. its shear strength.

The third condition must be guarded against in order to prevent disastrous failure. The usual safeguard is to carry out a stability analysis, for which the shear strength of the soil for the relevant conditions must be known. The analysis should ensure that the shear stresses in the soil are everywhere less than its shear strength by a suitable margin, which has to be both adequately safe and economically feasible.

12.1.3 Shear strength of soil

The term shear strength, as applied to soils, is not a fundamental property of a soil in the same way as, for instance, the compressive strength is a property of concrete. On the contrary, shear strength is related to the conditions prevailing in-situ, and can also vary with time. The value measured in the laboratory is likewise dependent upon the conditions imposed during the test and in some instances upon the duration of the test.

The aspects of shear strength dealt with in this chapter can be divided into four categories.

1. The shear resistance of free-draining non-cohesive soils (i.e. sands and gravels), which is virtually independent of time.
2. The drained shear strength of cohesive soils, which depends upon the rate of displacement being slow enough to permit full drainage to take place during shear.
3. The long-term or residual drained shear strength of soils such as overconsolidated clays, for which a slow rate of displacement and a large displacement movement are required.
4. The shear strength of very soft cohesive soils under undrained conditions, i.e. in which shearing is applied relatively quickly.

This chapter describes the use of the shearbox apparatus for the measurement of items 1–3, and the use of the laboratory vane apparatus for item 4. The triaxial apparatus, described in Chapter 13 (Section 13.6.2), is more satisfactory for other types of total stress shear strength measurements, and for most types of effective stress test. The latter, which involve the measurement of pore water pressures, will be covered in Volume 3.

The fall cone test is carried out on remoulded or undisturbed samples of soil, but when testing undisturbed soils, the results may be affected by soil anisotropy. For this reason and given its empirical nature, the test should be treated as an index test.

12.1.4 Principle of shearbox test

The shearbox test is the simplest, the oldest and the most straightforward procedure for measuring the 'immediate' or short-term shear strength of soils in terms of total stresses. It is also the easiest to understand, but it has a number of shortcomings which are discussed in Section 12.4.5.

In principle the shearbox test is an 'angle of friction' test, in which one portion of soil is made to slide along another by the action of a steadily increasing horizontal shearing force, while a constant load is applied normal to the plane of relative movement. These conditions are achieved by placing the soil in a rigid metal box, square in plan, consisting of two halves. The lower half of the box can slide relative to the upper half when pushed (or pulled) by a

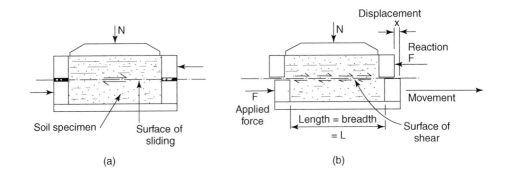

Figure 12.1 Principle of shearbox test: (a) start of test; (b) during relative displacement

motorised drive unit, while a yoke supporting a load hanger provides the normal pressure. The principle is shown in Figure 12.1.

The usual shearbox apparatus provides no control of drainage and no provision for measuring pore water pressure. It is therefore not suitable for carrying out undrained tests, and its usual application is restricted to drained tests in which effective stresses are equal to total stresses.

During the shearing process the applied shearing force is steadily increased. The resulting relative displacement of the two portions of the specimen, and the applied force, are both measured at suitable intervals so that a load–displacement curve can be drawn. The vertical movement of the top surface of the specimen, which indicates changes of volume, is also measured and enables changes in density and voids ratio during shear to be evaluated.

For convenience of presentation, the shearbox test is described under two categories. The first procedure, given in Section 12.5, is a 'quick' test which is applicable to free-draining soils (i.e. sands), whether dry or fully saturated. This is the procedure for which the traditional shearbox apparatus for a 60 mm square sample, commonly used in the UK and the USA, was originally designed. Details of the apparatus and its use are described in this section. Shearboxes 100 mm square are now also widely used. The same principle applies to the large shearbox, about 300 mm square or 12 inches square, which is described in Section 12.6. For free-draining granular soils, drained conditions apply in a 'quick' test.

The second category applies to soils that are not free-draining and therefore require more time to allow drainage to take place. The procedure is described in Section 12.7. Both categories of test are covered together in Clause 4.5 of BS 1377:Part 7:1990.

In order to carry out a slow 'drained' shear test, provision is made for the specimen to be consolidated before shearing and for further drainage to take place during shear at a suitably slow rate of displacement, so that the consolidated-drained shear strength parameters can be determined. The use of reversing attachments enables a specimen to be re-sheared a number of times in order to determine the drained residual shear strength. These aspects are covered in Section 12.7.

The environmental requirement of BS 1377:Part 7:1990 is that the laboratory area in which these tests are carried out should be maintained at a temperature that is constant to within ± 4°C. Apparatus should also be protected from direct sunlight, from local sources of heat, and from draughts.

12.1.5 Ring shear apparatus

The ring shear apparatus was developed to overcome certain disadvantages of the conventional shearbox in the measurement of residual shear strength. The apparatus accommodates a ring-shaped specimen, as shown in Figure 12.2. An unlimited rotational shear displacement can be applied to the specimen continuously without having to stop and reverse the shearing movement, while the area of contact on the shear plane remains constant.

The same environmental requirements as stated in Section 12.1.4 apply to ring shear tests.

12.1.6 Principle of vane test

In the vane test a four-blade cruciform vane is pushed into the soil and then rotated. The torque required to cause rotation of the cylinder of soil enclosing the vane is measured, which enables the undrained shear strength of the clay to be calculated. The principle is shown in Figure 12.3. A repeat test immediately after remoulding the soil by rapid rotation of the vane provides a measure of the remoulded strength, and hence the sensitivity. Details are given in Section 12.8. Reference is also made to the BS field vane from which the laboratory apparatus was derived.

The use of a small pocket shear vane tester is briefly described in Section 12.8.5.

12.1.7 Principle of fall cone test

For an undisturbed soil or a soil remoulded at a specified moisture content, the penetration of the fall cone using the apparatus for the determination of the liquid limit (see Volume 1 (third edition), Section 2.6.4) is inversely proportional to the soil shear strength. The geometry and mass of the cone are selected with reference to the anticipated shear strength, with heavier cones being used for soils of greater shear strength. The test is described in Section 12.10.

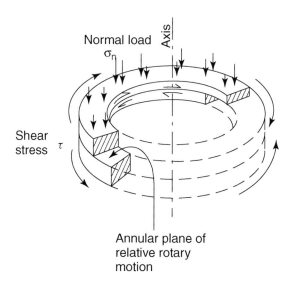

Figure 12.2 Principle of ring shear test (after Bishop et al., 1971)

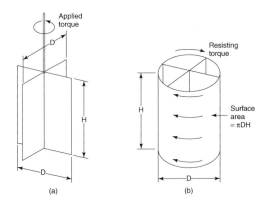

Figure 12.3 Principle of vane shear test: (a) vane blades; (b) cylinder of soil rotated by vane

12.1.8 Historical outline

Shearbox apparatus

The earliest known attempt to measure the shear strength of a soil was made by the French engineer, Alexandre Collin(1846). He used a split box, 350 mm long, in which a sample of clay 40 × 40 mm section was subjected to double shear under a load applied by hanging weights (see Figure 12.4). The earliest measurements in Britain were made by Bell (1915), who constructed a device which was to be the prototype for subsequent developments of the shearbox. Bell was the first to carry out and publish results of a really practical series of shear tests on various types of soil (Skempton, 1958).

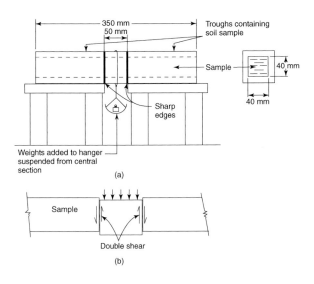

Figure 12.4 Shearbox apparatus devised by Collin (1846): (a) general arrangement; (b) forces on sheared portion of sample

In 1934 a simple shearbox with a single plane of shear was designed at the Building Research Station (BRS) (Cooling and Smith, 1936). In this apparatus the load was applied in increments (the 'stress control' principle) by progressively adding weights to a pan (see Figure 12.5). This required considerable care and judgement on the part of the operator in order to ascertain the load at which failure occurred.

A shearbox in its modern form was designed by Casagrande at Harvard University (Cambridge, Massachusetts [MA], USA) in 1932, but details were not published. At MIT (Cambridge, MA, USA) Gilboy (1936) developed a constant rate of displacement machine (the 'strain control' principle), using a fixed speed drive motor, which overcame the disadvantages of the BRS design. A further development using this principle was described by Golder (1942). Improvements to details of design were introduced by Bishop at Imperial College, London in 1946.

Most commercial shearbox machines are still based on the displacement control principle (see Figure 12.6) and today provide a wide range of displacement speeds, from a few millimetres per minute to about 10,000 times slower. Electronic control using thyristors provides steplessly variable speed control throughout a similar range.

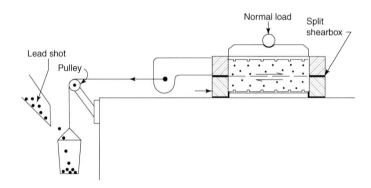

Figure 12.5 Principle of early type of controlled-stress shearbox

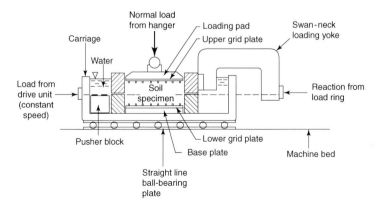

Figure 12.6 Arrangement of conventional 60 mm shearbox apparatus with displacement control

The stress control method has certain advantages in some long-term tests in which increments of stress must be applied very slowly, and in tests for the study of the effect of 'creep' under constant shear stress. However, for routine testing applications the displacement control method is the one now normally used.

The first shearbox tests to measure the shear strength parameters in terms of effective stress on a natural slip surface in a clay landslide were probably those carried out in 1963 in connection with the Walton's Wood landslide (Early and Skempton, 1972). They included the earliest known multi-reversal shearbox tests for the measurement of the residual strength of clays. From 1963 to 1966 the original author carried out similar tests on tectonic shear zones in heavily overconsolidated Siwalik clays of the Mangla Dam Project in West Pakistan (Binnie *et al.*, 1967).

Ring shear apparatus

An early ring shear apparatus for testing soils was designed by Casagrande and reported by Hvorslev (1939). Subsequent developments have been described by Bishop *et al.* (1971) and by Bromhead (1979), whose simplified apparatus uses a thin sample of remoulded soil. Use of the latter is outlined in Section 12.9.

The laboratory vane apparatus was originally designed by the Road Research Laboratory in the UK in about 1954. It was based on the use of the existing field vane apparatus, such as that described by Skempton (1948). One of the early uses of the laboratory vane was in the investigation of the relationship between undrained shear strength and moisture content in cohesive soils.

The main application of the laboratory vane now is for the measurement of the undrained shear strength of clays and peats which are too soft for the preparation of satisfactory test specimens. The apparatus is available either hand operated or motor driven. BS 1377:1990 includes the laboratory vane test in Clause 3 of Part 7, in addition to the field test in Clause 4.4 of Part 9.

Fall cone apparatus

The fall cone apparatus was first introduced in 1915 and is used for the determination of the liquid limit and also undrained shear strength of soil. The test was developed by the Geotechnical Commission of the Swedish State Railway and was conceived by the Secretary of the Commission, John Olsson. The test is widely used in Scandinavia and has been published as a European Standard DD CEN ISO/TS 17892-6:2004. This has been adopted as a British Standard, as there is no BS equivalent for this test.

12.2 Definitions

Force That influence which causes a change of state of motion of a body

$$(\text{Force}) = (\text{mass}) \times (\text{acceleration})$$

Normal force or *direct force* A force which acts normal (perpendicular) to a plane of section.

Shear force A force which acts tangential to a plane of section.

Stress Intensity of force, i.e. force per unit area.

Strain (linear) Change in length per unit length due to a stress, measured in the direction of the stress.

Shear stress Shear force per unit area.

Shear strain Angular distortion, measured in radians, due to the action of shear stresses.

Displacement Horizontal movement of one portion of a specimen relative to the other along the surface of sliding and in the direction of the applied force, in a direct shear test.

Shear resistance (of a soil) The resistance offered (by a soil) to deformation when it is subjected to a shear stress.

Shear strength (τ_f) The maximum shear resistance which a soil can offer under defined conditions of effective pressure and drainage. (Often used synonymously with peak strength.)

Undrained shear strength (c_u) The shear strength of a soil under undrained conditions, i.e. immediately after the application of stress and before drainage of water can take place.

Angle of shear resistance (ϕ') (in terms of effective stress) The slope of the line on a graphical plot relating shear strength on a surface of failure to the effective stress normal to that surface.

Apparent cohesion (c') (in terms of effective stress) That part of the shear strength which is independent of effective normal stress, i.e. the intercept of the shear strength envelope in terms of effective stress.

Dilatancy Expansion of a soil when subjected to shear stress.

Free-draining soil A soil in which water can move easily through the void spaces so that no excess pore pressure or suction develops as a result of the application of stress or deformation.

Critical voids ratio The voids ratio at which a granular soil neither contracts nor dilates when subjected to shear.

Peak strength (see Shear strength)

Residual strength The shear resistance which a soil can maintain when subjected to large shear displacement after the peak strength has been mobilised.

Failure The point at which continued shear deformation under a constant or decreasing shear stress begins.

Couple A combination of two numerically equal but directionally opposite parallel forces which can be balanced only by an equal and opposite couple.

Torsion The twisting moment due to the action of equal and opposite couples.

Torque The moment of a couple producing torsion, expressed as (force) × (distance apart of the two components).

Vane shear strength (τ_v) The shear strength of a soil as determined by applying a torque in the vane shear test.

12.3 Theory

12.3.1 Force

The usual definition of 'force' describes it as an influence which changes the state of motion of a body, i.e. which results in an acceleration. Unit force produces unit acceleration in a unit mass, hence 1 newton (N) is equal to one kilogram metre per second per second, or

$$1 \text{ N} = 1 \text{ kg m/s}^2$$

However, in problems of statics, and especially in soils, it is more relevant to think of force as an influence which causes deformation when applied to a body. The deformation may be

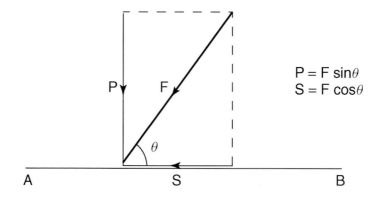

Figure 12.7 Resolution of a force into two perpendicular components

quite obvious, as when an elastic band is stretched by pulling, or undetectable except by using sensitive instruments, such as the compression of the legs of a chair when one sits on it.

Force is a vector quantity, that is, it acts in a certain direction as well as having a certain magnitude. A force can be represented on paper by a straight line, the orientation of which represents its line of action (with an arrow to indicate the direction) and the length of which represents its magnitude, drawn to a certain scale. This graphical representation enables a force to be resolved into two components by using the principle of the parallelogram of forces, (described in textbooks on mechanics or physics, e.g. Abbott (1969, Chapter 3).

A force can be resolved into two components at right angles to each other, as illustrated in Figure 12.7. The force F is resolved into two forces P and S acting normal to and along the surface represented by AB. The resolved components P and S are given by the equations

$$P = F \sin \theta \tag{12.1}$$

$$S = F \cos \theta \tag{12.2}$$

The component P is the 'normal' component of the force F (i.e. it acts normal to the surface) and the component S is the tangential or 'shear' component.

12.3.2 Stress

When an external force is applied to a body it sets up an internal force which provides an equal and opposite reaction. The intensity of this force, assumed to be distributed uniformly over the area of cross-section at right angles to the direction of the force, is known as the 'stress'. Stress is the internal force per unit area. The units are the same as used for pressure, but 'pressure' is the term usually applied to fluids.

The stress unit most often used for soils is the kilopascal (kPa), also known as kilonewton per square metre (kN/m²). In laboratory work forces are often measured in newtons (N) and areas in square millimetres (mm²), and the following conversions are worth remembering

$$1 \text{ N/mm}^2 = 1 \text{ MN/m}^2 = 1000 \text{ kN/m}^2$$
$$= 1 \text{ MPa} \quad = 1000 \text{ kPa}$$

There are two types of stress
    Normal or direct stress:
        compressive
        tensile
    Shear stress

Normal (direct) stress

Compressive stresses and tensile stresses act in a direction normal to the plane of cross-section being considered. They are referred to as normal stresses or direct stresses and may be positive or negative according to the sign convention used.

A compressive stress is set up in a body when it is subjected to compressive forces and resists the tendency of the forces to shorten its length. For instance, the column of bricks in Figure 12.8(a) is compressed by the downward force P applied at the top.

If the area of cross-section of the column is denoted by A_1, the stress resisting compression on any horizontal section XX (ignoring the weight of the bricks) is equal to P/A_1. The symbol σ is used for direct stress, so

$$\sigma = P/A_1 \tag{12.3}$$

A tensile stress is set up when a body is subjected to tensile forces and resists the tendency of the forces to pull it apart and increase its length, i.e. to stretch it. For instance, the wire in Figure 12.8(b) is being pulled downwards by the weight force W of the mass it is supporting. If the area of cross-section of the wire is denoted by A_2, the stress resisting extension or pulling apart on a horizontal section, such as YY, is equal to W/A_2.

The stress is acting in the opposite sense to that in Figure 12.8(a), i.e. it is 'pulling' instead of 'pushing'. It must therefore be given the opposite sign and is written

$$\sigma = -W/A_2$$

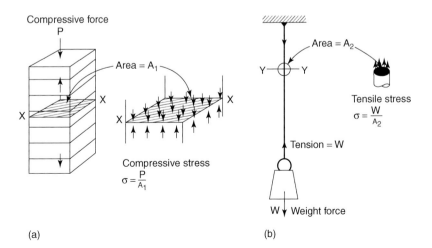

Figure 12.8 Direct stresses: (a) compressive stress in a column; (b) tensile stress in a wire

In the above examples, it is evident that the brick pillar (if the bricks are not cemented together) cannot sustain a tensile stress; and the wire cannot sustain a compressive stress. But a steel bar for instance can, within limits, sustain either compressive or tensile stresses.

Soils possess little or no resistance to tension and are usually analysed in terms of compression and shear stresses. It is convenient therefore to adopt the convention that compressive stresses are positive and tensile stresses negative. This is the opposite convention to that used in structural analysis, but it is of no significance provided that the chosen convention is used consistently. The sign convention applied to soils is summarised in Table 12.1.

Shear stress

Shear stresses act parallel to the plane being considered, and are set up when applied forces tend to cause successive layers to slide over each other. A shear stress in a body resists an angular change of shape, just as a normal stress resists a tendency either to compress or to elongate. The effect of shear can be seen for instance if a block of rubber placed on the table is pushed sideways (see Figure 12.9(a)). The rubber deforms and in doing so mobilises resistance to further shear deformation. On the other hand a pack of playing cards (see Figure 12.9(b))

Table 12.1 Sign convention used for soils

Physical quantity	Positive sense (+)	Negative sense (−)
Force	Compression	Tension
Stress	Compression	Tensile
Displacement	Compression or contraction	Extension or expansion
Strain	Compressive (shortening)	Tensile (lengthening, stretching)
Volume change	Decrease (consolidation)	Increase (swelling)
Pressure change	Increase	Decrease
Voids ratio change	Decrease	Increase

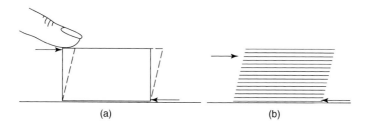

(a)   (b)

Figure 12.9 Illustrations of shear: (a) shear force distorting rubber block; (b) shearing action on pack of playing cards

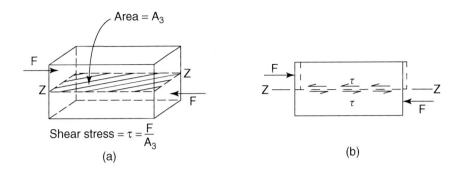

Figure 12.10 Shear stress: (a) shear stress in a block; (b) representation of shear stress on a surface

offers little resistance to shear and illustrates the relative sliding effect throughout its thickness which is characteristic of shearing action.

If the block in Figure 12.10(a) has a horizontal area of cross-sectional equal to A_3 and is pushed by a horizontal force F, the shear stress resisting sliding on a horizontal surface ZZ is equal to F/A_3. The symbol τ is used for shear stress, so

$$\tau = F/A_3 \qquad (12.4)$$

The equal and opposite shear stresses acting on either face of a horizontal section are usually represented by single-barb arrows as shown in Figure 12.10(b).

12.3.3 Strain

The deformation produced in a body due to the application of a force is called 'strain'. Strain is related to stress and in a perfectly elastic material strain is directly proportional to stress. Direct stresses produce linear strains and shear stresses produce shear strains.

In a body subjected to a longitudinal force which induces a direct stress in the longitudinal direction, the length measured in that direction will change. The ratio of the change in length to the original length is defined as the strain and is often multiplied by 100 to express it as a percentage.

In Figure 12.11(a) the cylindrical soil specimen of original length L and diameter D is subjected to the application of a compressive force P. The area of cross-section of the specimen is equal to $\pi D^2/4$ and is denoted by A. The compressive stress is therefore equal to P/A (positive), and this causes the specimen to decrease in length by δL The strain is therefore equal to $\delta L/L$, and is positive. Linear strain is denoted by the symbol ε, and is a dimensionless number. Therefore

$$\varepsilon = -\frac{\delta L}{L} \text{ or } -\frac{\delta L}{L} \times 100 \qquad (12.5)$$

Similarly the strain in the wire subjected to tension in Figure 12.11(b) is negative, and is written

$$\varepsilon = -\frac{\delta L}{L} \text{ or } -\frac{\delta L}{L} \times 100$$

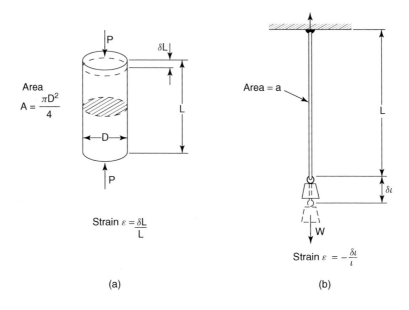

Figure 12.11 Longitudinal strain: (a) compression of clay cylinder; (b) extension of wire

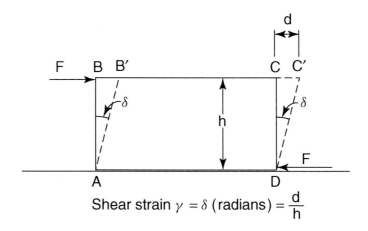

Figure 12.12 Shear strain

Shear strain is the angular deformation resulting from the application of shear forces. In Figure 12.12 the rectangular rubber block ABCD has been distorted into the shape AB'C'D by the shear forces F, so that the face AB has been rotated through an angle δ. The shear strain or angular strain (usually denoted by γ) is equal to the angle measured in radians and is a dimensionless number. For small displacements, $\gamma = d/h$. If the strains are uniform throughout the block, vertical lines marked on it initially, as in Figure 12.13(a), remain parallel and straight, but become inclined as in Figure 12.13(b). This type of deformation is produced by

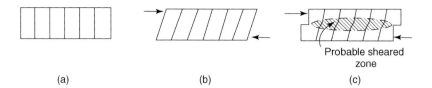

(a)   (b)   (c)

Probable sheared zone

Figure 12.13 Strains due to shear: (a) marked specimen before shearing; (b) specimen in simple shear; (c) strains in shearbox test

'simple shear'. However, this condition is very difficult to reproduce in a laboratory test on soils, although it has been done, for instance, at Cambridge University (Roscoe, 1953).

In the standard direct shear test in the shearbox, one portion of soil is constrained to slide on another, as in Figure 12.10(b), and the resulting strains are more like those shown in Figure 12.13(c). The soil subjected to shear is probably contained within the zone indicated by the dotted lines and the soil near the front and back edges of the box tends to reach failure before that near the middle. Thus the actual strain pattern due to shear is complex. However, in routine shearbox tests the measurement of strain is simplified by merely taking account of the linear displacement of one half of the box relative to the other. It is not correct to refer to the relative displacement as 'strain'.

12.3.4 Friction

Consider a block of weight W (force units) resting on a level table-top which is not perfectly smooth (see Figure 12.14(a)). The force of reaction N from the table on the block acts vertically upwards and is equal to W. If the block is pushed by a small horizontal force P (see Figure 12.14(b)), less than that required to move it, an equal and opposite force F will act on the block at the surface of contact, opposing the tendency to move. This force is due to friction between the block and the table-top. The resultant reaction R on the block from the table is obtained by combining the force vectors N and F as shown in Figure 12.14(c). The resultant R is inclined at an angle θ to the normal through the surface of contact.

As the force P is gradually increased, the frictional force F increases until it reaches its limiting value F_{max}, when the block just begins to move. Since the normal force N remains constant, the angle θ gradually increases as F increases until it reaches a maximum value ϕ when F reaches its maximum value F_{max}. The ratio F_{max}/N is known as the coefficient of friction between the block and the table, and is denoted by μ. The angle ϕ, which is the maximum obliquity of the reaction R, is known as the angle of friction. From Figure 12.14(d) it can be seen that

$$\tan \phi = \frac{F_{max}}{N} = \mu$$

(12.6)

If a number of measurements of F_{max} are made for various weights of block (i.e. for various values of N), a graph can be drawn relating F_{max} to N as shown in Figure 12.15. The points obtained will be on a straight line through the origin rising at an angle ϕ to the horizontal axis. This enables the angle of friction ϕ to be obtained experimentally.

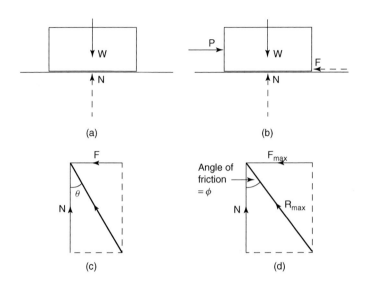

Figure 12.14 Friction and angle of friction

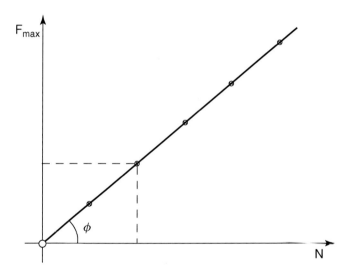

Figure 12.15 Measurement of angle of friction

12.3.5 Shear in soils

If one mass of soil can be made to slide on another of the same kind while a load is applied normal to the sliding surface, a test similar to that described above can be carried out to determine the frictional characteristics of the soil. This is the basis of the shearbox test which is used to measure the angle of internal friction or more correctly the 'angle of shear resistance' of the soil. In this sense the test is applicable only to granular soils and was originally intended for sands.

The principle of the shearbox is shown in Figure 12.1(a). A normal load N applied to the soil produces a vertical stress σ_n, where $\sigma_n = N/L^2$, and L is the length of side of the square box. A steadily increasing displacement, which causes an increasing shearing force F, is applied to one-half of the sample in a horizontal direction, while the other half is restrained by the load-measuring device. The shear stress induced on the predetermined plane of slip is equal to F/L^2. Unlike the block on the table, a horizontal displacement (see Figure 12.1(b)) of the soil in the bottom half of the box relative to that in the top half takes place gradually while the force F is increasing, as shown by OA in the load–displacement graph (see Figure 12.16). Eventually the point B is reached, at which the maximum shear stress (τ_f) which can be sustained on the surface of sliding is offered by the soil. This shear stress is the shear strength of the soil under the particular normal stress σ_n and the point B is known as the 'peak' of the shear stress–displacement curve. After the peak the shear resistance often falls off, as indicated by BC, and failure of the soil in shear has occurred.

Several tests, usually three, can be carried out on specimens of the same soil under different normal loads (denoted here by N_1, N_2, N_3) giving three different values of the normal stress σ_n. From each stress–displacement curve (Figure 12.17) the maximum shear stress τ_f can be read off, and plotted against the corresponding value of σ_n, as in Figure 12.18. This graph generally approximates to a straight line, its inclination to the horizontal axis being equal to the angle of shearing resistance of the soil, ϕ, and its intercept on the vertical (shear stress) axis being the apparent cohesion, denoted by c.

12.3.6 Coulomb's law

The general relationship between maximum shearing resistance, τ_f, and normal stress, σ_n, for soils was suggested by Coulomb in 1772 to be of the form

$$\tau_f = c + \sigma_n \tan \phi \tag{12.7}$$

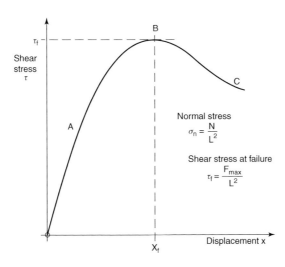

Figure 12.16 Shear stress–displacement relationship in shearbox test

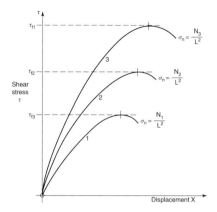

Figure 12.17 Shear stress–displacement curves for specimens tested under three different normal pressures

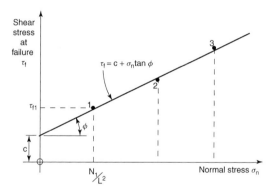

Figure 12.18 Maximum shear stress related to normal stress from shearbox tests (Coulomb envelope)

The Coulomb envelope represented by this equation is shown graphically in Figure 12.18, and is known as the 'failure envelope' (in terms of total stress). However, it is not generally correct to assume that the shear strength of soil is governed by the total normal stress of the failure surface. It has been established that soil behaviour is controlled by effective stress rather than total stress, and the Coulomb equation has to be re-defined in terms of effective stress. To determine the effective stress it is necessary to know the pore water pressure (u). The concept of effective stress, and measurement of pore pressure, will be covered in Volume 3. Stated briefly, for saturated soils the effective normal stress σ'_n is given by the equation

$$\sigma'_n = \sigma_n - u$$

and the Coulomb equation is then modified to the form

$$\tau_f = c' + (\sigma_n - u)\tan \phi' = c' + \sigma'_n \tan \phi' \tag{12.8}$$

in which c', ϕ' are the parameters expressing the shear strength in terms of effective stress.

This equation suggests that the shearing resistance of soils can be considered as consisting of two components.

Friction (denoted by tan ϕ'), which is due to the interlocking of particles and the friction between them when subjected to normal effective stress.

Cohesion (denoted by c'), which is due to internal forces holding soil particles together in a solid mass.

The friction component increases with increasing normal effective stress but the cohesion component remains constant. This interpretation is not strictly correct because ϕ' is not a true angle of friction in the physical sense, but it does provide a simple concept of the nature of shear strength in soils.

In the special case of fully saturated soils sheared under conditions which allow no drainage of pore water (i.e. for saturated clays), Coulomb's equation in terms of total stress is applicable. Under these conditions, $\phi = 0$ and Equation (12.7) becomes

$$\tau_f = c_u$$

The term c_u defines the shear strength of the saturated soil in terms of total stress for the undrained condition, and its value depends on the voids ratio (i.e. the moisture content).

12.3.7 Shear strength of dry and saturated sand

The shear strength of dry sand depends upon several factors, such as the mineralogical composition of the grains; their size, shape, surface texture and grading; the soil structure, i.e. packing of the grains; and the moisture content. For a particular sand in the dry state the only variable is the state of packing, which has an important influence on shear strength, as discussed below. The state of packing can be expressed in terms of density index (either descriptively or numerically), or voids ratio, or porosity, or dry density. These terms were explained in Volume 1 (third edition), Sections 3.3.2 and 3.3.5, and Table 3.4.

Experience has shown that shear strength results obtained on saturated sand are very similar to those for dry sand, provided that the sand remains saturated and that drainage takes place freely during shear. Sections 1–3 below relate equally to dry sand, or to fully saturated free-draining sand. In both cases effective stresses are equal to total stresses.

1. Dense sand

The state of packing of the grains in a dense sand (low voids ratio) is represented diagrammatically in Figure 12.19(a). If the sand is sheared along a plane such as XX, and if it is assumed that distortion and crushing of individual grains does not occur, grains lying just above the surface XX will be forced to ride up and over those lying just below when relative movement occurs. This causes an expansion, which can be measured by observing the upward movement of the top surface of the sand. The resulting increase in volume is known as dilatancy and in free-draining submerged sands results in additional water entering the soil structure.

The resulting shear stress–displacement curve is of the form marked (D) in Figure 12.20(a), and the corresponding volume change relationship with displacement is marked (D) in Figure 12.20(b). The small initial contraction is due to some bedding down of grains when shearing begins. The shear stress curve rises quite sharply to a peak and then falls

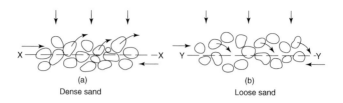

Figure 12.19 Effect of shear on grain structure in sands

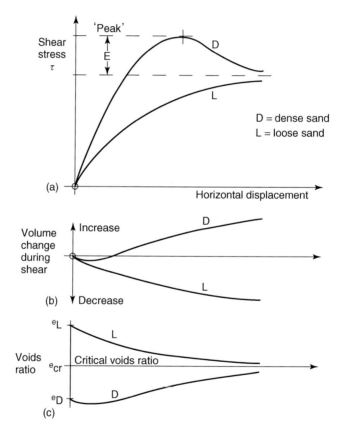

Figure 12.20 Shear characteristics of dense and loose sands (all plotted against displacement): (a) shear stress; (b) volume change; (c) voids ratio change

off to a somewhat lower value. The excess of the peak over the final value, denoted by E, represents the extra work which has to be put in to produce the vertical movement due to dilatancy.

After shearing, the grains adjacent to the shear surface are in a less dense state of packing than they were initially.

2. Loose sand

A loose state of packing of grains is shown diagrammatically in Figure 12.19(b). Shearing along a plane such as YY will result in a collapse of the relatively open structure as grains move downwards into void spaces. This causes a volume decrease (contraction), which can be measured as a downward movement of the top surface, and in free-draining submerged sands results in water being expelled from the soil structure.

The resulting shear stress–displacement curve, marked (L) in Figure 12.20(a), is less steep than curve (D) and does not have a pronounced peak. The corresponding volume change relationship with displacement is shown as curve (L) in Figure 12.20(b). After shearing, the grains adjacent to the shear surface are in a more dense state of packing than they were initially.

3. Comparison of loose and dense sands

Volume changes during shear for both states of packing are represented in Figure 12.20(c) in terms of voids ratio. Initial voids ratios are denoted by e_D (dense) and e_L (loose). At the end of the shearing displacement the voids ratio in each case approaches a common value e_{cr}, known as the critical voids ratio.

Any voids ratio e can be related to the limiting voids ratios (e_{max} and e_{min}) in terms of the density index (I_D), which is defined by the following equation (see Volume 1 (third edition), Section 3.3.5):

$$I_D = \frac{e_{max} - e}{e_{max} - e_{min}}$$

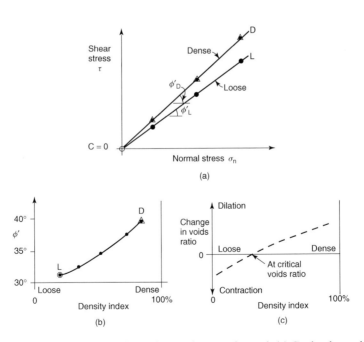

(a)

(b)

(c)

Figure 12.21 Effect of initial voids ratio on shear resistance of a sand: (a) Coulomb envelopes for 'dense' and 'loose' states; (b) value of ϕ related to relative density, (c) voids ratio change during shear related to initial density index

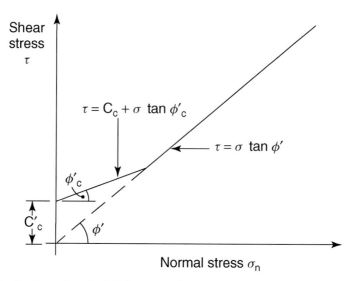

Figure 12.22 Coulomb envelope for lightly cemented sand

If three tests are carried out under three different normal stresses for each state of packing, the shear strength–normal stress relationships can be plotted as shown in Figure 12.21(a). The angle of shear resistance (ϕ'_D) for the dense state is greater than that (ϕ'_L) for the loose state. By performing additional tests at intermediate void ratios or density index, a relationship between ϕ' and density index can be obtained as in Figure 12.21(b). The corresponding changes of voids ratio up to the point of failure (i.e. maximum shear stress) can also be plotted against the density index (see Figure 12.21(c)). The intersection of this curve with the horizontal axis gives the density index corresponding to the critical voids ratio, because this is the density at which no change in voids ratio occurs due to shear.

Under large normal stresses the dilatancy of dense sand can be suppressed and a curved envelope is likely to be obtained. Imposing a linear relationship could give a spurious cohesion intercept, therefore it is important to limit such linearisation to the range of normal stress appropriate to the field conditions being analysed. The value of ϕ' depends on the effective normal stress when more than a narrow range of effective stress is considered.

4. Lightly cemented sands

Dry or fully saturated cohesionless sands give a zero cohesion intercept, but if there is a cementing agent between the grains a failure envelope similar to Figure 12.22 may be obtained. The first part AB of the envelope represents the cohesive effect of the cementing agent (apparent cohesion denoted by c_c). Above the point B the stress level is sufficient to cause breakdown of the cementing agent and the steep portion BC represents the behaviour as a normal granular (cohesionless) material. Re-shearing this type of sand reduces c_c to zero.

A similar effect, but with only a small cohesion intercept, is observed in moist or damp sands, but this is due to surface tension of the water between grains. The value of c_c is not reduced by re-shearing if conditions remain unaltered.

12.3.8 Shear strength of clays
Test conditions
The undrained strength of clays depends not only on the soil type and composition, in the sense described in Section 12.3.7 for sands, but also on factors related to the mineralogy, grain size and shape, adsorbed water, and water chemistry of the clay minerals present. Shear strength also depends to a great extent upon the initial moisture content of the clay, and the rate at which the soil structure can expel or take in water during a test. When a saturated soil is subjected to shear, excess pore water pressures are generated, and the rate at which these excess pressures can dissipate depends on the soil permeability.

Quick and slow tests
In a 'quick' test on a soil of low permeability, such as clay, the time period is not sufficient to allow excess pore pressures to dissipate and the soil is tested in the undrained condition. The shearbox apparatus, in which drainage cannot be entirely prevented, is not suitable for this type of test. The undrained shear strength, c_u, is usually determined by means of tests on cylindrical specimens, as described in Chapter 13. For saturated and soft clays the undrained shear strength can be measured directly using the vane apparatus (see Section 12.8). However, different methods of test, and different rates of testing, can give significantly different values of c_u.

The drained shear strength of clays, for derivation of the effective strength parameters, can be determined from 'slow' drained shearbox tests, as explained in Section 12.3.9 below.

12.3.9 Drained tests on clays and silts
Principle
Measurement of the drained shear strength of clays and silts is the same in principle as for sands, the only practical difference being the length of time required. Drained tests on low permeability soils take longer than tests on sands (often periods of several days) because of the longer time needed for drainage of excess pore water. The test method includes a procedure for measuring the rate of drainage during a consolidation stage, from which a suitable rate of shearing can be assessed empirically.

Tests are usually carried out on a set of specimens each at a different pressure. The soil is first allowed to consolidate under the selected normal pressure, until consolidation is completed and there is virtually no excess of pore pressure remaining (see Chapter 14, Section 14.3.2). Shear displacement is then applied slowly enough to allow the dissipation of any further pore water pressure (whether positive or negative) which may develop due to shear, the rate of displacement being determined from the consolidation stage. Under these conditions the effective stresses are equal to the applied stresses.

The shear strength envelope obtained from a set of drained tests is typically of the form shown in Figure 12.23. The envelope is approximately linear, the inclination to the horizontal axis being the angle of shear resistance in terms of effective stress, ϕ'. The intercept with the shear stress axis gives the apparent cohesion in terms of effective stress, denoted by c'.

Rate of displacement
The rate of displacement at which the specimen should be sheared in a drained test depends upon the drainage characteristics, i.e. the permeability of the soil and the thickness of the

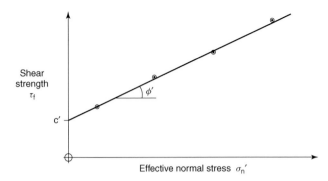

Figure 12.23 Typical shear strength envelope from a set of drained shearbox tests

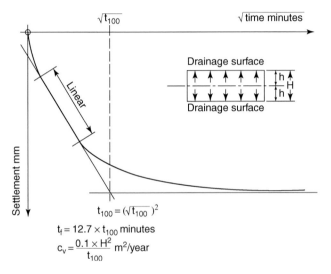

$$t_{100} = (\sqrt{t_{100}})^2$$

$$t_f = 12.7 \times t_{100} \text{ minutes}$$

$$c_v = \frac{0.1 \times H^2}{t_{100}} \text{ m}^2/\text{year}$$

Figure 12.24 Derivation of time to failure from consolidation curve. Example calculation:
If $\sqrt{t_{100}}$ min = 5.8, $t_{100} = (5.8)^2 = 3.6$ min

$$\therefore t_f = 12.7 \times 33.6 = 427 \text{ min}$$

If H = 21 mm

$$C_v = \frac{0.103 \times (21)^2}{33.6} = 1.35 \text{ m}^2/\text{year}$$

sample. Since the permeability is related to the coefficient of consolidation (see Chapter 14, Section 14.3.11), the consolidation stage of the test can provide the data for estimating a suitable time to failure. An empirical derivation is used, the principle of which is outlined as follows. (The theoretical background will be presented in Volume 3.)

From the consolidation of the specimen under the applied normal pressure a curve of settlement against square-root-time (minutes) is obtained, of the form indicated in Figure 12.24. A tangent is drawn to the early straight-line portion of the curve, in the same way as described in Section 14.3.7, method 2. This line is extended to intersect the horizontal line representing 100% consolidation, which often corresponds to the 24 hour reading. The point of intersection gives the value of $\sqrt{t_{100}}$ (see Figure 12.24), which when multiplied by itself gives the time intercept t_{100} (min) as defined by Bishop and Henkel (1962). The time required to failure, t_f, is related to t_{100} by the empirical equation

$$t_f = 12.7 \times t_{100} \text{ min} \tag{12.9}$$

(Gibson and Henkel, 1954).

The coefficient of consolidation, c_v, can be calculated from the equation

$$c_v = \frac{0.103 H^2}{t_{100}} \text{ m}^2 / \text{year} \tag{12.10}$$

where H is the specimen thickness (mm) and t_{100} is in minutes. For a standard specimen of height $H = 20$ mm, Equation (12.10) becomes

$$c_v = \frac{41}{t_{100}} \text{ m}^2 / \text{year} \tag{12.11}$$

A difficulty arises with this method if the consolidation curve does not resemble the theoretical curve, in that the initial portion up to about 50% consolidation is not linear. This may be due to the effect of bedding of the grid plate, or to the presence of air in the voids of the soil (i.e. partial saturation).

A method which gives a reasonable estimate of $\sqrt{t_{100}}$, from which c_v and the time to failure may be derived, is illustrated in Figure 12.25 and is as follows (Binnie and Partners, 1968). It requires a number of settlement readings to be taken in the later stages of consolidation.

Find the point C which is the earliest at which consolidation is substantially complete, i.e. beyond which the curve virtually flattens out. Make AB = 0.5C, and read off the value of $\sqrt{t_{100}}$ at the point B. Values of t_{100}, t_f and c_v are then calculated as described above.

Consolidation of free-draining soils takes place very rapidly, and consolidation readings against time are not practicable. The 'quick' procedure is then applicable, in which failure should occur within 5–10 minutes.

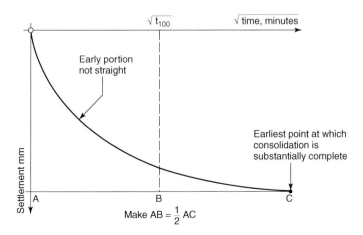

Figure 12.25 Derivation of time to failure from 'non-standard' consolidation curve

12.3.10 Residual shear strength

Meaning of residual strength

The effect of shearing a dense sand is described in Section 12.3.7, and the form of the shear strength displacement curve is indicated in Figure 12.20(a) by curve D. If shearing is continued after the peak point to the maximum displacement of the shearbox, a curve of the type shown in Figure 12.26(a) is obtained. The shear strength decreases rapidly from the peak value at first, but eventually reaches a steady-state (ultimate) value which is maintained as the displacement increases.

It was shown by Skempton (1964) that overconsolidated clays behave in a similar manner to dense sands when subjected to large shearing displacements under fully drained conditions (see Section 12.3.9). This requires shear strengths to be measured in terms of effective stress, not total stress. The shear strength which the clay ultimately reaches is known as the 'residual strength', which is often appreciably lower than the maximum or 'peak strength'.

Peak and residual envelopes

From a set of drained residual strength tests on three or more identical specimens, Coulomb envelopes can be drawn for both the peak and the residual strength conditions, as shown in Figure 12.26(b). Peak shear strength is represented by the equation

$$\tau_f = c' + \sigma'_n \tan \phi' \tag{12.12}$$

Residual strength is defined over the linear range by the equation

$$\tau_f = c'_r + \sigma'_n \tan \phi'_r \tag{12.13}$$

as represented by the full line in Figure 12.27. However, many tests suggest that the residual strength envelope is not linear at low effective stresses, and that it may pass through the origin as shown by the dashed curve in Figure 12.27 (see also the example in Figure 12.57). This gives $c'_r = 0$ and a value of ϕ'_r which depends upon the normal effective stress at low effective stresses.

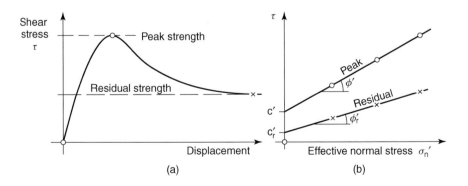

Figure 12.26 (a) Peak and residual shear strength; (b) Coulomb envelopes for peak and residual conditions (after Skempton, 1964)

The peak envelope may also show curvature. But in general, except when considering very low stresses, both envelopes can be represented adequately by straight lines over a limited stress range of interest.

Reversal shear box tests

The effect of a large displacement can be obtained in the ordinary shearbox apparatus by returning the split box to its starting position after completing the extent of its travel, and then shearing again. This process can be repeated a number of times until a steady (residual) value of shear strength is observed. A typical form of a set of shear stress–displacement curves after four traverses is shown in Figure 12.28.

The multi-reversal procedure does have certain shortcomings. There is often a small 'peak' on re-shearing after reversal, especially on the second run. Some loss of material, in the form of slurry, may occur from the shear surface. If the test is terminated too soon by not applying a sufficient number of reversals, the residual strength will not be reached. These

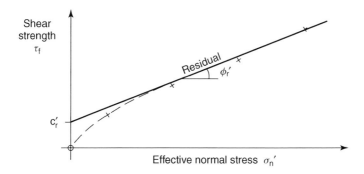

Figure 12.27 Typical form of residual strength envelope

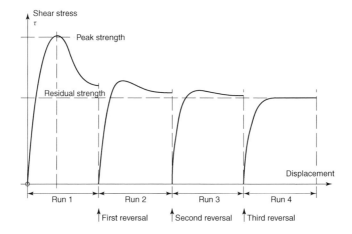

Figure 12.28 Shear stress–displacement curves from multi-reversal drained shearbox test

difficulties may be overcome by using the ring shear apparatus (see Section 12.9) in which displacement is applied continuously in one direction. However, for many purposes the reversal shearbox procedure described in Section 12.7.5 enables reasonably representative results to be obtained with the use of relatively simple apparatus.

Effect of stress history

The influence of overconsolidation on the shear strength–displacement relationship is illustrated in Figure 12.29(a), which represents direct shear tests extended to large displacements for a normally consolidated clay (NC) and for an identical clay which has been overconsolidated (OC). The preconsolidation effective stress for the OC clay is appreciably higher than the applied normal stress, which is the same for both tests.

There is little difference between the peak and residual strengths of the NC clay, although this difference tends to be greater for clays of higher plasticity index. The OC clay shows a much higher peak strength, at a smaller displacement, compared with the NC clay, followed by a marked decrease in strength to a residual value which is the same as that of the NC clay.

Changes in volume and in voids ratio during shear are indicated in Figures 12.29(c) and (d). Coulomb envelopes for the NC and OC peak strengths are shown by the full lines in Figure 12.29(b) and the residual strength envelope is shown by the broken line. This envelope is usually found to be slightly curved, implying that ϕ'_r is dependent on stress level.

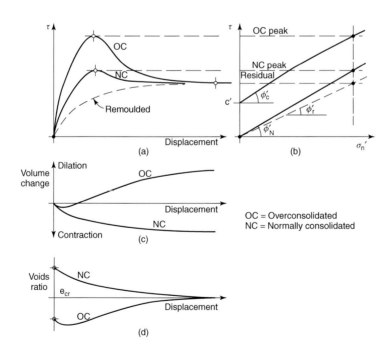

Figure 12.29 Influence of overconsolidation and remoulding on peak and residual shear strengths in a clay (after Skempton, 1964): (a) shear stress/displacement; (b) Coulomb envelopes; (c) volume change during shear; (d) voids ratio change during shear

A drained shear test on a fully remoulded specimen of the same clay would give a shear stress–displacement relationship as indicated by the dashed curve in Figure 12.29(a), requiring a large displacement to reach the residual strength without first giving a peak value.

12.3.11 Rotational shear (vane test)

Torque and torsion

The explanation given in this section continues the notes on force in Section 12.3.1. Two equal and opposite forces, acting in the same plane at a distance from each other as shown in Figure 12.30(a), produce what is termed a 'couple'. The magnitude of each force is denoted by F, the distance apart of their lines of action is d, and the moment of the couple is equal to $(F \times d)$. If F is measured in newtons, d in millimetres, the moment of the couple is expressed in newton millimetres (abbreviated Nmm).

Two equal and opposite forces forming a couple applied to one end of a rod are shown in Figure 12.30 (b). The effect of this couple is similar to turning on a water tap. The moment of the forces about the axis of the rod is known as the 'torque', T, and is equal to

$$\left(F \times \frac{d}{2} \right) \times 2 = F \times d$$

i.e. the moment of the couple. If the other end of the rod is resisted by an equal and opposite couple or torque, as in Figure 12.30(c), the rod is said to be subjected to 'torsion' under the twisting effect of the two opposing couples.

A torque may be resisted by a uniformly distributed shear stress, s, per unit area (N/mm²) acting around the curved surface of a cylinder attached to a rod, as shown in Figure 12.31. The area of the curved surface is πdh mm². Therefore the total circumferential force is equal to $s \times \pi dh$ newtons, and its moment about the axis of the rod provides the resisting torque T_r, where

$$T_r = s \times \pi dh \times \frac{d}{2}$$

$$= \frac{\pi d^2 hs}{2} \ \text{Nmm} \tag{12.14}$$

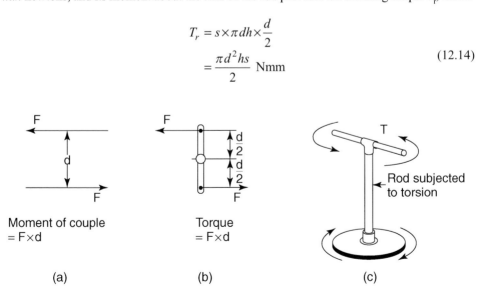

Figure 12.30 Rotational shear: (a) a couple; (b) torque; (c) torsion

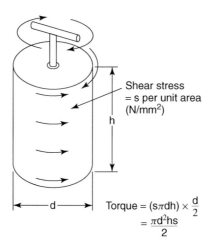

Figure 12.31 Shear stress on surface of cylinder resisting rotation

Vane shear test

The above principle is used in the vane shear test to determine the shear strength of soft clay by measuring the torque applied to cause failure. The apparatus consists of a cruciform vane of diameter D mm and height H mm, attached to the lower end of a rod of small diameter, as shown at d in Figure 12.58. The vane rotates a cylinder of soil in the manner shown in Figure 12.32(b) when torsion is applied to the rod. This is done by means of a calibrated spring device (the torsion head) at its upper end, which enables the applied torque to be measured.

So long as the applied torque is smaller than that required to shear the soil, it is resisted by an equal and opposite torque provided by the shear resistance of the soil acting on the surface of the cylinder of potential rotation. When the torque applied to the vane is increased to a value which is just sufficient to cause the cylinder of soil to rotate, it is assumed that the

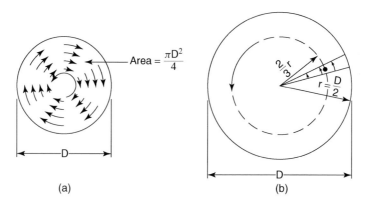

Figure 12.32 (a) Stresses on end surface of cylinder; (b) one small sector of end surface

maximum shear resistance, i.e. the shear strength of the soil, is reached simultaneously on all sliding surfaces.

The total resisting torque, T_r, is made up of two components, T_1 and T_2, where T_1 = torque provided by shear resistance on the cylindrical surface and T_2 = torque provided by shear resistance on each of the two circular end areas. Therefore

$$T_r = T_1 + 2T_2 \qquad (12.15)$$

Applying Equation (12.14) to the cylinder shown in Figure 12.3(a), the value of the torque T_1 is given by

$$T_1 = \frac{\pi D^2 Hs}{2} \text{ Nmm} \qquad (12.16)$$

The stresses acting on each of the circular end surfaces of the cylinder are indicated in Figure 12.32(a). Stresses are assumed to be uniformly distributed, therefore the total resisting shear force acting on each end is equal to $\pi D^2 s/4$ N. (The diameter of the rod can be neglected.) The length of the lever arm is not immediately obvious. The resisting torque is usually derived by the application of calculus, but the following method is easier to understand.

The circular end can be divided into a large number of small sectors, each approximating to an isosceles triangle of height r, where $r = D/2$. One such sector is shown in Figure 12.32(b). If the stress acts uniformly over the sector, the line of action of the resultant force will be through the centre of gravity of the triangle, i.e. at a distance of $(2r/3)$ from the apex. Putting all triangular sectors together, the total resultant force will act at a distance $2r/3$ (i.e. $D/3$) from the centre of the circle. The torque T_2 is therefore given by the equation

$$T_2 = \left(\frac{\pi D^2 s}{4} \right) \times \frac{1}{3} D$$
$$= \frac{\pi D^3 s}{12} \qquad (12.17)$$

From Equations (12.15)–(12.17) the total resisting torque is equal to

$$T_r = T_1 + 2T_2$$
$$= \frac{\pi D^2 Hs}{2} + 2\left(\frac{\pi D^2 s}{12} \right)$$
$$= \pi D^2 \left(\frac{H}{2} + \frac{D}{6} \right) s \text{ Nmm}$$

If the vane shear strength, τ_v, of the clay is to be measured in kPa, then $s = \tau_v/1000$ N/mm². Therefore

$$T_r = \frac{\pi D^2}{1000} \left(\frac{H}{2} + \frac{D}{6} \right) \tau_v \qquad (12.18)$$

The torque T (Nmm) applied to the shaft of the vane is proportional to the angular rotation q (degrees) of the torsion spring, i.e.

$$T = C_s\theta$$

where C_s is the spring calibration factor measured in Nmm per degree, and is provided by the manufacturer. At failure, $T = T_f$, and $\theta = \theta_f$

$$T_r = C_s\,\theta_f = \frac{\pi D^2}{1000}\left(\frac{H}{2}+\frac{D}{6}\right)\tau_v \tag{12.19}$$

Therefore, τ_v can be determined from the measurements made in the vane test by rearranging Equation (12.19) as follows:

$$\tau_v = \frac{1000\,C_s\theta_f}{\pi D^2\left(\dfrac{H}{2}+\dfrac{D}{6}\right)}\,\text{kPa}$$

Putting $\pi D^2\,(H/2 + D/6) = K$ (12.20)

$$\tau_v = \frac{1000\,C_s\theta_f}{K}\,\text{kPa}$$

where K (mm^3) is a factor depending only on the dimensions of the vane.
   The dimensions of a typical laboratory vane are

$$D = 0.5\ \text{in} = 12.7\ \text{mm and } H = 0.5\ \text{in} = 12.7\ \text{mm}$$

$$\therefore \frac{H}{2}+\frac{D}{6} = 6.350+2.117\ \text{mm} = 8.467\ \text{mm}$$

and the value of K from Equation (12.20) is

$$\pi\times(12.7)^2\times8.467 = 4290\ \text{mm}^3$$

i.e.
$$\tau_v = \frac{1000\ C_s\theta_f}{4290} = \frac{C_s\theta_f}{4.29}\ \text{kPa} \tag{12.21}$$

in which C_s is expressed in N mm/degree and θ_f is measured in degrees. Equation (12.21) holds good only for a vane of the stated dimensions. A different relationship must be worked out, using Equation (12.20), for other sizes of vane.
   The rate of rotation of the vane is not critical provided that it is neither excessively fast nor extremely slow. If the vane is rotated too fast, viscous effects may lead to a result which is too high. If the testing time is unnecessarily prolonged, partial drainage may take place during shear in some soils. The constant angular rate of rotation of the torsion head (which exceeds that of the vane itself) specified in the BS is 6°/min to 12°/min for both the laboratory vane test (Part 7, Clause 3.3.7) and the field vane test (Part 9, Clause 4.4.4.1.4). A motorised drive, with suitable reduction gearing, is more convenient than hand operation.

12.4 Applications

12.4.1 Applications of shear strength parameters

Many stability problems in soils are concerned with a hypothetical limiting condition in which the mechanism of failure involves the sliding of a body of soil relative to the main soil mass. The surface of slip along which relative movement is assumed to take place may be plane or curved and it is assumed that the soil along the whole of the slip surface is at a state of failure, i.e. its maximum shear strength has been mobilised. In practice it is necessary to ensure that this condition will never occur. For this reason, and to limit deformations to within tolerable limits, a suitable factor of safety is applied to ensure that the shear stress in the soil is nowhere greater than a certain proportion of its maximum shear strength.

In cases where it can be assumed that the water content of the soil does not change under load, an analysis in terms of total stresses, based on undrained shear strength, is sometimes appropriate. Examples with simplified illustrations are indicated in simple terms below.

1. Bearing capacity of footings and foundations for structures on saturated homogeneous clays, immediately after construction. The soil beneath a foundation, if loaded to failure, is assumed to fail by shear in the manner indicated in Figure 12.33(a).
2. Earth pressure on a retaining wall, for the conditions prevailing immediately after construction (see Figure 12.33(b)).

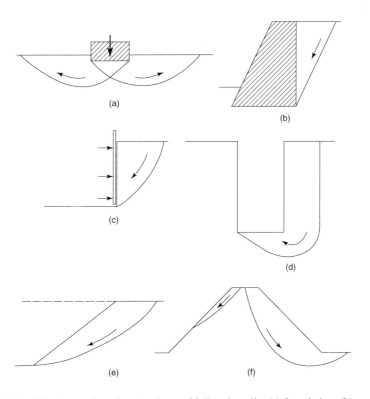

(a)

(b)

(c)

(d)

(e)

(f)

Figure 12.33 Simplified examples of mechanisms of failure in soils: (a) foundation; (b) retaining wall; (c) bracing in excavation; (d) deep excavation; (e) cutting; (f) embankment or earth dam

3. Earth pressure against bracing in temporary excavations (see Figure 12.33(c)).
4. Safeguard against heave of the bottom of temporary oven excavations in clay (see Figure 12.33(d)).
5. Stability of the side slopes of cuttings immediately after excavation (see Figure 12.33(e)).
6. Short-term stability of embankments and earth dams during construction (see Figure 12.33(f)).

In short-term stability problems such as these, the value of the undrained shear strength, c_u, could be applicable in the appropriate circumstances, but this has to be assessed on the basis of experience.

For the analysis of long-term stability of retaining walls, embankments, and earth dams the drained shear strength parameters c', ϕ', are required. The long-term stability of slopes and cuttings in OC clays is based on the residual shear strength parameters c'_r, ϕ'_r, (Skempton, 1964; Skempton and La Rochelle, 1965; Symons, 1968).

12.4.2 Use of standard shearbox

The shearbox apparatus was developed originally for the determination of the angle of shear resistance, ϕ', of recompacted sands. It provides the most direct means of relating ϕ' to the voids ratio, e, and of determining the critical voids ratio (or critical density) of dry sands or of saturated sands which do not contain fine material in sufficient quantity to impair the drainage characteristics.

The shearbox is not suitable for determining the immediate undrained shear strength of cohesive soils, but for these materials the triaxial test is more satisfactory.

One of the main applications of shearbox testing is the measurement of the residual shear strength of OC clays as an extension to the procedure for measuring peak drained strength.

12.4.3 Use of large shearbox

The usual size of a large shearbox is 12 in (305 mm) square, requiring a specimen about 150 mm thick, and is suitable for soils containing particles up to 37.5 mm. Triaxial testing of these materials is impracticable unless exceptionally large equipment is available.

The shear strength of gravelly soils is rarely a critical factor for foundations, but it is significant in the design of embankments or earth dams which incorporate gravel fill material. Shear strength can also be used as a means of classifying road construction materials and granular sub-bases (Pike, 1973; Pike *et al.*, 1977).

Other materials besides gravels containing particles up to 37.5 mm or even occasional 50 mm particles can also be tested in a large shearbox where triaxial testing would be impracticable. These materials include shale, industrial slag, brick rubble, and colliery spoil. Provided that the material is free-draining, the quick testing procedure may be used to determine the angle of shear resistance; otherwise slow (drained) tests are necessary.

This apparatus can also be used for testing large block samples of undisturbed soil. It has been used by the present author to measure the shear strength along surfaces of discontinuity, such as fissures or shear zones, in large samples of OC clay; and the shear strength of clay laminations present in some sandstones.

12.4.4 Applications to miscellaneous materials

In addition to its application to soils, the shearbox apparatus can be used for the measurement of the frictional resistance of other engineering materials. Some examples are as follows:

Friction between soil and rock

Friction on a joint surface in rock

Friction between soil and manufactured materials such as concrete, fabric matting, reinforcing materials used in reinforced earth construction, components of ground anchor systems

Bond strength of adhesives and cementing agents

Friction between materials and components used in laboratory testing, e.g. latex rubber and silicone grease on stainless steel

In most of the above applications the property measured is the angle of friction, or coefficient of friction. The strength of a bonding or cementing agent would show up as an apparent cohesion.

12.4.5 Limitations and advantages of the shearbox test

There is some uncertainty in the interpretation of results obtained from shearbox tests for providing a failure criterion for the soil. Soil behaviour in direct shear tests is discussed in more detail by several authors in the Sixth Géotechnique Symposium in Print (1987). The method outlined in this chapter has been generally used for obtaining a soil failure criterion. The main limitations and disadvantages of the shearbox test are summarised below.

Limitations

1. The soil specimen is constrained to fail along a predetermined plane of shear
2. The distribution of stresses on this surface is not uniform
3. The actual stress pattern is complex and the directions of the planes of principal stresses rotate as the shear strain is increased
4. No control can be exercised over drainage, except by varying the rate of shear displacement
5. Pore water pressures cannot be measured
6. The deformation which can be applied to the soil is limited by the maximum length of travel of the apparatus
7. The area of contact between the soil in the two halves of the shearbox decreases as the test proceeds. A correction to allow for this was proposed by Petley (1966), but its effect is small. It affects the shear stress and normal stress in equal proportions, and the effect on the Coulomb envelope is usually negligible, so it is generally ignored

Advantages

Notwithstanding the above limitations, the shearbox apparatus has certain merits for routine shear strength testing, as summarised below.

1. The test is relatively simple to carry out
2. The basic principle is easily understood
3. Preparation of recompacted test specimens is not difficult

4. Consolidation is relatively rapid due to the small thickness of the test specimen
5. The principle can be extended to gravelly soils and other materials containing large particles, which would be more expensive to test by other means
6. Friction between rocks and the angle of friction between soils and many other engineering materials can be measured
7. In addition to the determination of the peak strength at failure the apparatus can be used for the measurement of residual shear strength by the multi-reversal process

12.4.6 Typical values of angle of shear resistance

Values of ϕ' for grains of pure quartz, over the range of possible states of packing depending upon particle shape and grading, are given in Table 12.2. Typical ranges of values of ϕ' for dry non-cohesive soils are indicated in Table 12.3.

The presence of certain mineral particles such as mica can decrease the ϕ' value of sands and silts. Gravels consisting of relatively soft particles which are susceptible to crushing, give a lower value than those with hard particles.

12.4.7 Application of vane test

The laboratory vane enables the low shear strengths of very soft soils (i.e. less than 20 kPa) to be measured, which would be very difficult to do by other means.

It may sometimes be important to know the value of a low shear strength to a reasonable degree of accuracy. For instance, if an embankment is to be built on a stratum of very soft soil, the shear strength would indicate the maximum safe bearing pressure which it could

Table 12.2 Values of ϕ' for quartz grains (taken from Terzaghi and Peck, 1967)

Particle shape and grading	*Degrees*	
	Loosest	*Densest*
Rounded, uniform	28	35
Angular, well-graded	34	46

Table 12.3 Typical values of ϕ' for dry non-cohesive soils (taken from Lambe and Whitman, 1979)

Type of soil and grading	*Degrees*			
	Loose		*Dense*	
	Rounded	*Angular*	*Rounded*	*Angular*
Sand:				
uniform fine to medium	30	35	37	43
well graded	34	39	40	45
Sand and gravel	36	42	40	48
Gravel	35	40	45	50
Silt	28–32		30–35	

sustain initially, and hence the thickness of embankment which could be placed as the first stage. Subsequent consolidation would increase the shear strength and construction could proceed in stages based on the shear strength criterion.

The shear strength of soft strata near the surface may be required for the estimation of 'negative skin friction' due to adhesion on piles taken down to a firmer stratum below (Lambe and Whitman, 1979). Remoulded strength may be significant in areas of soft deposits which have been extensively disturbed, such as during the driving of piles.

The vane shear test can be used to provide a relationship between shear strength and moisture content of clays extended to moisture contents greater than those at which it is practicable to prepare conventional test specimens. It has been used to investigate the relationship of shear strength to soil moisture suction (Lewis and Ross, 1955). The laboratory vane, suitably adapted if necessary, can also provide data on the shear strength of some grouting materials, and on the gelling or setting time required to obtain a required strength.

12.5 Small shearbox: rapid test (BS 1377:Part 7:1990:4, and ASTM D 3080)

12.5.1 General

The apparatus described here is typical of that which is commercially available in the UK for routine testing, and the procedures follow BS practice. The most common type of apparatus accommodates a 60 mm square specimen, 20 or 25 mm high, and is described in detail. Some shearbox machines can also provide for other sizes up to 100 mm square. The same apparatus is also used for 'slow' drained and residual strength tests (see Section 12.7), for which many of the procedural details are similar. Use of a much larger apparatus, for specimens 12 inches (305 mm) square or larger, is outlined in Section 12.6.

The procedure described in this section is the rapid test for determination of the shearing resistance of dry or free-draining saturated sands. Usually a set of three similar specimens from the same sample of soil, all placed at the same dry density, are tested under three different normal pressures, enabling the relationship between the measured shear stress at failure and the applied normal stress to be obtained. Since no excess pore water pressure can develop in these soils, effective stresses are equal to the measured total stresses and the shear strength can be expressed as the angle of shearing resistance, ϕ'.

The normal pressures applied to specimens in a set of tests should generally 'bracket' the maximum stress likely to occur in the ground. Normal pressures of about 50%, 100% and 150–200% of this value are often appropriate, but these are suggested only as a general guide and the pressures used should be decided by the engineer. There is no 'standard' set of pressures.

Slow drained tests on soils that are not free-draining, including shearbox tests for measurement of residual strength, are described in Section 12.7.

Before starting a set of tests the following test conditions need to be defined:

Size of test specimens

Dry density (or porosity or void ratio) at which reconstituted specimens are to be prepared

Orientation of undisturbed test specimens

Number of specimens to be used for the set

Normal pressures to be applied

12.5.2 Apparatus

The shearbox machine comprises a drive unit, shearbox assembly, shearbox carriage, load hanger and other items detailed below which may be supported on a bench or mounted in a steel-framed stand. A typical shearbox machine with a counterbalanced lever arm is shown in Figure 12.34.

When fully loaded with hanger weights the whole apparatus could weigh up to 200 kg, so a substantial floor mounting is necessary. The machine should be bolted down.

The component parts of the shearbox apparatus and ancillary items required for the test are listed below.

1. Shearbox carriage, watertight, running on ball or roller bearings.
2. Shearbox assembly, comprising
    (a) Shearbox body, in two halves (Figure 12.35), the upper half fitted with a 'swan-neck' yoke, the whole box being rigid enough to resist distortion under load. The points of application of the horizontal shearing forces must be in line with the plane of separation of the two halves of the box (see Figure 12.6).

    The two halves can be temporarily fixed together by means of two clamping screws (marked (C) in Figure 12.35) using the holes marked C, threaded in the lower half. Two lifting screws (L) enable the upper half of the box to be lifted slightly (see Section 12.5.6, stage 8) using the threaded holes at the opposite corners of the upper half, marked L

    (b) Lower pressure plate (base plate), retained by small lugs in the lower half of the box which fit into four recesses
    (c) Upper pressure plate (load pad), with spherical seating and ball bearing
    (d) Upper and lower porous plates

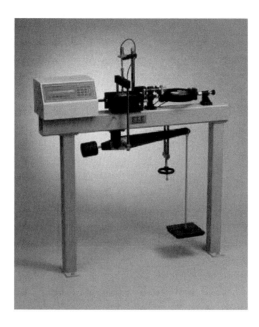

Figure 12.34 Typical shearbox apparatus for 60 × 60 mm or 100 × 100 mm specimens

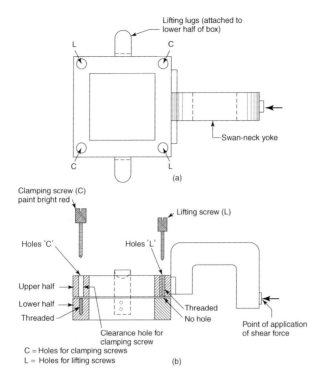

Figure 12.35 Details of 60 mm shearbox: (a) plan view; (b) section through box

Figure 12.36 Component parts of 60 mm shearbox, with wood 'pusher'

(e) Upper and lower grid plates, plain or perforated. The grid plates enable the shearing forces to be transmitted uniformly along the length of the sample. Their use is not mandatory in the BS, where they are referred to as spacer plates, but the author favours the school of thought that they should be used for dry or free-draining sands. Solid plates should be used only for dry sands

The above components are shown in Figure 12.36 and identified in Figure 12.37. A cross-section of the assembled box is shown in Figure 12.6.

3. Loading yoke and weight hanger, for applying the normal pressure to the specimen.

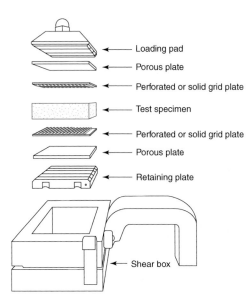

Figure 12.37 Assembly of 60 mm shearbox

4. Lever-arm loading system, for extending the range of normal pressures. Beam ratios of 5:1 or 10:1 are usual. The procedure for verifying the effect of the lever arm is described in Section 12.5.6, stage 7(b). Some machines incorporate a counterbalanced lever-arm and a hanger support jack.
5. Calibrated slotted weights for the load hanger or lever system. A suitable set comprises
    9 no. × 10 kg
    1 no. × 5 kg
    2 no. × 2 kg
    1 no. × 1 kg
    totalling 100 kg altogether.
6. Calibrated force measuring device (load ring or load cell (see Section 8.2.1, subsection on Conventional and electronic measuring instruments) for measuring the horizontal shear force. A ring suitable for most purposes is of 2 kN capacity, but a ring of 4.5 kN or perhaps up to 10 kN capacity may be required for measuring very high shear strengths.
7. Electric motor and multi-speed drive unit, typically providing 24 speeds ranging from 5 mm/min to about 0.0003 mm/min; or infinitely variable speed control. The full displacement of 10 mm can be obtained in a period ranging from 2 min to about 3 weeks. The quick test is usually carried out in about 5–10 min, which requires a nominal rate of displacement of about 1 mm/min.

    A speed chart gives the rate of displacement for every combination of gear settings. The motor is reversible.

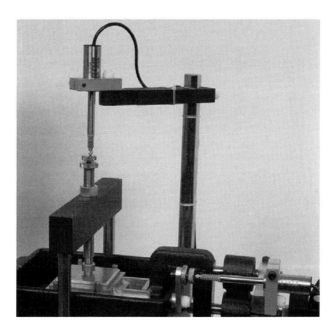

Figure 12.38 Shearbox fitted with transducers for measuring vertical displacement and relative horizontal displacement

8. Loading jack or worm reduction unit, which pushes the shearbox carriage forwards. It may be connected to the drive unit by a sprocket and chain, or by enclosed gears. The drive unit can be disengaged to enable rapid manual adjustment of the jack to be made by means of a handwheel.
9. Tailstock unit, which provides the reactive force, fixed to bed of machine, to which the load ring is attached and which provides linear adjustment. The stem of the load ring which bears against the yoke of the split box is supported in a ball-bearing sleeve.
10. Micrometer dial gauge or linear transducer (see Section 8.2.1), 12 mm travel reading to 0.002 mm, for measuring vertical movement of the top of the specimen, together with supporting post and mounting bracket. The gauge should preferably read anti-clockwise, i.e. reading increasing as the stem moves downward.
11. Micrometer dial gauge or linear transducer, 12 mm travel reading to 0.01 mm for measuring horizontal displacement. This should be mounted so as to measure the *relative* movement of the two halves of the box, such as by fixing it to the front end of the carriage so that the stem bears against a bracket fixed to the swan-neck yoke (see Figure 12.38, in which a displacement transducer is so mounted).

    If the gauge measures the displacement of the carriage only, the deflection of the load ring must be subtracted to obtain the relative movement of the two halves of the specimen. The relationship between dial reading and linear displacement (mm) should be ascertained before hand.
12. Specimen cutter, 60 mm square and 20 mm or 25 mm deep, with smooth internal surfaces and sharp externally chamfered cutting edges.
13. Wood pusher for removing specimen from cutter (shown in Figure 12.36).

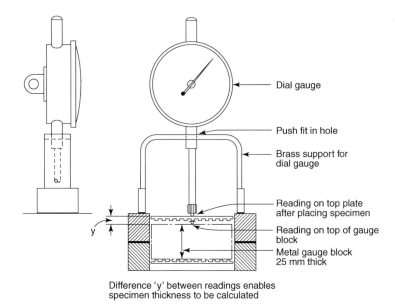

Difference 'y' between readings enables
specimen thickness to be calculated

Figure 12.39 Specimen thickness comparator for 60 mm shearbox (from design by Imperial College, London)

14. Jig for holding specimen cutter in place while extruding a sample into it from a 100 mm diameter sampling tube.
15. Cutting tools and straight edge for preparing specimen.
16. Flat glass plate, such as is used for the liquid limit test (see Volume 1 (third edition), Section 2.6.4).
17. Watch glass or metal tray, large enough to hold the specimen cutter.
18. Tamping rod with square end about 15 mm across, for tamping soil into the shear box. Alternatively, the Harvard compaction apparatus could be used if available (see Volume 1 (third edition), Section 6.5.10).
19. Stop-clock or seconds timer.
20. Balance and measuring instruments (steel rule, vernier callipers, depth gauge) for weighing and measuring the test specimen and specimen cutter.
21. Dial gauge, reading to 0.01 mm, and supporting stand, for accurate measurement of the thickness of specimens formed directly in the box. A small stand made at Imperial College specially to fit the 60 mm square shearbox is shown is Figure 12.39, and is used in comparison with a metal block of exactly 25 mm thickness. Alternatively a vernier depth gauge reading to 0.1 mm may be used.
22. Apparatus for determining moisture content (see Volume 1 (third edition), Section 2.5.2).
23. Engraving tool (see Chapter 9, Figure 9.14) fitted with a suitable tamping foot. This device has been found to be effective as a small vibrator for compacting sand to a high density in the shearbox.
24. Silicone grease or petroleum jelly.

12.5.3 Preparation of apparatus

The following notes indicate the general principles to be observed. Detailed instructions provided by the manufacturer of the equipment should be followed.

Ensure that the shearbox is clean and dry, especially the surfaces of the two halves which adjoin. Apply a thin film of silicone grease to the inside surfaces of the box and to the mating surfaces. Assemble the two halves, and tighten the clamping screws. Verify the internal measurements using vernier callipers. The lengths of the sides are denoted by L_1 and L_2 (mm), each measured to an accuracy of 0.1 mm.

Determine the mean depth from the top surface of the upper half of the box to the top of the baseplate (h_1) to 0.1 mm. Measure the thickness of the porous plates (denoted as t_p) to 0.1 mm with vernier callipers or a micrometer and determine the mean depth from the top surface of the upper half of the box to the top of the upper porous plate (h_2). The thickness of the specimen (H_0) is given by the difference in the depths less the thickness of the porous plates:

$$\text{e.g. } H_0 = h_1 - (h_2 + t_p)$$

The essential dimensions required to determine specimen thickness are shown in Figure 12.40. Where grid plates are used as spacers the thickness of the lower grid plate must also be determined and included in t_p.

See that the inside of the carriage is clean and that it runs freely on its bearings, which should be centrally positioned under the box.

Wind back the handwheel on the tailstock far enough to enable the load ring to be fitted in place, and to allow sufficient clearance for the shearbox and swan-neck yoke. Ensure that the load ring is securely fixed to the tailstock assembly, and that the loading stem can move freely in its sleeve. Check that the dial gauge is securely fixed to the bracket on the load ring and that the stem is in contact with the anvil at the loading end.

Place the baseplate, grooves uppermost, in the shearbox. Lift the shearbox with the lugs provided (see Figure 12.35), and lower it into position in the carriage. The lower half of the box should fit firmly against the spacer block at the driving end of the carriage. If reversing facilities are fitted to the machine, it may not be necessary to couple these for a quick test in one direction.

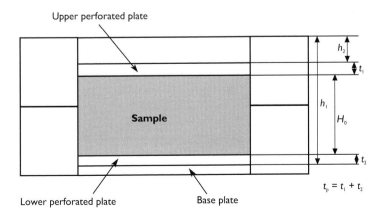

Figure 12.40 Dimensions required for determination of specimen thickness

Adjust the tailstock and drive unit to the correct starting position, using the handwheels, so that the carriage will have at least 10 mm movement in the forward direction. The shearbox is then ready to receive the specimen. Alternatively, if it is more convenient to place the sample in the box on the bench, turn the drive unit backwards slightly by hand so that there is enough clearance to take out the shearbox and to replace it after inserting the sample.

Check that the gears of the motor unit are set to the required rate of displacement, or that the required rate is selected. For a rapid test a rate of 1 mm/min is usually suitable, so that failure occurs within 5–10 min. If the machine incorporates a chain drive, check that the chain-drive sprockets are tight on their shafts and that the chain is neither too slack nor over-tight. Ensure that the reversing switch is in the correct position for forward movement if the gearbox is of the type in which alternate positions change the direction of travel.

12.5.4 Preparation of test specimens of cohesionless soils

The procedure depends upon the type of soil, and the condition in which it is to be tested. The maximum size of particles present in significant quantity should not exceed one-tenth of the specimen height. For a specimen 20 mm or 25 mm high, particles retained on a 2 mm sieve should be removed. The soil should not contain a significant quantity of fine material passing a 63 μm sieve. This is to avoid segregation of fine particles if placed dry, and to ensure that the free-draining condition is maintained if saturated. If the amount of fines is large enough to impede free drainage of water the rapid procedure is not applicable.

1. Dry sand: general

Sand is usually tested at a specified porosity or voids ratio, from which the dry density can be obtained. A test specimen is prepared by placing or compacting the material directly into the shearbox. Because of the large mass of the shearbox itself, the mass of sand used (*m* grams) is best obtained by weighing out a known dry mass, weighing the dried sand left over after compaction, and calculating the mass used (*m*) by difference.

The lower (solid) grid plate is placed on top of the baseplate in the shearbox, grid uppermost and with the ribs at right angles to the direction of shear. Unperforated grid plates can be used with dry sand because the question of drainage does not arise.

Details of the procedures used for placing the sand are given separately below for dense and loose states. After placing, carefully level off the top surface of the sand to within about 5 mm of the top of the box without disturbing the whole specimen. A levelling template as shown in Figure 12.41 facilitates this operation. Retain any unused sand for weighing, as referred to above.

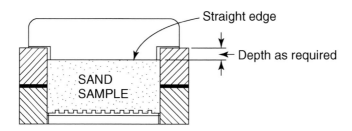

Figure 12.41 Levelling template for preparing sand specimen

Place the top porous on the surface of the sand, and press it down evenly to bed into the sand with the top surface level. Measure the distance from the top of the box to the back of the perforated plate at the mid-point of each side or at the four corners. A depth gauge or a dial gauge comparator of the type shown in Figure 12.39 is desirable to obtain accurate measurements. The average measurement gives the distance h_2.

The height of the specimen H can then be determined from the equation

$$H_0 = h_1 - (h_2 + t_p)$$

as given in Section 12.5.3.

The volume of the specimen in the box, V, is equal to $L^2 \times H/1000$ cm^3 and the density is equal to

$$\frac{m}{V} \text{ Mg/m}^3 \text{ or } \frac{1000\,m}{L^2 \times H} \text{ Mg/m}^3$$

2. Dry sand: dense

Dry sand can be placed in the shearbox at a reasonably high density by slow pouring at high velocity, i.e. from a relatively high drop (Kolbuszewski, 1948). A drop of about 450 mm is probably enough, for practical reasons. If the sand can be subjected to vibration a higher density should be obtainable. An electric engraving tool (see item 23, Section 12.5.2) may be used for vibration. Compaction of dry sand with a small hand tamper will have little effect.

When pouring, place the shearbox on a tray so that all the surplus poured sand can be retained for weighing.

3. Dry sand: loose

A low density may be obtained by pouring the sand very rapidly into the shearbox from a small height. Use a tray to retain the surplus poured sand. When levelling the top surface, the excess sand should be carefully scooped off, avoiding a scraping action as far as possible. The top grid plate should be placed very carefully and bedded down with minimum pressure. Jolting and bumping of the shearbox when placing in the machine should be avoided because loose sand is very sensitive to sudden shock.

4. Dry sand: medium density

For an intermediate density, the sand can be placed in the shearbox in three layers, each layer subjected to a controlled amount of tamping. The exact amount of compaction necessary can only be determined by trial. The mid-height of the middle layer should be about level with the plane of shear.

5. Saturated sand

Saturate the sand by boiling a known dry mass in water for about 10 minutes and allowing it to cool; or by placing the sand and water mixture under vacuum to remove air bubbles.

The shearbox is placed in the carriage, which is then partly filled with water. Saturated sand is poured into the box and tamped or gently vibrated, depending upon the density required. It is impracticable to obtain a low relative density by this method, but medium to high densities can be achieved.

This method is not suitable if fine material (silt or clay) is present because of segregation. An alternative procedure is to place the mixed soil in a damp state and compact it as necessary, then pour water carefully into the carriage so that it percolates slowly upwards through the specimen.

Drainage will take place rapidly from a 'clean' sand. To facilitate drainage the perforated grid plates are used instead of the solid plates, together with a porous plate behind each grid plate. The additional thickness t_3 of the perforated grid plate has to be taken into account when calculating the specimen height H, i.e.

$$H = B - (h_2 + t_p + t_3) \text{ mm}$$

assuming that the measurement x' is made before placing the upper porous plate.

12.5.5 Procedural stages

1. Prepare and check apparatus
2. Prepare test specimen; procedure depends on soil type, as explained in Section 12.5.4
   Dry sand: general (1)
   Dry sand: dense (2)
   Dry sand: loose (3)
   Dry sand: medium density (4)
   Saturated sand (5)
3. Assemble apparatus
4. Fit load hanger
5. Set vertical dial gauge
6. Add water (if appropriate)
7. Apply normal stress
8. Lift top half of box
9. Final checks
10. Shear
11. Remove load
12. Drain box
13. Remove shearbox
14. Remove specimen
15. Repeat stages 2–15 using at least two other specimens
16. Calculate
17. Analyse data (see Section 12.5.7)
18. Report results see (Section 12.5.7)

12.5.6 Test procedure

1. *Preparation of apparatus*
   See Section 12.5.3.
2. *Preparation of test specimen*
   (See Section 12.5.4, whichever of sub-sections 1–5 is relevant to the type of soil and the required condition).

3. *Assembly of apparatus*

Place the shearbox in position in the carriage if it has been removed, taking care not to jolt the box if it contains loose sand. Check that the top grid plate is correctly in position and that there is a small clearance all around its edge. If perforated plates are used, place the top porous plate on top of the grid plate. Place the load pad on top, again ensuring that there is a small all-round clearance.

Adjust the worm drive unit and the tailstock by hand, if necessary, so that contact is just made at all five contact points indicated in Figure 12.42. A slight rotation of the handwheel on the worm drive unit should produce a small deflection of the load ring. Ensure that the worm drive is positioned so that it can give at least 12 mm of forward movement.

Adjust the load ring to the zero load position and set the dial gauge to zero or to a convenient zero reading.

Mount the horizontal displacement dial gauge or transducer on its bracket with the stem bearing on the moving bracket, ensuring that the gauge has sufficient travel available in the right direction. Set the gauge to zero or to a convenient initial reading.

4. *Fitting load hanger*

Place the ball bearing (if one is used) in the spherical seating on the load pad. Lift the load hanger and place it gently so that the recess under the yoke registers with the ball bearing or hemispherical bearing surface. A small normal stress will be induced in the specimen due to the weight of the hanger, but the resulting settlement is not usually measured because the vertical-movement dial gauge cannot normally be fitted until the hanger is in place. The specimen height after the addition of the hanger is usually taken as the datum from which subsequent vertical movements are measured.

This limitation can be overcome if the simple attachment shown in Figure 12.43, devised at Imperial College, is fitted to the load pad. The dial gauge can then be mounted and set to an initial zero reading before placing the load hanger in position, and the settlement due to the weight of the hanger can be measured.

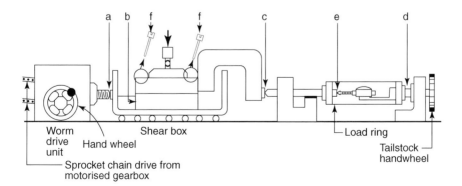

Figure 12.42 Points of contact to check before starting shearbox test: (a) worm drive to carriage; (b) pusher block to lower half of box; (c) swan-neck yoke to load ring stem; (d) load ring to tailstock; (e) stem of dial gauge to load ring, and note especially (f) remove clamping screws

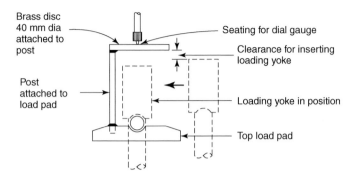

Figure 12.43 Dial gauge seating arrangement to load pad (from design by Imperial College, London)

5. *Setting vertical dial gauge*

Mount the vertical movement dial gauge or transducer on its support and swing it into position so that the stem rests on the stud or knurled screw at the centre of the hanger yoke or on the load pad attachment. Adjust the height of the gauge so that it gives a convenient zero reading (say 1000 divisions) in the middle range of its travel. It can then indicate either downward (positive) or upward (negative) movements.

6. *Adding water*

If water is to be added to the soil in the shearbox it should be poured steadily into the space between the carriage and the shearbox, so that water can penetrate upwards through the specimen, thereby displacing much of any air present in the voids. Any vertical movement resulting from inundation should be recorded.

7. *Application of normal stress*

Calculate the amount of weight required on the hanger to give the desired normal stress on the specimen.

(a) *Loading yoke only* For normal stresses within the capacity of the loading yoke weights, the calculation is as follows:

If the mass of the loading yoke with weight hanger is denoted by W_h kg, and the weights it supports by W_1 kg, the total mass supported by the specimen is $(W_h + W_1)$ kg = W kg. This is equal to a force of 9.81 W N. The pressure or stress on the specimen is therefore equal to $9.81W/L^2$ N/mm², where L is the length of the side of the square shearbox in mm.

Therefore

$$\sigma_n = \frac{9.81\,W}{L^2} \times 1000 \ \text{kN/m}^2$$

$$\text{or} \quad W = \frac{\sigma_n L^2}{9810} \ \text{kg}$$

$$\text{and} \quad W_1 = \frac{\sigma_n L^2}{9810} - W_h \ \text{kg} \tag{12.22}$$

For the standard 60 mm^2 shearbox, Equation (12.22) becomes

$$W_1 = (0.367\ \sigma_n - W_h)\ \text{kg}$$

and for a 100 mm^2 shearbox

$$W_1 = (1.02\ \sigma_n - W_h)\ \text{kg}$$

Place the weights gently on the hanger, starting with the heaviest at the bottom, and the smallest on top. Settlement of a sand specimen will stop after a very short time.

(b) *Lever arm loading* If the normal stress to be applied is greater than can be achieved with the hanger weights alone, use of the lever arm is necessary.

The following symbols are used in calculating the required weights, all masses being in kilograms (see Figure 12.44).

W_h = mass of loading yoke combined with its hanger
W_1 = mass placed on yoke hanger
W_2 = mass placed on beam hanger
W_j = mass of beam hanger suspended from lever arm beam
W_b = mass of lever-arm beam
W = total load (kg) applied to specimen
a = distance from fixed fulcrum to pivot point of beam
b = distance from fulcrum to centre of gravity of beam
c = distance from fulcrum to hanger suspended from beam

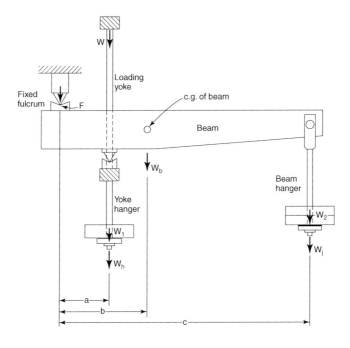

Figure 12.44 Details of lever arm for applying high normal pressures to shearbox specimen

Taking moments about the fulcrum F, the moment of the total load W is equal to the sum of the moments of the separate components, i.e.

$$Wa = (W_h + W_1)a + W_b b + (W_j + W_2)c$$

$$\therefore W = W_h + W_1 + W_b \frac{b}{a} + (W_j + W_2)\frac{c}{a}$$

The force applied to the specimen is equal to 9.81 W N.

The ratio b/a requires that the position of the centre of gravity be obtained. This can be done accurately enough by finding the point at which the beam just balances, for instance on the edge of a spatula blade clamped in a vice or held firmly on the bench. The ratio c/a is the nominal beam magnification ratio (usually 5 or 10), but should be verified. These ratios and the constant masses of the yoke, beam and beam hanger should be recorded and displayed, together with the pressures produced on the specimen by the available range of weights

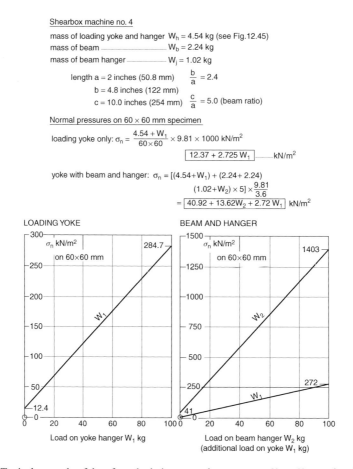

Shearbox machine no. 4

mass of loading yoke and hanger W_h = 4.54 kg (see Fig.12.45)
mass of beam .... W_b = 2.24 kg
mass of beam hanger .... W_j = 1.02 kg

length a = 2 inches (50.8 mm) $\frac{b}{a}$ = 2.4
b = 4.8 inches (122 mm)
c = 10.0 inches (254 mm) $\frac{c}{a}$ = 5.0 (beam ratio)

Normal pressures on 60 × 60 mm specimen

loading yoke only: $\sigma_n = \frac{4.54 + W_1}{60 \times 60} \times 9.81 \times 1000$ kN/m²

$\boxed{12.37 + 2.725\,W_1}$ .... kN/m²

yoke with beam and hanger: $\sigma_n = [(4.54+W_1) + (2.24 + 2.24)$

$(1.02+W_2) \times 5] \times \frac{9.81}{3.6}$

$= \boxed{40.92 + 13.62W_2 + 2.72\,W_1}$ kN/m²

LOADING YOKE

BEAM AND HANGER

Load on yoke hanger W₁ kg

Load on beam hanger W₂ kg
(additional load on yoke W₁ kg)

Figure 12.45 Typical example of data for calculating normal pressure on 60 × 60 mm shearbox specimen

both with the yoke hanger and the lever arm hanger. A set of data relating to a particular machine is given in Figure 12.45. If a counterbalanced beam is fitted, only the weights need to be taken into account and the calculation of normal stress is simplified. If pound weights are used on the hangers, multiply their value in pounds by 0.4536 to obtain the value in kilograms.

Fix the beam carefully into position, adjust it to be level, fit the weight hanger, then start the timer. Carefully add the weights to the hanger, bearing in mind that the effect on the specimen of each weight and of any sudden shock is magnified 5 or 10 times (depending on the lever ratio). The downward movement of the hanger will be in the same ratio to the vertical movement of the specimen, so some adjustment may be necessary to maintain the beam near to the horizontal.

Weights may also be added to the loading yoke hanger, if necessary, to provide a finer adjustment of normal stress than would be possible using the lever arm alone.

The relationships between W (the total load, in kg, applied to the specimen) and σ_n for four sizes of shearbox are summarised in Table 12.4.

8. *Lifting top half of box*

Remove the clamping screws which lock the two halves of the shearbox together (see Figure 12.35). If this is overlooked, the test will merely be an attempt to shear them (which might succeed). If the tops of the screws are painted bright red it will provide a visual reminder.

Insert these screws, or separate screws if provided, into the other two holes (marked L in Figure 12.35) and screw them down until contact with the lower half of the box can just be felt. Rotate them both together a further half-turn, so as to separate the two halves of the box by about 0.5 mm. The whole of the horizontal shear force can now be transmitted through the specimen itself. Retract the screws and remove them.

Lifting the upper half of the box by a half-turn of the screws is a nominal amount, but if sand is being tested the amount of lift should ideally be slightly more than the diameter of the largest particle. This is to prevent crushing of grains between the two halves of the box which could cause additional friction. For example, if the material contains medium sand up to 600 μm, a lift of say 0.8 mm would be appropriate. If the pitch of the screw thread is 0.8 mm, this would require one whole turn. The amount of lift must be kept to within a reasonable limit, say 1 mm at the most.

9. *Final checks*

Before proceeding to the shearing stage, the following items should be checked:
Contact made at all contact points (see Figure 12.42(a)–(e))

Table 12.4 Load on specimen (W) related to normal stress (σ_n)

Dimensions of shearbox (mm)	σ_n for given W (kg) (kPa)	W for σ_n = 100 kPa (kg)
60 × 60	2.725 W	36.7
100 × 100	0.981 W	102
300 × 300	0.109 W	917

Clamping screws and lifting screws removed (see Figure 12.42(f))
Dial gauges set correctly
Load pad not tilted or jamming
Machine speed and reversing switch correctly set
All dial gauge readings recorded in the initial or 'zero' positions
Timer wound up, set at zero

10. *Shearing*

Switch on the motor and simultaneously start the timer. At regular intervals of the displacement dial reading, record the readings of the load dial, the vertical movement dial and the time. A suitable interval for taking readings initially is every 10 divisions (0.1 mm) of the displacement dial gauge, but if the load ring reading increases rapidly, additional readings should be taken. At least 20 sets of readings should be taken up to the maximum reading of load ('peak' shear strength). For brittle specimens such as dense sands, which show a rapid increase in load, sets of readings should be taken at regular intervals of load ring reading instead of displacement. If the load is seen to change slowly, the number of recorded readings may be reduced to one every 20 or 50 divisions of the displacement gauge.

Shearing should be continued until the maximum stress or 'peak' point has been clearly defined, that is until at least four consecutive readings indicate a decrease in shear force. If a peak is not observed, shearing should continue until the full length of travel of the box has been reached. A steady rapid increase in load ring reading will indicate that the permitted travel has been exceeded, unless a travel limit switch is fitted.

If possible, plot a rough graph of load ring reading against displacement as the test proceeds.

11. *Removal of load*

At the end of the shear test switch off the motor, wait until it has completely stopped, and switch on in the reverse direction until the drive unit has returned to its starting position. Unless the reversing coupler has been engaged (and this is not usually done for a quick test) the specimen will remain in its sheared position.

12. *Draining the box*

If water was added to the carriage, remove it by siphoning into a beaker or by sucking it into an empty plastic wash bottle. Allow to stand for about half an hour so that the porous plates can drain. Record any further movement of the vertical dial gauge.

13. *Removal of shearbox*

Take the weights off the load hanger and if the lever arm hanger was used remove the hanger and beam. Swing the vertical movement dial gauge into a safe position. Lift off the hanger yoke and place it in its resting position. Take off the load pad and upper porous plate if used.

Lift out the shearbox, using the lifting lugs, and place on the bench. Clean out the carriage.

14. *Removal of specimen*

Remove the top grid plate and tip the specimen into a small metal tray which has previously been weighed. Brush out the inside of the box, and the grid plate, so that all the material is transferred to the tray without loss. If the sand was tested dry, weigh the tray and soil.

If the sand was not tested in the dry state, dry the soil in the oven overnight, allow to cool in a desiccator and weigh. The moisture content is not usually important for a granular soil but the mass of soil provides a check on the mass initially used and therefore on the density and voids ratio.

The whole specimen may then be used for particle density and particle size tests, if required.

15. *Repeat tests*

To obtain a set of three points on the Coulomb envelope, repeat stages 1–14 on two additional identical specimens under different normal pressures. A set of three specimens is usual, but additional specimens may be tested if required.

The values of the normal pressures should be related to the stress levels in the particular application and should be selected to provide a reasonable spread of points (see Section 12.5.1). There is no 'standard' set of pressures.

16. *Calculations*

Density and moisture content The calculation of the initial density of the specimen, ρ, is given under item 1 of Section 12.5.4.

If the application of the normal stress causes a settlement of y mm before shearing, the consolidated density ρ_c is given by the equation

$$\rho_c = \frac{H}{H-y}\rho \tag{12.23}$$

or if y is small compared with H, it is approximately

$$\rho_c = \rho\left(1 + \frac{y}{H}\right)$$

Normal stress The calculation of the normal stress is given in stage 7.

Shear stress If the load ring calibration, in newtons per division (N/div), is denoted by C_R, the shear stress t corresponding to a load dial reading R divisions is given by the equation

$$\tau = \frac{C_R R}{L^2} \times 1000 \text{ kPa} \tag{12.24}$$

where L is the length of side of the square shearbox (mm). The continual change in area of contact between the two halves of the specimen is not taken into account in calculating either the shear stress or the normal stress, because the changes almost cancel out.

For a 60 × 60 mm shearbox, the shear stress is given by

$$\tau = \frac{C_R R}{3.6} \text{ kPa}$$

and for a 100 × 100 mm shearbox

$$\tau = \frac{C_R R}{10} \text{ kPa}$$

Shear Box Test Data

Location	Mawne Hill			Sample No	30 / 8	A
Operator	D. P. R			Date	24 . 6 . 80	
Type of test	Quick			Nominal size	60 × 60 mm	
Soil description	Light brown fine to medium sand (dry)					
Type of specimen	COMPACTED					
Specimen preparation	Compacted by rodding in 3 layers					

INITIAL MEASUREMENT	Length L	60.0 mm	Area A	3612 mm²	Specific gravity
	Breadth B	60.2 mm	Volume V_O	145.6 cm³	
	Height H_O	40.3 mm	Bulk density ρ	1.69 Mg/m³	ASSUMED
	Mass m	246.2 g	Dry density ρ_D	1.69 Mg/m³	G_S
	Moisture w	0 %	Voids ratio e_O	0.568	2.65

SHEARING

	AFTER CONSOLIDATION
Machine No 2 Load ring No R 225	Settlement 0.018 mm Height H_1 40.1 mm
Mean calibration C_R 0.936 N/div	Dry density ρ_{D1} 1.70 Mg/m³
Stress factor C_T 0.259 kN/m² per div.	Voids ratio e_1 0.559
Rate of displacement / mm/min	Normal stress σ_n 36 kN/m²

Date	Time	Horizontal displacement mm	Load dial reading divs.	Horizontal load N	Shear stress τ kN/m²	Vertical movement dial reading μm	Vertical movement expansion– settlement + mm	Remarks
24/6	1527	0	0	0	0	5000	0	
		0.25	31.5	29.5	8.2	5000	0	
		0.50	58.0	54.3	15.0	4996	+0.004	
		0.75	68.0	63.6	17.6	4987	+0.013	
		1.00	97.5	91.3	25.3	5016	−0.016	
		1.25	112	105	29.0	5053	−0.053	
		1.50	119	111	30.8	5084	−0.084	
		1.75	121	113	31.4	5123	−0.123	
		2.00	122	114.2	31.6	5160	−1.60	max.τ
		2.50	121.5	113.7	31.5	5235	−0.235	
		3.00	118	110	30.6	5312	−0.312	
		3.50	114	107	29.5	5377	−0.377	
		4.00	110	103	28.5	5424	−0.424	
		4.50	105	98.3	27.2	5455	−0.455	
		5.00	102	95.5	26.4	5470	−0.470	
		5.50	98.5	92.2	25.5	5491	−0.491	
		6.00	99.0	92.7	25.7	5504	−0.504	
		6.50	98.0	91.7	25.4	5518	−0.518	
	1538	7.00	95.0	88.9	24.6	5520	−0.520	test stopped

Figure 12.46 Typical set of shearbox test data for one specimen of dry sand

If the load ring calibration is reasonably constant so that a mean value of C_R can be used, it is convenient to first calculate the load ring 'stress factor', C_T (kPa per division), where

$$C_T = \frac{C_R}{L^2} \times 1000 \qquad (12.25)$$

For a 60×60 mm shearbox, $C_T = C_R/3.6$ and for a 100×100 mm shearbox, $C_T = C_R/10$. It is then only necessary to multiply each load ring reading (after subtracting the zero reading if there is one) by C_T to obtain the shear stress.

A typical set of shear box test data for one specimen, with the shear stresses calculated, is shown in Figure 12.46.

12.5.7 Analysis of results

17. (a) *Shear stress and volume change*

Plot the calculated values of shear stress (kPa) as ordinates against displacement (mm) along the horizontal axis. On the same horizontal scale, plot the observed vertical movements during shear. All the curves from a set of three tests can be plotted on the same axes, as in Figure 12.47.

From the graphs read off the maximum shear stress for each specimen (the 'peak' value, i.e. shear stress at failure) and the corresponding displacement and vertical movement. Tabulate the data as shown in Figure 12.47, together with the normal stress applied to each specimen.

Alternatively, if a mean value of the load ring calibration can be assumed, load ring readings can be plotted directly against displacement. The reading corresponding to the 'peak' of each graph is determined, from which the relevant maximum shear stresses are the only values which need be calculated.

18. (b) *Coulomb envelope*

On a separate graph sheet, plot the shear stress at failure, τ_f, against the corresponding normal stress, σ_n, for each specimen, as in Figure 12.48. The horizontal and vertical scales must be the same for this graph. Draw the line of best fit through the three points. If the soil is granular and non-cohesive, the line should pass through the origin ($c' = 0$), which provides a fourth point. This line is the failure envelope or the Coulomb envelope.

Measure the angle of inclination ϕ' of the failure envelope to the horizontal axis to the nearest $0.5°$. Read off the cohesion intercept c' (kPa) on the vertical axis if relevant.

19. *Report results*

From a set of shearbox tests on three or more specimens, the following data are reported:

Angle of shearing resistance, ϕ', to the nearest $0.5°$
Apparent cohesion, c', to two significant figures
Dimensions of test specimen
Description of sample
Whether specimen was undisturbed or recompacted, and how prepared
Initial bulk density, moisture content, dry density
Bulk density and dry density at start of shear
Voids ratios and relative densities (sands only)

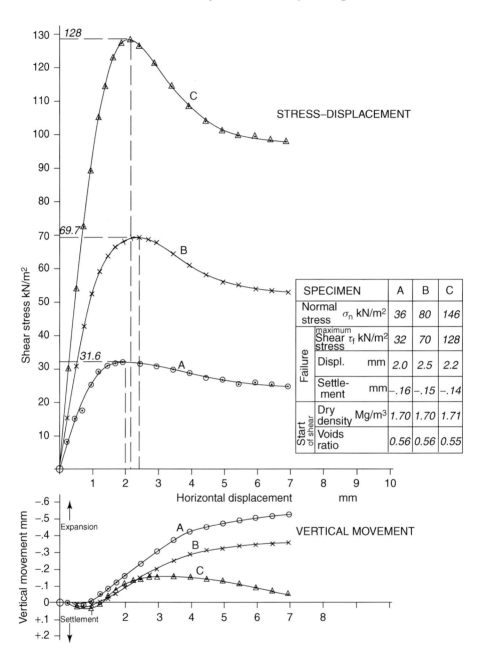

Figure 12.47 Typical graphical results from a set of shearbox tests on dry sand (readings for specimen A are shown in Figure 12.46)

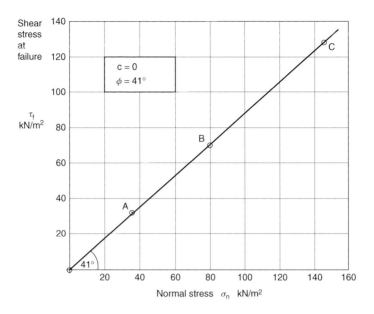

Figure 12.48 Coulomb envelope for set of shearbox tests on dry sand shown in Figure 12.47

Whether water was added for test
Rate of displacement
Tabulated data summarising each test (see Figure 12.47)
Shear stress–displacement and vertical movement–displacement curves (see Figure 12.47)
Failure envelope (see Figure 12.48)
The test is described as a 'rapid shearbox test'.

12.6 Large shearbox: rapid test (BS 1377:Part 7:1990:5)

12.6.1 General

The large shearbox referred to here is a typical commercial machine designed for testing specimens up to 12 in (305 mm) square, although larger machines have been built. The apparatus was originally designed for testing free-draining materials containing particles up to coarse gravel size (37.5 mm), and this is the main application which is described (Bishop, 1948). Preparation and testing of undisturbed samples such as very stiff clays or soft rocks are described in Sections 12.7.3 and 12.7.4.

An early shearbox of this size uses a lever arm and hanging weights for applying the normal pressure. A more recent design, shown in Figure 12.49, incorporates a hydraulic loading system.

12.6.2 Apparatus

The large shearbox apparatus consists essentially of components similar to those of the standard shearbox (see Section 12.5.2) except that they are on a larger scale and therefore

Figure 12.49 Large shearbox machine with hydraulic loading (photograph courtesy of Soil Mechanics Ltd)

more difficult to handle. The shearbox itself, for instance, when filled with a specimen for testing could weigh about 60 kg and is not easy to manhandle into place.

Items which differ from those listed in Section 12.5.2 are as follows:

Load ring (see Section 8.2.1): 50 kN capacity is normally required, but depends upon the load capacity and size of the box.

Dial gauges (see Section 8.2.1): For horizontal movement: 50 mm travel, reading to 0.01 mm.

For vertical movement: 25 mm travel reading to 0.01 mm.

Four such gauges can be mounted, one at each corner of the box, so that any tilt of the top loading plate can be measured.

Normal loading system: A lever-arm loading system may be limited to 400 kPa, but a hydraulic system can provide up to 1 MN/m^2.

Motor unit: A powerful motor is necessary, and travel limit switches are desirable safeguards against damage. A multi-speed gearbox can provide a wide range of displacement speeds.

The component parts of the large shearbox shown in Figure 12.49 can be seen in Figure 12.50. Additional items required for tests on granular materials are as follows:

1. Miscellaneous tools: shovel, large scoop, straight edge.
2. Riffle box, suitable for medium to coarse gravels.
3. Electric vibrating hammer, with a square tamping foot. For the size of specimen to be compacted, a hammer of at least twice the power of that specified for the BS vibrating hammer compaction test (see Volume 1 (third edition), Section 6.5.9) is desirable, preferably mounted in a rig with a hydraulic ram. A unit used by Pike (1973) at the TRRL had the following characteristics:
Power consumption of hammer, 1.5 kW
Operating frequency, about 14 Hz

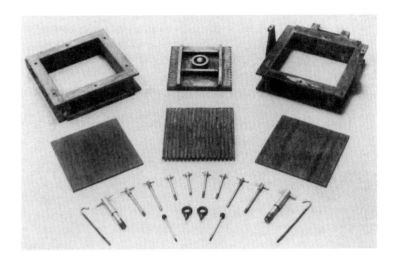

Figure 12.50 Component parts of large shearbox

Force applied through hydraulic ram, about 4.5 kN
Tamping foot, 300 mm square
With a smaller hand-held vibrating hammer, a smaller tamping foot (e.g. 100 mm²)
would be more suitable.

12.6.3 Preparation of samples: coarse grained cohesionless materials

The specimen to be tested is taken from a known mass of material, and the total residue is weighed afterwards so that the mass of specimen can be obtained by difference. The BS specifies that any particles larger than one-tenth of the specimen height should be removed from the original sample, and their proportion by dry mass determined. However, it is recognised that inclusion of some larger particles is sometimes unavoidable, and up to 15% of oversize material is permitted, up to a maximum particle size of 20 mm. In practice, for a specimen height of about 150 mm, this means that there should be no more than 15% by dry mass of particles retained on a 14 mm sieve, all of which pass a 20 mm sieve. If the largest particles present are few in number, care should be taken to ensure that they are evenly distributed within the test specimen, by hand placing if necessary.

The material is compacted into the shearbox already in place in the machine, if possible, to avoid the difficulties of subsequent handling. Compaction should normally be done in three layers, with the plane of shear near the mid-height of the middle layer. Normally the criterion will be to achieve a desired dry density, rather than to apply a specified compactive effort. The difficulty of achieving even the BS 'light' compactive effort by hand ramming is illustrated by the following calculation.

Volume of 12 in shearbox with specimen in (152.4 mm) thick

$$= \frac{304.8 \times 304.8 \times 152.4}{1000} \text{ cm}^3 = 14158 \text{ cm}^3$$

Volume of BS compaction mould = 1000 cm³
Ratio of volumes = 14.16
= ratio of masses at equal densities
Number of blows per layer (equivalent to 27 in the compaction mould) = 27 × 14.16 = 382 in the shearbox.

Clearly the use of a vibrating hammer is preferable to a tiring and time-consuming hand compaction procedure. However, for dry materials, compaction using a hand tamping rod may be adequate.

If a hand-held vibrator is used, with a tamping foot smaller than the surface area of the shearbox, the material around the edge of the box should be compacted first, working systematically over the inner area afterwards. Several passes of the vibrator may be necessary and the total time should be based on experience or trial compaction. At least two minutes per layer may be needed to equate with BS 'light' compaction.

If the specimen contains a high proportion of coarse material, and is short of fines, a layer of medium sand a few millimetres thick may be spread over the top surface to make it easier to bed down the top grid plate uniformly.

12.6.4 Test procedure

The procedure for carrying out the test and for calculating, plotting and reporting results, is similar in principle to that described in Sections 12.5.5–12.5.7. Detailed procedures will depend upon the type of equipment used.

It should be indicated on the test data sheets that the large shearbox apparatus was used and the size of specimen and method of preparation should be stated.

12.7 Drained strength and residual strength shearbox tests

12.7.1 General

Measurement of the drained shear strength of clays is based on the principles of effective stress, which will be covered in Volume 3. The procedure outlined below, using the shearbox equipment already described, is included here in order to complete the account of the use of this apparatus in routine testing of soils. Procedural details relating to use of the apparatus are the same as described in Sections 12.5.3 and 12.5.6. Measurement of pore water pressure is not necessary in this type of test, neither is it practicable in the ordinary shearbox. Significant changes in pore water pressure during shear are prevented by allowing drainage from the specimen, and by applying a sufficiently slow rate of displacement. This can be assessed by using the empirical method described in Section 12.3.9, which is based on the requirement that at least 95% dissipation of excess pore pressure should take place.

The residual shear strength of clays is measured by extending the drained test well beyond the point at which the maximum (peak) strength occurs. Displacement is continued to the limit of travel of the shearbox, which is then returned to the starting position so that the specimen can be re-sheared. This process is repeated a number of times until a constant value of shear resistance (the residual strength) is reached.

A large shearbox can be used for drained and residual strength tests if the nature and size of the sample make it necessary. The principle is the same as for the standard shearbox test, but consolidation and shearing times are likely to be very much longer because of the greater thickness of soil which must be drained.

12.7.2 Apparatus

The apparatus required for a drained test for measurement of peak shear strength is the same as that described in Section 12.5.2. A gearbox providing a wide range of displacement speeds is necessary. Built-in micro-switches which automatically stop the motor at the limits of forward and reverse travel provide a desirable safeguard against damage due to over-running. A dial gauge mounted to measure directly the relative displacement of the two halves of the box (see Figure 12.38) is necessary in order to monitor the actual rate of relative displacement.

Figure 12.51 Additional fittings to shearbox apparatus for residual strength tests: (a) link between drive-rod and outer box; (b) packer between outer box and lower half of shearbox; (c) link between swan-neck and load ring stem; (d) load ring and anchorage must sustain tension

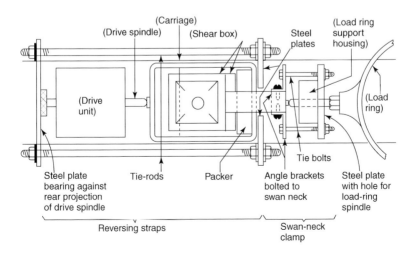

Figure 12.52 Reversing attachments for older shearbox machines (present author's design)

For measurement of residual strength the shearbox must be fitted with a means of reversing back to the starting point, while the normal load remains on the specimen, after the full travel has been reached. A spring-return device was described by Marsh (1972), but the drive motor running in reverse is generally used to provide the return travel. This requires the inclusion in the apparatus of two linkages to transmit tension, a rigid packer between the outer box and the lower half of the split box, and a load ring which can sustain tension (see Figure 12.51). It is not essential to measure the shear force applied to the specimen during the reverse travel, although some machines are fitted with a load ring which can measure in both compression and tension.

Older shearbox machines which were not fitted with reversing links can be provided with tie-rods and straps, and a clamp, in the manner indicated in Figure 12.52. A packer in front of the lower half of the split box is also required.

12.7.3 Preparation of test specimen

Specimens of cohesive soil for shearbox tests may be cut from undisturbed samples, or may consist of remoulded or compacted disturbed soil. The same requirements for the maximum size of particles apply as stated at the beginning of Section 12.5.4. Undisturbed specimens for the large shearbox are usually hand-trimmed from block samples.

1. Cohesive soil: undisturbed

Undisturbed specimens of cohesive soil are prepared by using the square specimen cutter, as described in Section 9.2.2 (from a U-100 tube sample) or Section 9.3.1 (from a block sample). The specimen is transferred to the shearbox by using a wood pusher or by pressing down on the top grid plate with the thumbs as shown in Figure 12.53.

When trimming stiff clays by hand a sharp blade should be used, taking care not to shatter the specimen. If hard fragments of gravel size are present these may cause difficulties in specimen trimming. The presence of one or more coarse particles within the zone of shear could invalidate the results, especially in a small specimen.

Specimens of peat should initially be of greater thickness than is normally used, to allow for the likelihood of very large settlements taking place when consolidated. It may be

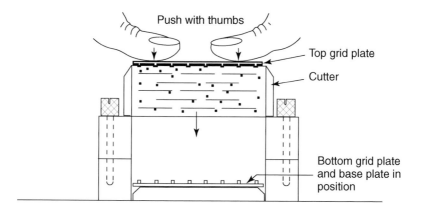

Figure 12.53 Pushing cohesive specimen out of cutter into shearbox

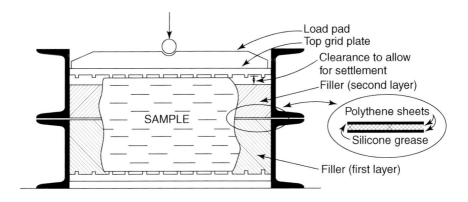

Figure 12.54 Irregular block sample set up in large shearbox

necessary to add further material to the top of the specimen after partial consolidation in order to ensure that the surface of shear remains well within the boundaries of the consolidated specimen.

2. Cohesive soil: recompacted or remoulded

A specimen may be recompacted or remoulded directly into the shearbox at the appropriate moisture content, as described in Section 9.5.3. If enough material is available, it is better to first compact the soil into a mould (such as a compaction mould) and then use the square specimen cutter as for an undisturbed sample (see Section 9.5.5).

3. Block samples for large shearbox

The preparation of block samples of cohesive materials for the large shearbox is described in Section 9.3.4.

Block samples that are smaller than the shearbox can be placed in the middle of the box and the surrounding space filled with a rapid setting filler material such as stiff Polyfilla. The sample should first be protected by several coats of paraffin wax, brushed on, to prevent absorption of water from the filler. The filler should be placed in two layers, the first exactly up to the level of the plane of shear. This should be levelled off smooth, allowed to set and then covered with two layers of polythene sheet cut to fit around the sample. The polythene sheets are separated by a layer of silicone grease, to eliminate any shear resistance due to the presence of the filler material. The second layer of filler is placed on top of the polythene, taking care not to trap any air, and brought up to about 5 mm below the trimmed top surface of the sample. This will ensure that the top grid plate rests only on the specimen, not on the filler, and that the whole of the normal load is transmitted through the soil (see Figure 12.54). Grid plates are shown in direct contact with the soil to provide a horizontal force along the top surface, as the soil projects above the top of the surrounding matrix, which is deliberately mixed to ensure that it is weaker than the soil.

The area of specimen on which the normal and shear stresses are assumed to act can be determined after the test when the two halves of the specimen are separated. The outline of the specimen can be marked on a sheet of tracing paper laid on the sheared surface. The area of the outline can be calculated by dividing the figure into rectangles and triangles; or by laying on a sheet of graph paper and counting squares; or by using a planimeter.

12.7.4 Drained test procedure

The procedure described below is for the measurement of the peak and residual drained strengths of a cohesive soil. In a single-stage drained test in which the peak strength only is required, shearing may be terminated as soon as the maximum shear resistance has been established, or when the limit of travel of the shearbox has been reached.

The first part of the test is very similar to that described in Section 12.5.6 for the rapid shearbox test, but it is important to obtain a complete set of readings from the consolidation stage.

Setting up

The apparatus is prepared, and the specimen set up, as described in stages 1–6 of Section 12.5.6. Water is added to the outer container immediately before applying weights to the hanger.

Consolidation

Apply the normal load to the specimen, as described in Section 12.5.6, stage 7, and at the same instant start the timer. Record the readings of the vertical movement dial gauge in the same way as for an oedometer consolidation test (see Section 14.5.5, stage 14). The time intervals convenient for square-root-time plotting (0.25, 1, 2.25, 4 min, see Table 14.11) may be used.

Plot settlement readings against square-root-time (minutes), allowing consolidation to continue until settlement is virtually completed. This may take 24 hours.

Estimation of rate of displacement

Determine the value of t_{100} by the method described in Section 12.3.9, and illustrated in Figure 12.24. If the linear portion of the settlement curve is not evident, use the alternative procedure indicated in Figure 12.25, also described in Section 12.3.9.

Calculate the minimum time to failure, t_f, from the relationship $t_f = 12.7 t_{100}$.
As an example, if $\sqrt{t_{100}}$ is found from the graph to be 5.8, then

$$t_{100} = 5.8^2 = 33.6 \text{ min}$$

and $t_f = 12.7 \times 33.6 = 427$ min, i.e. about 7 h

$$\text{If } H = 21 \text{ mm}$$
$$c_v = \frac{0.103 \times 21^2}{33.6} = 1.35 \text{ m}^2/\text{year}$$

Estimate the displacement at which the peak strength is likely to be mobilised. This is usually based on experience, and a general guide is given in Table 12.5. If in doubt use a low estimate which will err on the safe side. For this example it is assumed that the peak strength will be reached at a displacement of 3 mm. Therefore, the rate of displacement should not be greater than $3/427 = 0.00703$ mm/min, and this is represented by the line OA in Figure 12.55.

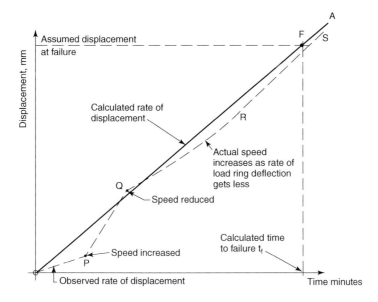

Figure 12.55 Graphical monitoring of rate of displacement

Table 12.5 Typical displacements for peak shear strength in 60 mm shearbox

Type of soil	Shear box displacement for peak strength (mm)
Loose sand	5–8
Dense sand	2–5
Plastic clay	8 (typical limit of travel)
Stiff clay	2–5
Hard clay	1–2

If the nearest machine speed on the lower side is 0.0070 mm/min this would be a suitable speed at which to run the test.

Shearing: single stage test

The specimen is sheared in the same way as described in Section 12.5.6, using the rate of displacement obtained as above. As the test proceeds the load ring reading (which is approximately proportional to shear stress) should be plotted against cumulative horizontal displacement, in the manner shown in Figure 12.28. In addition, the actual rate of displacement should be monitored by plotting observed displacements against time as shown in Figure 12.55. If the observed points fall below the line OA, the rate of displacement is less than the calculated value; if above the line, the rate of displacement is too fast and should be reduced. Soon after the start of the test, when the load ring is deforming as the shear force builds up, the actual rate of relative displacement may be less than the machine speed (curve OP).

It is then permissible to increase the machine speed, provided that the rate of strain continues to be monitored. When the plot of observed points reaches the line OA (point Q) the speed should be reduced, giving a curve such as QRS. The aim should be to maintain an average rate of displacement no greater than the calculated value represented by the line OA.

After passing the peak condition, continue shearing and taking readings so as to define the 'peak' clearly. If there is no defined peak, continue shearing until the full travel of the apparatus is reached.

Completion of test

Removal

When the maximum stress is clearly defined, or the limit of travel is reached, stop the motor and return the drive unit to its starting position. Drain the water from the carriage, remove the hanger weights, hanger and yoke, and lift out the shearbox, as in stages 11–13 of Section 12.5.6.

The moisture content and the form of the sheared surface are usually significant. Pull the upper half of the shearbox upwards from around the specimen, at the same time pressing down on the grid plate with the thumbs (see Figure 12.56(a)). Invert the lower half of the box over a small tray and push the specimen out of the box into the tray by pressing on the base plate (see Figure 12.56(b)). Any soil adhering to the box should be removed and added to the specimen.

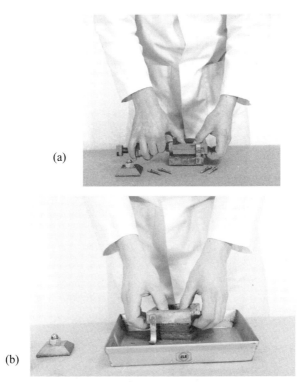

Figure 12.56 Removing cohesive specimen from shearbox: (a) pulling away upper half of box; (b) pushing out of lower half of box

Weigh the tray complete with specimen and two grid plates, or else slide off the grid plates and add any adhering soil to the specimen before weighing (m_2).

To examine the surface of shearing, separate the two halves of the box before pushing the specimen out by sliding them horizontally in the same relative direction as in the shear test. The two sheared faces can then be sketched or photographed. If the moisture content is to be measured, do not leave these surfaces exposed too long especially in the vicinity of photographic floodlamps.

Measurement of moisture content

Dry the specimen in the oven overnight or long enough to achieve constant mass. Allow to cool in a desiccator and weigh in the tray (m_3). The final moisture content w_2 is given by the equation

$$w_2 = \frac{m_2 - m_3}{m_3 - m_T} \times 100\%$$

where m_T is the mass of the tray or the mass of tray plus the two grid plates if these remained with the specimen when weighed wet and dry.

The initial moisture content is obtained from trimmings if taken from an undisturbed specimen. This may be verified from the dry mass at the end of the test, which is the same as at the beginning if no material has been lost.

$$\text{Intial mass of Specimen} = m$$
$$\text{Final dry mass} = m_s - m_T$$
$$\therefore \text{Initial moisture content} = \left\{ \frac{m - (m_3 - m_T)}{(m_3 - m_T)} \right\} \times 100\%$$

If the accepted initial moisture content is denoted by $w_0\%$, the initial dry density r_D is equal to

$$\frac{100\rho}{100 + w_0} \quad \text{Mg/m}^3$$

Calculation and plotting

Calculate the density, normal stress and shear stresses on the specimen, as described in stage 16 of Section 12.5.6. Determine the maximum shear stress (the 'peak' value) representing failure.

If a set of similar specimens has been tested at different normal stresses, plot each maximum shear stress against the normal stress and draw the Coulomb envelope. Determine the effective stress parameters c' and ϕ' and report the results as described in Section 12.5.7.

Multi-stage test

A multi-stage test in which a single specimen is sheared successively under two or more different normal pressures appears to be economically attractive. However, this procedure is open to the objection that each additional consolidation stage displaces the previously formed shear surface downwards and out of alignment with the plane of shear of the apparatus, so that a new shear surface has to be developed.

12.7.5 Residual strength test procedure

Initial stages

Specimen preparation, setting up, consolidation and estimation of the rate of displacement are as described in Section 12.7.4. The first stage of shearing is also as described above, but displacement is continued beyond the point of maximum shearing resistance until the limit of travel of the apparatus is reached.

Reversing

When the limit of travel is reached, stop the motor and ensure that the linkages provided for reverse travel are securely connected. Return the shearbox to its starting position by one of the following reversal procedures, without removing the vertical force on the specimen.

    (a) Reverse the motor so that the drive unit returns the lower half of the shearbox to its original alignment with the upper half. The motor speed for this operation should be such that the time taken for the reverse travel is about the same as the time from the start of shearing to the peak shearing force. Readings of shear force are not necessary during reverse travel, and have no significance.

    (b) Reverse the direction of travel by using the hand-winding facility within a period of a few minutes, until the original alignment is reached. Allow to stand for at least 12 h before starting the next stage, to allow re-establishment of pore pressure equilibrium.

    (c) By hand winding, apply five to ten rapid backward and forward traverses within a few minutes. This will establish a shear plane within the specimen. Return to the original alignment and allow to stand for at least 12 h as in (b).

After reversing, record the reading of the vertical displacement gauge. Check that the readings of the horizontal displacement gauge and the load ring have returned to their initial values. If they have not, then either make adjustments or record the new zero readings.

Re-shearing

The second shearing traverse is carried out in the same way as the first, except that the rate of displacement should be equal to that used for the reverse travel referred to in procedure (a) above. Displacement should be measured cumulatively, so that the second run begins from the point at which the first run ended, as shown in Figure 12.28.

Subsequent shearing stages

Further reversals and re-shearing are applied as described above, until a constant residual value of shear resistance has been achieved, as shown by the plot of load ring reading against cumulative displacement (see Figure 12.28).

Completion of test

After completion of the final shearing run, stop the motor. Remove the specimen, examine and describe the surfaces of shearing, and determine the moisture content, all as described in Section 12.7.4.

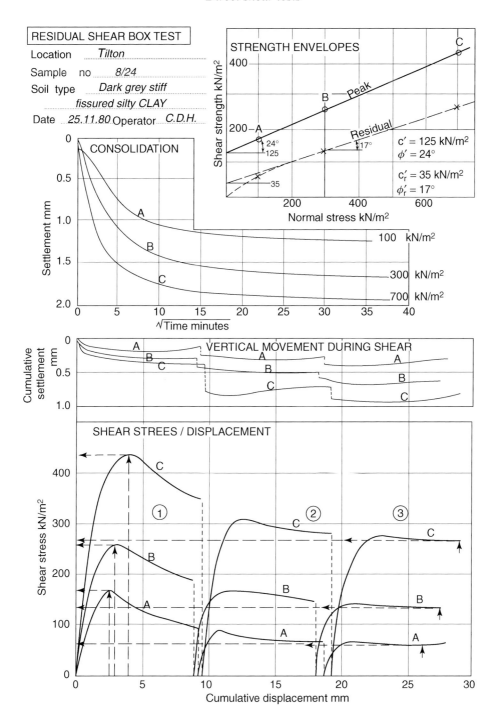

Figure 12.57 Graphical results from typical set of residual shearbox tests

From the load–displacement curves, derive the peak and the residual strengths as indicated in Figure 12.28. Coulomb envelopes for peak and residual drained conditions can be derived from a set of three tests (see Figure 12.57). The drained shear strength parameters c', ϕ' (peak) and c'_r, ϕ'_r (residual) can then be obtained.

12.7.6 Cut-plane method

The number of shearbox travels required to bring a clay to its residual condition may be reduced by slicing the specimen in two along the plane of shear, a procedure sometimes referred to as the 'cut-plane method'. If the specimen is sliced before setting up initially, allowance must be made for consolidation under the normal load to ensure that the cut will be at the level of the plane of shear of the apparatus for the shearing stage. This difficulty can be overcome if the specimen is consolidated first, then removed for slicing and carefully replaced before re-applying the normal load. If the intact peak strength is to be determined the specimen can be removed after the first travel of the shearbox and sliced along the resulting shear plane.

An alternative procedure which obviates removal of the specimen is to apply up to ten rapid back and forth travels of the shearbox after completing the first run (as in reversal procedure (c) in Section 12.7.5). A disadvantage is the almost inevitable loss of fine material from the shear surface, and it is the fine fraction that has an important influence on residual strength.

The measurement of shear strength along a fissure or other surface of discontinuity in a soil or rock can be regarded as a test on a naturally occurring 'cut plane'. A peak strength appreciably higher than the residual value is not usually to be expected if the surface is correctly aligned in the shearbox. Careful allowance must be made for settlement due to consolidation when setting up.

Cut-plane tests, and tests on natural discontinuities, were carried out by the original author during 1963–1966 on Siwalik clays at Mangla, West Pakistan. The procedure is also referred to by Bishop (1971), and Townsend and Gilbert (1973).

12.8 Vane shear test (BS 1377:Part 7:1990:3, and ASTM D 4648)

12.8.1 General

The laboratory vane test described here is similar in principle to the field vane test (Clause 4.4 of Part 9 of BS 1377) but is on a smaller scale, being designed for direct measurement of the shear strength of soil samples in the laboratory. The ASTM refers to it as the 'miniature vane shear test'. The vanes on the field apparatus may be up to 150 mm long and 75 mm wide, but the usual laboratory apparatus has vanes measuring 12.7×12.7 mm. Larger vanes, up to about 25 mm long, are available for measuring very low shear strengths.

Experience has shown that the results of laboratory vane tests on saturated clays can be compatible with the results of unconfined compression tests. The vane apparatus is particularly suitable for soils such as soft, sensitive clays having shear strengths of 20 kPa or less, from which it would be extremely difficult to prepare undisturbed specimens for other types of test. Soft samples can be tested in the sampling tube with the minimum of disturbance and this is the application which is described. However, the test can also be carried out on soft remoulded soils, for instance in a compaction mould.

Another type of apparatus using the same principle is the pocket shearmeter, a small hand-held device described in Section 12.8.5.

12.8.2 Apparatus

1. The laboratory vane apparatus is self-contained and consists essentially of the following components (see Figure 12.58). It may be fitted with a drive motor, but the following description is for hand operation:
   (a) Frame, stand and baseplate
   (b) Vane mounting assembly
   (c) Handle for raising and lowering the vane assembly by means of the square-thread lead screw
   (d) Vane, with four blades, typically 12.7 mm wide and 12.7 mm long
   (e) Handle for rotating vane head, which applies torque to the vane shaft
   (f) Graduated scales, marked in degrees, one on the fixed vane head, the other rotating with the vane (see Figure 12.59)
   (g) Vertical shaft attached to knob fitted with pointer carrier on friction sleeve
   (h) Set of calibrated springs (usually four) of different stiffness, to allow for a range of soil strengths
   (i) Calibration chart for the springs. Calibration curves for one particular set of springs are shown in Figure 12.60, but each spring must be individually calibrated before first use. A procedure for calibrating these springs is given in ASTM D 4648

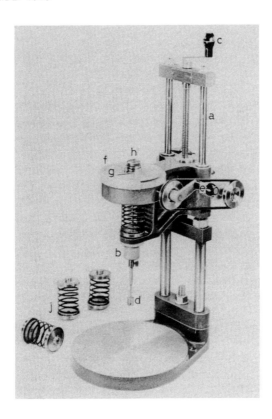

Figure 12.58 Laboratory vane apparatus

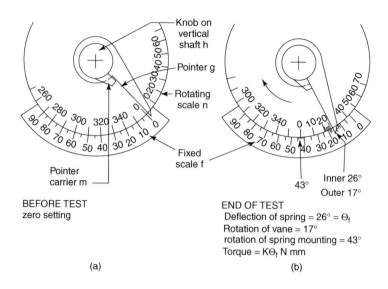

Figure 12.59 Angular scale on laboratory vane apparatus: (a) details; (b) example illustrating readings after a test

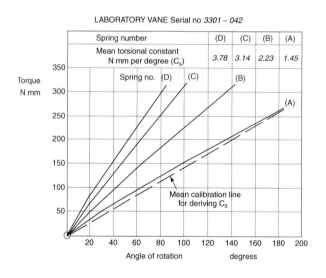

Figure 12.60 Typical example of calibration curves for torsion springs of laboratory vane apparatus

The dimensions of the frame and base are large enough to accommodate a standard compaction mould or a CBR mould containing the test specimen. For testing a sample in a long container, such as a U-100 tube or a piston tube, the frame can be swivelled through 180° so that the vane can hang over the edge of the bench, where the sampling tube can be

Figure 12.61 Laboratory vane test in U-100 sample tube

clamped in a suitable position (see Figure 12.61). Counterbalance weights must be added to the baseplate.
2. A means of supporting and clamping a 100 mm diameter sampling tube close to the edge of the bench
3. Drying oven and other equipment for the determination of moisture content
4. Small tools such as spatulas, trimming knives, steel rule
5. Stop-clock or timer

The principle of operation of the apparatus is as follows. If the vane is prevented from rotating as the handle (e) is turned, the inner graduated scale (n) (see Figure 12.59) rotates through the same angle as the upper end of the spring but the vertical shaft (h) and pointer (h) remain stationary. The pointer therefore reads zero on the outer scale (f), and indicates the torsional rotation, or 'twist', of the spring on the inner scale (n), from which the torque applied to the spring can be calculated. The pointer remains in this position when the torque is reduced. If the vane is not fully restrained, the reading indicated by (g) on the outer scale (f) gives the angle of rotation of the vane (d), while the reading on the inner scale (n) enables the torque to be calculated as before. The rotation of scale (n) relative to scale (f) is equal to the sum of these rotations, and merely indicates the total rotation of the drive unit. Therefore at the end of a test:

Pointer reading on inner scale gives *torque;*
Pointer reading on outer scale gives *vane rotation.*

12.8.3 Procedural stages

1. Prepare sample in tube
2. Clamp tube in position
3. Select spring
4. Prepare apparatus
5. Adjust scales
6. Insert vane
7. Measure shear strength
8. Remould
9. Measure remoulded strength
10. Remove vane
11. Repeat stages 5–10 (four more times at different positions)
12. Measure moisture content and density
13. Calculate
14. Report results

12.8.4 Test procedure

1. Preparation of sample

The following procedure relates to a test on an undisturbed sample contained in a U-100 sampling tube, but the same principle applies to other types of sample. The sample should not be jacked even part way out of the tube if this can be avoided, and it should neither be jolted nor tilted once the end seal has been removed.

Clamp the sample tube securely with its axis vertical and with the end to be tested uppermost. Remove the end cap, wax seal and any packing material. Trim the sample inside the tube so that its upper end is flat and perpendicular to the tube axis. Measure the distance from the end of the tube to the sample.

2. Clamping tube in position

Carefully adjust the position of the sample tube, keeping its axis vertical, to the position already prepared for it for the test. Clamp the tube adjacent to the edge of the bench, with the horizon that is to be tested a little above bench level, so that the vane can reach to the central axis of the tube (see Figure 12.61).

If its weight is supported by a stool, the tube can be held against the bend by tightly wrapping a length of cord or wire two or three times round it, and securing each end to a G-clamp clamped on the edge of the bench at either side (see Figure 12.61). A piece of foam rubber or plastic material interposed between the bench and the tube will help to keep it steady. Alternatively, the apparatus may incorporate a clamp for a U-100 tube in the baseplate.

3. Selection of torsion spring

The torsion spring to be used should be selected after examining the sample and assessing its range of probable shear strength. A general guide for a typical set of springs, based on the descriptive terms given in Volume 1 (third edition), Table 7.1, is given in Table 12.6.

Record the number of the spring used.

Table 12.6 Typical torsion springs for laboratory vane

General descriptive term for strength	Suggested spring reference*	Probable maximum shear stress (kPa)
Very soft	(A) (weakest)	20
Soft	(B)	40
Soft to firm	(C)	60
Firm	(D) (stiffest)	90

* Not necessarily manufacturer's identification marks

4. Preparation of apparatus

The apparatus is assembled and the torsion spring fitted in accordance with the manufacturer's instructions. The torsion spring must be fitted the right way round.

Fit the shaft of the vane into its socket, and tighten the fixing screw. To check that the spring and vane are correctly fitted, hold the vane between thumb and finger to prevent rotation and turn the torsion drive handle a little way with the other hand. The tendency of the vane to rotate should be felt.

Slacken the nut on the base of the pillar unit, swivel the unit through 180° so that the base and torsion head are on opposite sides of the pillars, and re-tighten the fixing nut. *Keep hold of the pillars until counterbalance weights have been added to the baseplate* (see Figure 12.61).

Raise the vane assembly by rotating the handle (c) (see Figure 12.58) so that the vane clears the top of the U-100 tube. Move the apparatus into position so that the vane is directly over the central axis of the tube (position 1, Figure 12.62).

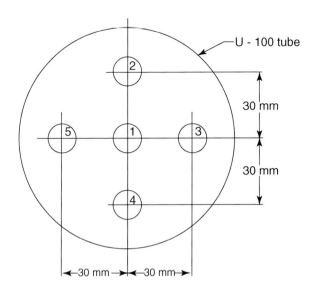

Figure 12.62 Vane test locations in U-100 tube

5. Adjustment of angular scales

Move the pointer (g) on the dial (see Figure 12.59) so that it is in contact with the carrier (m) on top of the vane shaft. Hold the knob (h) and rotate the carrier until the pointer is brought to the zero position on the inner scale (n). Rotate the handle to bring the pointer to zero on the outer scale (f). Check that any backlash in the torque mechanism has been taken up.

6. Inserting vane

Check that the vane is directly above the centre of the sample. Wind down the vane assembly by turning handle (c) (see Figure 12.58) until the vane just touches the end surface of the sample. Measure and record the level of the top of the vane mounting with reference to a fixed point such as the underside of the cross-head. The exact depth to which the lower end of the vane penetrates the sample can then be measured. Alternatively, if the pitch of the lead screw is known the vane penetration can be determined by counting the number of revolutions of the handle (c).

Wind down the vane steadily until the required penetration has been achieved. Usually the top of the vane should not be less than four times the blade width below the surface, giving a minimum cover of about 50 mm for the standard vane. Record the depth of penetration.

7. Measurement of shear strength

Record the initial readings of the pointer on both angular scales to the nearest half-degree.

Rotate the handle (e) (see Figure 12.58) clockwise at a steady rate, to apply torque to the vane. At the same time start the clock. The torsion head (spring mounting) should be rotated at a constant rate of 6°/min to 12°/min, which requires a very slow rotation of the handle. A suitable motorised drive is preferable. While the soil is resisting the applied torque, the pointer reading on the inner scale increases steadily until the maximum shear resistance of the soil is mobilised. At this point failure occurs and the torque decreases, but the pointer remains in the position indicating the maximum angular deflection of the spring, from which the vane torque at failure can be calculated. Stop turning the handle, stop the clock and record the angular scale readings to the nearest 0.5°. Record the time taken to reach failure.

If the spring deflection reaches 100°, or the upper limit imposed by the manufacturer, discontinue the test and repeat with a stiffer spring.

8. Remoulding

Rotate the vane rapidly through two complete revolutions, so as to remould the soil in the sheared zone.

9. Measurement of remoulded strength

Record the readings on the angular scales immediately after remoulding. Without further delay, rotate the vane as in stage 7 until failure is observed again and record the scale readings for the 'remoulded' test.

10. Removing the vane

Raise the vane steadily by turning the handle (c). As the vane emerges from the sample, press gently around it to prevent excessive disturbance due to tearing of the surface. Wipe the blades carefully to remove adhering soil. To avoid the possibility of damage to the vane and shaft, wind the assembly up until it is well clear of the top of the sample tube.

Move the apparatus away from the edge of the bench by about 30 mm, so that the vane is directly over the position marked 2 in Figure 12.62. This location allows reasonable clearances between positions 1 and 2, and between position 2 and the wall of the tube.

11. Repeat tests

Repeat stages 5–10 at position 2 (see Figure 12.62) and again at positions 3–5. Between each new location 2, 3, 4 and 5, either the sample tube may be rotated 90° or the vane apparatus can be moved. The distances between the centres of each test location should not be less than 2.3 times the width of the blade.

On completion of the set of five tests, clean the vane and remove it with its shaft from the mounting. Take out the spring. Replace all components in their places in the box provided.

12. Measurement of moisture content and density

The soil sample may now be extruded for examination and index testing. As it is extruded (tested end foremost) measure from the end surface so as to ascertain the location of the zone in which the vane tests were done. From this zone take specimens for moisture content determination, preferably one from each vane test position.

If, in addition, the density is required, it is better not to rely on measurements made after extrusion, but to measure and weigh the sample in the tube before extruding for moisture content measurements, as described in Volume 1 (third edition), Section 3.5.3. (The tested end of the sample has already been trimmed and measured; this should now be done at the other end.) After completing these measurements, the sample may be extruded for moisture content and index tests.

13. Calculations

The angular deflection reading indicated by the pointer (g) on the inner scale (n) after each shear test gives the relative angular deflection (θ_f degrees) of the ends of the spring at failure. The applied torque, M (N mm) is equal to $C_s \times \theta_f$, where C_s is the calibration factor (N mm/degree) for the spring being used, obtained from the calibration data. If curves similar to those in Figure 12.60 have been drawn, the torque M can be read directly from them.

If the usual vane (12.7 × 12.7 mm) was used, the vane shear strength of the soil (τ_v) is calculated from the equation

$$\tau_v = \frac{M}{4.29} \quad kN/m^2$$

(derived from Equation 12.21).

If a vane of different dimensions was used, the above relationship becomes

$$\tau_v = \frac{1000 \ M}{K} \quad kN/m^2$$

where the value of K (mm³) is defined by

$$K = \pi D^2 \ (H/2 \ + D/6)$$

(see Equation 12.20).

Calculate the average value of all five undisturbed vane shear strengths (τ_v) and of all five remoulded vane shear strengths (τ_{vr}). If one result differs appreciably from the others of the set (e.g. by more than 20%) it should be discarded. Also calculate the average angular strain at failure for each type of test.

Calculate the sensitivity S_t of the soil, where $S_t = \tau_v/\tau_{vr}$.

Calculate the average moisture content from the several locations at which measurements were made and calculate the overall density if measured.

14. Report results

Report the average undisturbed and remoulded vane shear strengths, in kPa, to two significant figures. Include the highest and lowest measured values.

Report the corresponding angular strains at failure to the nearest degree and the average time to reach failure.

Also report the sensitivity of the soil to two significant figures and the moisture content and bulk density if measured. Results of index tests (e.g. liquid limit, plastic limit) should be included if measured.

Report that the laboratory vane test was used in accordance with BS 1377:Part 7:1990:Clause 3 stating the size of vane. Indicate the horizon within the sample tube at which the tests were carried out, so that the result can be related to the depth below ground surface.

12.8.5 Pocket shearmeter

The pocket shearmeter, or pocket vane, shown in Figure 12.63 operates on a similar principle to the laboratory vane apparatus, but is applied to the surface and rotates a relatively thin disc of soil. It can be used on site, for instance on the sides of pits, trenches and in the laboratory on the ends of tube samples or on the faces of block samples. The instrument should be regarded as an aid to the visual classification of soil in the zone inspected, and not

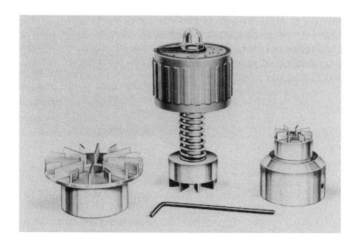

Figure 12.63 Pocket shearmeter

as a substitute for other methods of measuring shear strength for design purposes (see also Volume 1 (third edition) Section 7.5.6).

The shear strength of the zone of soil tested is measured by pushing the vanes into the soil and turning the knob until a maximum reading is achieved on the dial. This is calibrated to read directly in shear strength units.

The standard vane can be used for measuring shear strengths up to 100 kPa. In addition, a larger vane can be fitted, which gives greater sensitivity when measuring shear strengths below 20 kPa, and a smaller vane is available for extending the range up to 250 kPa.

The manufacturer's detailed instructions should be followed when using this instrument.

12.9 Ring shear test (BS 1377:Part 7:1990:6)

12.9.1 General

The ring shear test procedure described here makes use of the relatively simple apparatus designed by Professor Bromhead at Kingston University (London) (Bromhead, 1979). The test is designed only for the measurement of residual strengths of cohesive soils, for which small specimens of remoulded soil are adequate.

The ring shear principle enables virtually unlimited deformation to be applied to a laboratory test specimen without the disadvantages of the reversal process in a conventional shearbox. In the Bromhead apparatus, specimens that have been remoulded at the plastic limit or wetter can be tested. Shearing takes place along a surface close to the upper loading platen. The rate of displacement is not critical because the fully drained condition is bound to be reached eventually.

The test specimen is an annulus of 5 mm thickness with an outer diameter of 100 mm and an inner diameter of 70 mm. It is assumed that the normal stress and the shear stress on the surface of shear are both uniformly distributed across the plane of relative rotary motion when the residual condition is reached.

Before starting a test the following test conditions need to be defined:

- The moisture content at which the soil is to be remoulded for preparing test specimens
- The number of specimens to be tested as a set, or the number of stages to be applied to a single specimen and the re-loading sequence
- The normal effective pressures to be applied
- The procedure to be followed for forming the shear plane

12.9.2 Apparatus

Ring shear apparatus

The remoulded specimen used in this apparatus is confined radially between concentric rings, and vertically between porous annular discs with relatively rough surfaces. Vertical pressure can be applied to the specimen through the upper porous annulus by means of a lever-arm arrangement (usually counterbalanced) using hanger weights. The cell which contains the specimen is removable, and can be submerged in a water bath during the test. The lower part of the cell is rotated by a motorised drive unit while the upper part is restrained by a matched pair of calibrated load rings or load cells (see Section 8.2.1), which enable the restraining torque to be determined.

Angular rotation is measured by means of a scale graduated at 1° intervals. A dial gauge or displacement transducer (see Section 8.2.1) is fitted for measurement of vertical deformation.

The general arrangement is shown diagrammatically in Figure 12.64 and a typical ring shear test apparatus is shown in Fig 12.65.

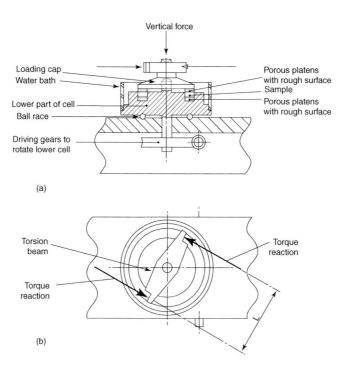

Figure 12.64 General arrangement of ring shear apparatus: (a) cross-section; (b) plan showing torque reactive forces from load rings on torsion beam (graduated scale of degrees not shown) (reproduced from Figure 4 of BS 1377:Part 7:1990)

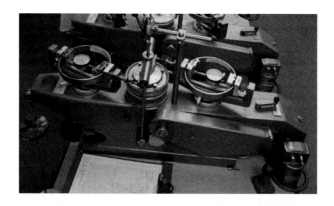

Figure 12.65 Typical ring shear apparatus (photograph courtesy of Soil Mechanics Ltd)

Ancillary items

Other items of equipment required are as follows:

Apparatus for determining moisture content (see Volume 1 (third edition), Section 2.5.2)
Balance, readable to 0.01 g
Spatula
Evaporating dish
Flat glass plate (as used for liquid limit tests)
Stop-clock or seconds timer
BS test sieve, 1.18 mm aperture

Measurements and checks

Before the apparatus is first used, verify the dimensions of the specimen cavity by measuring to the nearest 0.1 mm, using a vernier calliper.

Determine the value of the displacement factor (mm/division or mm/digit) for each load ring. If the load rings are fitted with dial gauges the factor is typically 0.002 mm/division but this should be verified. The mean of the two values is denoted by F.

Measure the distance between the points of application of the load rings on the torsion beam, to the nearest mm (L) (see Figure 12.64(b)).

12.9.3 Preparation and assembly

Test specimen

The material used for the test should contain no particles retained on a 1.18 mm sieve. If necessary remove any oversize particles, either by hand, or by wet sieving. The soil should not be dried before testing.

The test is prepared and set up as outlined below.

1. Take about 400 g of soil passing a 1.18 mm sieve, and thoroughly remould it to a uniform consistency. Determine the moisture content.
2. If any water has to be added to bring the soil to the desired moisture content, mix thoroughly in the water and leave the soil in a sealed container for at least 24 hours to allow the added water to permeate through the sample.
3. Weigh the empty ring shear cell to the nearest 0.1 g.
4. Mix the soil again and place the remoulded soil with a kneading action into the annular cavity of the cell, ensuring that no air is entrapped, using the fingers or a spatula or a short length of wood dowel. Level off the surface of the specimen flush with the top of the confining ring using the blade of the spatula. 'Smearing' of the soil surface in this way is not detrimental, since only the residual strength is to be measured. Clean off any surplus soil.
5. Weigh the cell with specimen to the nearest 0.1 g.
6. Use a representative portion of the excess soil to determine the initial moisture content of the specimen.

Assembly of apparatus

7. Place the cell on the frame of the test apparatus.
8. Apply grease to the cell centering spindle and place the upper platen in position on top of the specimen.

9. Fill the water bath with water and leave for one hour to allow the specimen to become saturated.
10. Place the counterbalanced loading yoke in position on the upper platen and apply just enough downward force to ensure that the yoke is properly seated.
11. Mount the vertical deformation gauge or transducer and secure it so that the stem is bearing correctly on the loading yoke.
12. Set the gauge or transducer to a convenient initial zero reading, which is recorded.

12.9.4 Test procedure

Consolidation

1. Apply weights to the load hanger to produce the required vertical stress (σ'_n kPa) on the specimen. At the same instant start the timer. When testing a soft clay, or when applying a high vertical stress, the specimen should be consolidated in stages so as to avoid squeezing out of soft soil.
2. Record and plot readings of vertical deformation against square-root-time, and derive a value for t_{100}, as described in Section 12.7.4 for the shearbox test. From that value calculate the time to failure, t_f, and the maximum rate of displacement for the test (denoted here by v mm/min), as in Section 12.7.4.
3. Calculate the corresponding maximum rate of angular displacement (degrees per minute), which is equal to $57.3v/r$, where r (mm) is the mean radius of the test specimen. For a specimen of the dimensions stated in Section 12.9.1

$$r = \frac{100 + 70}{2 \times 2} = 42.5 \text{ mm}$$

4. The alternative to the above procedure is to assume that a rate of angular displacement of about 0.048 degrees per minute is applicable. This has been found to be satisfactory for a wide range of soils, and the validity of the assumption can be verified during shearing (step 11 below). The small thickness of the specimen means that consolidation readings might not be of sufficient resolution to enable a reliable value of t_{100} to be determined, and a reasonable assumed rate of displacement is then the only practicable course. Bromhead (1992) prefers to omit the consolidation measurements altogether, and to start the shearing soon after applying the vertical stress, because consolidation then takes place simultaneously with the mobilisation of the forces applied by the load rings.

Final checks and adjustments

5. Set up the force measuring rings and carefully align and secure them so that they bear correctly on the torsion beam and at right angles to it. Take up any backlash in the drive unit by adjusting the hand wheel.
6. Form the shear plane by rotating the lower ring (either by hand drive, or by using the motor) by one to five revolutions within a period of about two minutes. However, if this relatively rapid rate of shearing is likely to cause extrusion of soil, then Bromhead (1992) recommends applying this initial shearing slowly; an overnight period is suitable because no readings need be taken.

7. Remove any torque indicated by the load rings by adjustment of the handwheel. Allow the specimen to stand for a period of not less than t_{100} (as determined in the consolidation stage), or for at least one hour if t_{100} was not derived, to allow excess pore pressures to dissipate

8. Set the drive unit of the machine to give a rate of angular displacement of not greater than that calculated in step 3 above.

Shearing

9. Record the initial readings of the load rings, the vertical deformation gauge and the angular scale of rotation (degrees).

10. Start the machine and at the same instant start the timer. Record readings of the following at regular intervals of angular rotation:
    Both load rings
    Vertical deformation (millimetres)
    Elapsed time (minutes)
    Angular rotation (θ degrees)

The intervals should be such that at least 20 sets of readings are obtained during the test.

11. For an immediate assessment of the strain rate sensitivity of the soil, the drive motor can be switched off. If the specimen then holds the applied torque, i.e. the load ring readings do not decrease over a period of about one hour, the rate of displacement is deemed to be satisfactory. A significant loss of torque, i.e. decrease in load ring readings, indicates that the rate of angular displacement was too fast and the test should be repeated with a fresh specimen.

12. Continue shearing until it can be seen from the load ring readings, or from the graphical plot (see Section 12.9.5 item 8), that the residual state has been reached, then stop the machine.

13. Reduce the torque applied to the specimen to zero before increasing the vertical stress for the next consolidation stage.

14. After applying the new vertical stress, repeat the procedures for consolidation, initial shearing and shear test (steps 2–4 and 6–13) to give at least three stages under vertical effective stresses which cover the required range of stress. Alternatively, three separate specimens may be tested, each under a different vertical stress.

Removal

15. After completion of the final shearing stage (step 12), reduce the applied torque to zero and siphon the water out of the water bath. Allow to stand for about 10 min.

16. Remove the vertical stress, take off the top platen and examine the shear surface. Record any significant features by means of sketches or photographs.

17. Take representative portions of the test specimen for determination of the final moisture content.

Figure 12.66 presents some test data on a worksheet similar to the format suggested by BS 1377:1990.

Contract:	EXAMPLE RESULT							Date:	10/11/09

Sample Description		*Stiff grey mottled light brown CLAY*							
Sample Preparation Procedure:		*<1.18 mm material remoulded by kneading*							
Machine No:		*1*							
Distance between points of application of force L (mm)		*154.11*							

Initial Conditions: / **Specimen Dimensions (mm):**

Initial Conditions:				Specimen Dimensions (mm):				
Mass of wet soil + cell	g	*1848.83*	Inside diameter	*69.56*	Inside radius r_1		*34.78*	
Mass of cell	g	*1807.64*	Outside diameter	*99.56*	Outside radius r_1		*50.00*	
Mass of wet soil	g	*41.19*	Mean radius	r	mm		*42.38*	
Moisture content from trimmings %		*31*	Height	H	mm		*5.25*	
Density	Mg/m³	*1.94*	Volume cm³ $\dfrac{2\pi r (r_2 - r_1) H}{1000}$				*21.28*	
Dry Density	Mg/m³	*1.48*	Particle Density Mg/m³ measured/assumed*				*2.7*	
Degree of saturation %		.						

Shear test								Stage No 1
Single stage/Multiple stage*		Run no.		*1*		Normal stress	kPa	*50*
Force Device			A	B		Average		
Mean Calibration	N/division		*0.0923*	*0.0924*		*0.09235*		
Displacement Factor F mm/division			*0.002*	*0.002*		*0.002*		

Time	Elapsed time (hh:mm:ss)	Force device reading			Angular displ. θ deg.	$D = \dfrac{\theta r}{57.3}$ mm	$d = \dfrac{(A+B)Fr}{L}$ mm	$D_1 = D - d$ mm	Shear stress τ kPa	Vertical deformation mm
		A	B	Average						
15:00:00	00:00:00	0	0	0.0	6.43	0	0	0	0	0
15:25:00	00:25:00	139	147	143.0	7.63	0.8875	0.1573	0.7302	11.75	0
15:54:47	00:54:47	130	143	136.5	9.06	1.9452	0.1501	1.7950	11.22	0
16:26:53	01:26:53	128	139	133.5	10.6	3.0842	0.1468	2.9373	10.97	-0.01
16:55:50	01:55:50	124	135	129.5	11.99	4.1123	0.1424	3.9698	10.64	-0.01
17:28:45	02:28:45	122	132	127.0	13.57	5.2809	0.1397	5.1412	10.52	-0.02
17:59:35	02:59:35	122	129	125.5	15.05	6.3755	0.1380	6.2374	10.31	-0.03
18:28:32	03:28:32	120	127	123.5	16.44	7.4036	0.1358	7.2677	10.23	-0.03
18:58:20	03:58:20	120	124	122.0	17.87	8.4612	0.1342	8.3270	10.03	-0.04
19:29:23	04:29:23	118	122	120.0	19.36	9.5632	0.1320	9.4312	9.94	-0.05

Comments: *See following sheets for remaining readings*		Operator	OP	Contract No	GEO/15555
				Lab Ref	S9541
		Input By	AN	BH/TP No	BH1
				Sample No	1
		Date	10/11/2009	Type	D
				Depth (m)	4.00
	Ring Shear Test BS 1377 : Part 7 : 1990 : 6	Checked By	LM	Date	13/11/2009

Figure 12.66 Typical worksheet for part of one stage of ring shear test

12.9.5 Calculation and plotting

General

1. Calculate the mass of the test specimen, m (grams), from the difference in masses determined in steps 3 and 5 of Section 12.9.3

2. Calculate the volume of the specimen from the equation

$$V = \frac{\pi(r_2^2 - r_1^2)h}{1000} \ \text{cm}^3$$

where r_2, r_1 are the outside and inside radii (half diameters), respectively, (mm) h is the height of the specimen (mm)

For a specimen of 100 mm and 70 mm external and internal diameters (50 and 35 mm radii), and height 5 mm, the volume V is 20.03 cm³. Hence calculate the specimen density, and then the dry density using the initial moisture content.

3. Calculate the vertical stress applied to the specimen as follows. Determine the equivalent load W (kg f) applied to the specimen, as described in Section 12.5.6 (item 7) for the shearbox apparatus. If the beam and weight hanger are counterbalanced, only the mass of the applied weights is multiplied by the beam factor.

Calculate the plan area (A) of the specimen from the equation

$$A = \pi(r_2^2 - r_1^2) \ \text{mm}^2$$

Calculate the vertical stress σ'_n (kPa) from the equation

$$\sigma'_n = \frac{9810 \ W}{A} \ \text{kPa}$$

For a specimen of the dimensions referred to above, the area A is equal to 4006 mm², and then

$$\sigma'_n = 2.45 \ W \ \text{kPa}$$

Shear stage

The following calculations apply to the test data taken from each set of readings during the shearing stage.

4. Calculate the apparent average linear displacement, D (mm), from the equation

$$D = \frac{\theta r}{57.3}$$

where θ is the measured angular displacement (degrees) and r is the mean radius (mm)

$$\left(r = \frac{r_1 + r_2}{2} \right)$$

5. Calculate the average linear displacement of the upper platen, d (mm), from the equation

$$d = \frac{(A+B)Fr}{L}$$

where A and B are the readings of the two load rings (divisions or digits); F is the mean displacement factor for the load rings (see Section 12.9.2); L is the distance shown in Figure 12.64(b) (see Section 12.9.2)

6. Calculate the corrected average linear displacement D_1 (mm) (i.e. the displacement of the lower part of the cell relative to the upper part) from the equation

$$D_1 = D \times d$$

7. Calculate the average shear stress, t (kPa) on the plane of shear from the equation

$$\tau = \frac{0.239(A+B)LR_f}{(r_2^3 - r_1^3)} \times 1000 \text{ kPa}$$

where R_f is the mean load ring factor (N/division) (the coefficient 0.239 equals 3/4 π).

8. Plot the calculated values of the shear stress, t, as ordinates against the calculated values of average relative displacement, D_1, as abscissae. This should be done as the test proceeds.

9. From the graph for each test run determine the residual shear stress, t_r (kPa)

10. Plot each value of τ_r as ordinate against the corresponding vertical effective stress, σ'_n (kPa) as abscissa, both to the same linear scale. Draw the line of best fit through the points and the origin, and determine its angle of slope. This gives the value of the residual strength, ϕ'_r, on the assumption that the cohesion intercept is zero

A typical set of results from a three-stage ring shear test is given in Figures 12.66–12.68. Figure 12.66 shows specimen data, and test data for the initial part of the first shear stage. Graphical plots from the three consolidation and shear stages are given in Figure 12.67, including the data from Figure 12.66. Figure 12.68 summarises the specimen and test data, and includes a plot of shear stress against normal stress for the three shear stages.

Reporting results

From a set of tests the following data are reported;

Dimensions of the test specimen

Initial and final moisture contents

Rate of angular displacement (or average linear displacement) applied during the shear test

Tabulated values of the applied normal stress, the corresponding residual shear stress, and the angular displacement at the end of each shearing stage

Graphical plot of residual shear stress against applied vertical effective stress, showing the derivation of the angle of shear resistance ϕ'_r

The angle of residual shear resistance, to nearest 0.5°

Graphical plots of mean shear stress, and change in specimen thickness, against average linear displacement (usually plotted cumulatively) for each shear stage

If appropriate, graphical plots of the change in specimen thickness against square-root time for each consolidation stage, showing the derivation of t_{100}

Whether the test was carried out on one specimen in stages (and if so, details of the re-loading sequence), or on separate specimens for each vertical stress.

The test method, i.e. the ring shear test on remoulded soil, in accordance with Clause 6 of BS 1377:Part 7:1990.

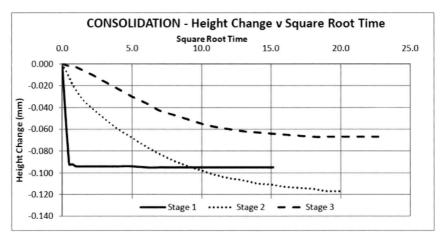

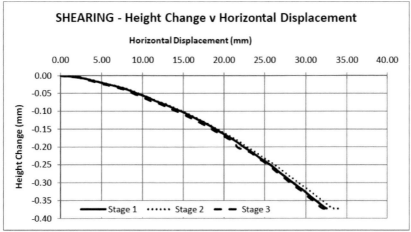

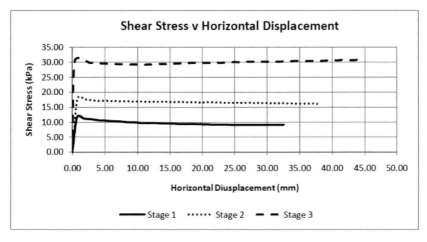

Figure 12.67 Graphical plots of output from ring shear test (data courtesy of Geolabs Limited)

Borehole No:	BH1	Description:
Sample No:	1	Stiff grey mottled light brown CLAY
Depth:	4.00 - 5.40m	

Specimen Details

Natural moisture content	%	31
Preparation		<1.18 mm material remoulded by kneading
Particle density	Mg/m³	2.70 (assumed)
Inner radii	mm	34.8
Outer Radii	mm	50.0
Initial height	mm	5.3
Initial moisture content	%	31
Initial wet density	Mg/m³	1.94
Initial dry density	Mg/m³	1.48

Consolidation Stage

Stage Number		1	2	3
Applied normal effective stress	kPa	50	100	200
Duration	day(s)	1	1	1

Shearing Stage

Applied normal effective stress	kPa	50	100	200
Duration	day(s)	1	1	1
Residual Conditions:				
Rate of angular displacement	degs/min	0.048	0.048	0.048
Residual shear stress	kPa	9.1	16.1	30.8
Final mean linear displacement	mm	32.6	38.0	44.5

Final Conditions

Final moisture content	%	42

Shear Strength Parameters

Angle of Residual Shear Resistance, Φ'_R	deg	9

Notes:

After consolidation, specimen pre-sheared by a revolution (360°) and then allowed to equalise before each shearing stage.

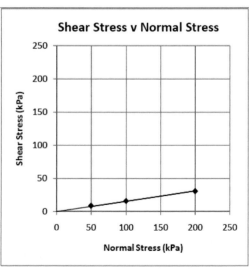

Fig 12.68 Summary report of ring shear test results given in Figures 12.66 and 12.67

12.10 Fall cone test

12.10.1 General

The fall cone test utilises the apparatus for the determination of the liquid limit by cone penetrometer, but the geometry and mass of the cone used varies according to the anticipated shear strength. The description of the test given below is based on the British Standard DD CEN ISO/TS 17892-6.

12.10.2 Apparatus

The cone apparatus should provide a mechanism for the instantaneous release of the cone such that it can fall freely in the vertical direction into the soil specimen from a fixed point. The apparatus requires a means of raising or lowering the cone and adjustment so that the tip of the cone just touches the surface of the specimen before the cone is released.

It should be fitted with a scale or other means of reading penetration capable of reading from 5 mm to 20 mm to a resolution of ± 0.1 mm.

A typical fall cone apparatus is shown in Figure 12.69.

Fall cones

The cones have a tip angle of 30° or 60°, with masses which vary cover the whole range of shear strengths as set out in Table 12.7. The geometry of the 60° cone is shown in Figure 12.70 and the following requirements must be met for both types of cone:

- An index line is required for manual readings
- The mass of the cone shall be within 1% of the nominal mass
- Tip angles shall be within 0.2° of the nominal tip angle
- Deviations from the geometrical tip shall not exceed the following tolerances:
  Manufacturing tolerance (a): 0.1 mm
  Tolerance due to wear (b): 3 mm
  Cone height (*h*) shall not be less than 20 mm

Figure 12.69 Fall cone apparatus (photograph courtesy of Geonor AS)

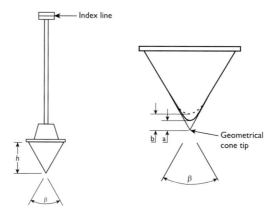

Figure 12.70 Essential requirements of a typical fall cone

Table 12.7 Size and geometry of cones for fall cone test

Penetration (mm)	Apex angle (°)	Mass (g)	Shear strength factor cf (kPa)**	Undrained shear strength (kPa)
5–20	60	10	26.5	0.063–1
5–15	60	60	159	0.67–6
15–25*	30	80	628	1–3
5–15	30	100	785	4.5–40
5–15	30	400	3139	18–250

* Cone as used in liquid limit test in required set of cones
** Derivation of shear strength factor is given in Section 12.10.4, Calculation of results

Table 12.8 Dimensions and tolerances of feeler gauges for checking cone tip

Cone angle	30°	60°
Plate thickness (mm)	1.75 ± 0.1	1.0 ± 0.1
Hole diameter (mm)	1.50 ± 0.2	1.50 ± 0.2

Cones with obvious wear or scratches should be replaced, and also when the point can no longer be felt when brushing lightly with the finger with the tip pushed through a perforated feeler gauge. The gauge consists of a metal plate with a hole conforming to the dimensions and tolerances given in Table 12.8.

Ancillary apparatus
- a cutting ring to prepare a specimen of at least 50 mm diameter and 25 mm thickness in the manner describes in Sections 9.2.2 and 9.3.1, unless an undisturbed sample is used
- wire cutter
- a mixing cup of at least 55 mm diameter and 30 mm depth
- spatula

12.10.3 Procedural stages

1. prepare test specimen from tube OR prepare sample in tube OR prepare remoulded specimen at required moisture content
2. Prepare apparatus
3. Release cone and measure penetration to nearest 0.1 mm
4. Repeat stages 1–3 at least twice more for undisturbed samples; or until two successive test yield identical values for remoulded specimens
5. Calculate
6. Report results

12.10.4 Test procedure

1. Preparation of sample

Samples tested in a sample tube are prepared by extruding disturbed material from the top of the sample tube and cutting such material off with a wire cutter to provide a level even surface.

Alternatively, a specimen of 50 mm diameter is extruded and prepared by trimming to provide parallel plane ends. The height of the test specimen should be at least 5 mm more than the anticipated cone penetration, but a height of 25 mm should normally comply with this requirement.

Any coarse material (described as gravel sized or shells in the standard, i.e. > 2 mm) should be removed when preparing a remoulded specimen and this should be noted. The remaining soil is thoroughly remoulded in such a way as to avoid air bubbles being mixed into the sample. The mixing cup is filled with soil without entrapping air and the soil surface is levelled flush with the brim of the cup using a spatula or a straight edge. Any surplus sample is placed in a spare cup.

2. Preparation of apparatus

Lock the cone in position and check the zero or record the initial scale reading to the nearest 0.1 mm (d_0).

Place the specimen with the smooth and level surface below the cone and lower the supporting assembly with the cone locked in the raised position until the cone just touches the soil.

3. Release the cone promptly, taking care not to jerk the apparatus when an automatic release is not provided.

Read the scale after 5 s. If the penetration exceeds 20 mm the test should be repeated using a lighter or blunter cone.

Determine the penetration (*i*) from the scale reading (d_1) or from the difference between scale readings ($d_1 - d_0$) mm. Record the penetration and type of cone used.

4. When repeating stages 1–3, position the test points no closer to the perimeter than 7 mm, and the distance between the outer boundary of two test points should exceed 14 mm.

If further material from an undisturbed sample is needed, extrude a thickness of 1.5 times the cone penetration and prepare a fresh surface.

Remoulded specimens are mixed and levelled again after each test.

5. Calculation of results

Calculate the average penetration (i) mm for undisturbed specimens. For remoulded specimens i is the common value from two successive tests.

The undrained shear strength of the specimen under test conditions is given by the expression

$$c_u \text{ (or } c_{ur}) = cg\frac{m}{i^2}$$

where c_u is the undrained shear strength of the undisturbed soil specimen in its tested state in kPa; c_{ur} is the undrained shear strength of the remoulded soil, in kPa; c is the constant, dependent on the state of the soil and the tip angle of the cone: ($c = 0.8$ for cones with 30° tip and $c = 0.27$ for cones with 60° tip); g is the acceleration due to gravity (= 9.81 m/s^2); m is the mass of the cone, in grams; and I is the cone penetration, in mm.

This expression can be simplified to

$$c_u = \frac{c_f}{i^2}$$

where $c_f = c \times g \times m$.

The calculated undrained shear strength is corrected by applying the empirical factor μ

$$c_u \text{ (or } c_{ur}) = cg\frac{m}{i^2}$$

where $\mu = \left(\dfrac{0.43}{w_L}\right)^{0.45}$ and $1.2 \geq \mu \geq 0.5$

w_L is the liquid limit of the soil expressed as a fraction.

7. Test report

The following data are reported.

- Condition of test specimen (undisturbed/remoulded) and method of sampling undisturbed samples
- Mass and tip angle of the cone used
- Cone penetration values obtained for undisturbed soil, and average penetration
- Calculated undrained shear strength in kPa to two significant digits

References

Abbott, A. F. (1969). *Ordinary Level Physics*, second edition, Chapter 3. Heinemann, London.

ASTM D 3080-04. Standard test method for direct shear test of soils under consolidated drained conditions. American Society for Testing and Materials, Philadelphia, PA, USA

ASTM D 4648-05. Standard test method for laboratory miniature vane shear test for saturated fine-grained clayey soil. American Society for Testing and Materials, Philadelphia, PA, USA.

Bell, A. L. (1915). The lateral pressure and resistance of clay, and the supporting power of clay foundations. *Proc. Inst. Civ. Eng.*, Vol. 199, 233–272.

Binnie, G. M., Gerrard, R. T., Eldridge, J. G., Kirmani, S. S., Davis, C. V., Dickinson, J. C., Gwyther, J. R., Thomas, A. R., Little, A. L., Clark, J. F. F. and Seddon, B. T. (1967). *Proc. Inst. Civ. Eng.*, Vol. 38, Paper No. 7063, Part I, 'Engineering of Mangla'.

Binnie & Partners (1968). Private communication to author

Bishop, A. W. (1948). A large shearbox for testing sands and gravels. *Proc. 2nd. Int. Conf. Soil Mech. and Found. Eng.*, Rotterdam, Vol. 1.

Bishop, A. W. and Henkel, D. J. (1962). *The Measurement of Soil Properties in the Triaxial Test*, 2nd edition. Edward Arnold, London.

Bishop, A. W. (1971). The influence of progressive failure on the choice of the method of stability analysis. *Géotechnique*, Vol. 21, No. 2 (Technical Note).

Bishop, A. W., Green, G. E., Garga, V. K., Andresen, A. and Brown, J. D. (1971). A new ring shear apparatus and its application to the measurement of residual strength. *Géotechnique*, Vol. 21, No. 4.

Bromhead, E. N. (1979). A simple ring shear apparatus. *Ground Engineering*, Vol. 12, No. 5.

Bromhead, E. N. (1992). *The Stability of Slopes* (second edition). Blackie, London and Glasgow.

Collin, A. (1846). (Translated by W. R. Schriever, 1956) *Landslides in Clays*. University of Toronto Press.

Cooling, L. F. and Smith, D. B. (1936). The shearing resistance of soils. *Proc. 1st Int. Conf. Soil Mech. and Found. Eng. 1936*, Vol. 1.

CEN ISO/TS 17892-6. Geotechnical investigation and testing – Laboratory testing of soil – Part 6: Fall Cone Test. European Committeee for Standardisation, Brussels.

Early, K. R. and Skempton, A. W. (1972). Investigations of the landslides at Walton's Wood, Staffordshire. *Q. J. of Eng. Geol.*, Vol. 5, No. 1, pp. 19–41.

Gibson, R. E. and Henkel, D. J. (1954). Influence of duration of tests on "drained" strength. *Géotechnique*, Vol. 4, No. 1.

Gilboy, G. (1936). 'Improved soil testing methods'. *Engineering News Record*, 21st May 1936.

Golder, H. Q. (1942). An apparatus for measuring the shear strength of soils. *Engineering*, 26th June 1942.

Hansbo, S. (1957). A new approach to the determination of the shear strength of clays by the fall-cone test'. *Proc. Roy. SGI*, Vol, 14, pp. 7–48.

Hvorslev, M. J. (1939). Torsion shear tests and their place in the determination of the shearing resistance of soils. *Proc. Am. Soc. Testing Mat.*, 99, pp. 999–1022.

Kolbuszewski, J. (1948). An experimental study of the maximum and minimum porosities of sands. *Proc. 2nd Int. Conf. Soil Mech. and Found. Eng.*, Rotterdam, Vol. 1.

Lambe, T. W. and Whitman, R. V. (1979). *Soil Mechanics, S.I. Version*, Wiley, New York.

Lewis, W. A. and Ross, N. F. (1955). An investigation of the relationship between the shear strength of remoulded cohesive soil and the soil moisture suction. Unpublished Research Note No. RN/2389/WAL.NFR. Transport and Road Research Laboratory, Crowthorne, UK.

Marsh, A. D. (1972). Determination of residual shear strength of clay by a modified shearbox method. Report LR 515, Transport and Road Research Laboratory, Crowthorne, UK.

Petley, D. J. (1966). The shear strength of soils at large strains. Unpublished PhD thesis, University of London.

Pike, D. C. (1973). Shearbox tests on graded aggregates. Report LR 584, Transport and Road Research Laboratory, Crowthorne, UK.

Pike, D. C., Acott, S. M. and Leech, R. M. (1977). Sub-base stability: A shearbox test compared with other prediction methods. Report No. LR 785. Transport and Road Research Laboratory, Crowthorne, UK.

Roscoe, K. H. (1953). An apparatus for the application of simple shear to soil samples. *Proc. 3rd Int. Conf. Soil Mech.* I: pp. 186–191, held in Zurich, Switzerland.

Sixth Géotechnique Symposium in Print (1987). The engineering application of direct and simple shear testing. *Géotechnique*, Vol. 37, No. 1.

Skempton, A. W. (1948). Vane tests in the alluvial plain of the River Forth near Grangemouth. *Géeotechnique*, Vol. 1, No. 2.

Skempton, A. W. (1958) Arthur Langtry Bell (1874–1956). and his contribution to soil mechanics. *Géotechnique*, Vol. 8, No. 4.

Skempton, A. W. (1964). Long term stability of clay slopes. Fourth Rankine Lecture, *Géotechnique*, Vol. 14, No. 2.

Skempton, A. W. and La Rochelle, P. (1965). The Bradwell slip: A short term failure in London Clay. *Géeotechnique*, Vol. 15, No. 3.

Symons, I. F. (1968). The application of residual shear strength to the design of cuttings in overconsolidated fissured clays. Report No. LR 227, Transport and Road Research Laboratory, Crowthorne, UK.

Terzaghi, K. and Peck, R. B. (1967). *Soil Mechanics in Engineering Practice*. Wiley, New York.

Townsend, F. C. and Gilbert, P. A. (1973). Tests to measure residual strengths of some clay shales. *Géotechnique*, Vol. 23, No. 2 (Technical Note).

Chapter 13

Undrained compression tests

13.1 Introduction

13.1.1 Scope

This chapter covers the measurement of the undrained shear strength of soils by means of axial compression tests on cylindrical specimens in which no drainage, i.e. no change of water content, is permitted during the test. These relatively simple 'quick' tests, in which undrained conditions apply, enable the undrained shear strength of a soil to be determined either when unconfined or when subjected to a defined confining pressure. Total stresses only are measured, which in general does not enable shear strength parameters to be derived from a set of tests on similar specimens. Derivation of shear strength parameters should normally be made only in terms of effective stress, not total stress.

However there is one important exception which is relevant in certain circumstances and that is the case of saturated soils in undrained conditions. For situations in which these conditions apply, measurement of total stresses might be all that is necessary.

The section on theory (see Section 13.3) follows on from Chapter 12, Section 12.3. It deals with the concept of principal stresses and the Mohr circle of stress for the graphical representation of direct and shear stresses on any plane within a specimen subjected to axial compression and a lateral confining pressure. This leads to the Mohr–Coulomb analysis of test data for deriving the undrained shear strength, c_u, from triaxial tests.

Some general items of equipment required for triaxial tests are described in Chapter 8, together with notes on their use and calibration. Items required specifically for uniaxial and triaxial compression tests are covered in this chapter.

13.1.2 Types of test

Compression tests described in this chapter are divided into two categories.

> Unconfined, or uniaxial, compression tests
> Triaxial compression tests

The first is a special case of the second, and requires simpler apparatus. Both are covered in BS 1377:Part7:1990, and in ASTM Standards.

Unconfined compression tests

The tests described here are as follows:

1. Standard laboratory test using a load frame, either hand-operated or machine driven, which can accommodate a wide range of specimen sizes (see Section 13.5.1).

2. Test with the traditional portable autographic apparatus, which can be used either in the laboratory or in the field, and is normally used for 38 mm diameter specimens (see Section 13.5.2). Although this apparatus is now rarely used in the UK, it is manufactured in India and the test has been retained while it remains in the British Standard.

3. Measurement of the remoulded strength of clays, for the determination of their sensitivity (see Section 13.5.3).

Triaxial compression tests

Triaxial tests which are described in Section 13.6.3 relate to testing of a single specimen, which can vary in size from 38 mm to 110 mm in diameter and includes the American 1.4 in and 2.8 in diameters and the European 35 mm and 79 mm diameters. The equipment listed in Section 13.6.2 and the drawings and photographs presented in Section 13.6.3 reflect the procedures required for 38 mm diameter specimens, which can obviously be extended to apply to other specimen sizes.

The other common British size for which procedures are given (see Section 13.6.4) is 100 mm (4 in) diameter, for specimens taken direct from standard U-100 or UT-100 sampling tubes, or rotary core samples. Reference is also made to triaxial tests on specimens 150 mm diameter and larger.

Other procedures described in this chapter are as follows:

Multi-stage triaxial tests (see Section 13.6.5).
Triaxial tests using free ends (lubricated ends) (see Section 13.6.6).
Triaxial tests under high confining pressures (see Section 13.6.7).
Preparation of specially orientated test specimens (see Section 13.6.8).
Preparation of recompacted specimens for triaxial test, including dry and saturated sand specimens (see Section 13.6.9)

13.1.3 Principle of test

General

A cylindrical specimen of soil is subjected to a steadily increasing axial load until failure occurs. In the unconfined test the axial load is the only force or stress which is applied (see Figure 13.1). In the triaxial test the specimen is first subjected to an all-round confining pressure, which is then maintained constant as the axial load is increased (see Figure 13.2). In either case the rate of loading is such that failure occurs within a relatively short time, usually between 5 and 15 min.

Drainage

No drainage of pore water from the specimen is permitted either during the application of confining pressure or during axial loading. These tests are therefore referred to as 'undrained' tests and no change in moisture content takes place. Because the test duration is short, and to distinguish them from slower tests in which pore water pressures are measured (to be described in Volume 3), they are often referred to as quick-undrained (QU) tests.

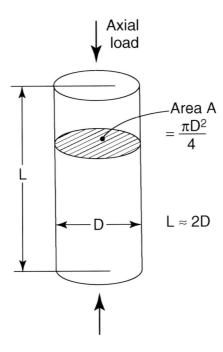

Figure 13.1 Principle of uniaxial (unconfined) compression test

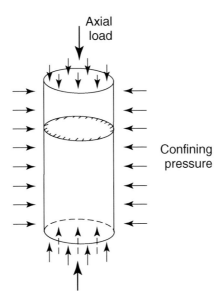

Figure 13.2 Principle of triaxial compression test

Specimen proportions

Test specimens are normally of a height: diameter ratio of 2:1, in British practice, and a ratio of 2:1 to 2.5:1 is permissible in ASTM Standards. If the ratio is much less than 2:1, results can be influenced by end restraints unless 'free ends' (see Section 13.6.6) are used. If the ratio is greater than 2.5:1, instability leading to buckling can occur and the specimen does not fail in true compression.

Rate of testing

A constant rate of compression (strain control) is applied, usually up to 2% of the specimen length per minute. Stress control is not usually appropriate for these tests. Rates of strain from about 0.3% to 10% per minute make little difference to the results. The rate of strain is not very critical, but should normally be such that failure is reached within a period of about 5–15 min. Brittle soils need a slower rate of strain than plastic soils to conform to this requirement.

Failure criteria

Failure normally implies the condition in which the specimen can sustain no further increase in stress, i.e. the point at which it offers its maximum resistance to deformation in terms of axial stress. Allowance has to be made for the bulging or 'barrelling' of the specimen as it is compressed, and for the effect of the rubber membrane used in a triaxial test.

In very plastic soils in which the axial stress does not readily reach a maximum value, failure is deemed to have occurred when a certain axial strain (typically 20%) has been reached.

Types of failure

Three main types of failure are recognised.

Plastic failure, in which the specimen bulges laterally into a 'barrel shape' without splitting, as in Figure 13.3(a)

Brittle failure, in which the specimen shears along one or more well-defined surfaces, as in Figure 13.3(b)

Failure in a manner intermediate between 1 and 2, as in Figure 13.3(c)

The mode of failure is a significant feature in the description of the soil properties.

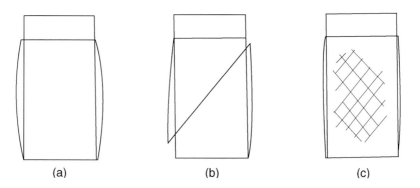

(a)   (b)   (c)

Figure 13.3 Modes of failure in compression test specimens: (a) plastic failure (barrelling); (b) brittle failure (shear plane); (c) intermediate type

Use of test data

In an unconfined compression test the axial load at failure and the corresponding axial strain are frequently the only numerical data that are recorded. If readings are taken during the test or if an autographic apparatus is used, a load–strain curve can be obtained.

In triaxial compression tests, it is usual to plot load–strain or stress–strain curves. Two or more tests on identical specimens, using different confining pressures, enable the Mohr circles to be plotted, from which the undrained shear strength parameter, c_u, can be derived at different stress levels. Wherever possible, three specimens are commonly tested in this way as a set.

13.1.4 Purpose

Some of the purposes for which the undrained shear strength parameters of soils are required are referred to in Section 12.1.2. Compression tests, and particularly triaxial tests, are more versatile and in many ways more reliable procedures for measuring these parameters than direct shear tests, and can be used with most types of soil. Triaxial tests can also represent more realistically the stress conditions prevailing in the ground.

13.1.5 Historical development

Unconfined compression test

The earliest cylindrical compression test apparatus for soils in Britain was probably that constructed by C. J. Jenkin at the Building Research Station (BRS) in 1932. It was based on a suggestion by Jurgensen in the USA, and was described by Cooling and Smith (1936). The specimen size was 0.75 in diameter and 1.5 in long.

In 1940 Cooling and Golder at the BRS designed a portable apparatus with a choice of springs of various strengths, and incorporating autographic recording, for 1.5 in diameter specimens. (Cooling and Golder 1940; Transport and Road Research Laboratory, 1952). The apparatus, although rarely used today, is described in Section 13.5.2 and is almost exactly the same in principle. It was designed for carrying out quick tests on virtually saturated soils, for which $\phi = 0$, and was intended for testing specimens on site immediately after sampling. The apparatus also became useful in the laboratory where it could be used in place of a load frame for these tests.

In its early form this machine was fitted with shallow cone seatings and the specimen ends were recessed with a matching conical profile. It was believed that coned ends reduced the tendency of a plastic soil specimen to become barrel-shaped when compressed, but they also tended to cause splitting of more brittle specimens. Flat end platens are now standard and are consistent with the triaxial test procedure.

In the laboratory it is possible to carry out an unconfined compression test in an unpressurised triaxial cell, but simpler apparatus is also available which enables the test to be carried out without the use of a cell, on virtually any load frame, using a wide range of specimen sizes.

Triaxial apparatus

A triaxial compression machine was designed in Britain by C. J. Jenkin and D. B. Smith by 1934. Axial load was applied by a spring and lateral pressure was developed in a brass cylinder. Specimens were 1 in diameter.

In 1940 an apparatus was constructed at the BRS for testing 1.5 in and 2.8 in diameter specimens. It made use of a lever-arm weighbridge for applying the axial load, similar to the principle which had been developed in the USA, and was therefore a stress-controlled machine. The confining pressure cell was a transparent cylinder, to enable failure of the specimen to be observed.

In 1943 this device was replaced by a hand-operated machine with a worm drive, using the same basic principle as that used in strain-controlled machines today. These early machines were used almost entirely for QU tests, but from about 1948 multi-speed drive units were fitted to enable tests of longer duration to be carried out. Bigger machines and cells were developed to accommodate larger diameter specimens, and the design of cells of all sizes was improved to provide positive seals and to make for easier handling.

Typical load frames available today are of 10 kN, 50 kN, 100 kN and 500 kN capacity. All may be fitted with multi-speed drive units, some with steplessly variable drive. Triaxial cells with Perspex walls are available for specimen sizes of 35–150 mm diameter. Steel cells for higher confining pressures, and for specimens up to 250 mm diameter and 500 mm high, can be obtained. Much larger cells, for samples up to 1 m diameter, have been specially constructed where the need has arisen, for instance for testing rockfill materials for a dam. (Marschi *et al.*, 1972.)

13.2 Definitions

Unconfined compression, uniaxial compression Longitudinal or axial compression of a specimen which is not subjected to a lateral or confining pressure.

Unconfined compressive strength (q_u) The compressive strength at failure of a specimen subjected to unconfined (uniaxial) compression. For a saturated clay it is equal to twice the undrained shear strength, i.e. $q_u = 2c_u$.

Triaxial compression Axial compression of a specimen which is subjected to a constant all-round lateral pressure.

Quick-undrained (QU) triaxial compression A triaxial compression test in which no change in water content of the specimen is allowed. The test is usually completed within 5–15 min.

Principal planes The three mutually perpendicular planes in a body subjected to stress, on which the shear stresses are zero.

Principal stresses The normal stresses which act on the principal planes.

Major principal stress (σ_1) The largest of the three principal stresses.

Minor principal stress (σ_3) The smallest of the three principal stresses.

Intermediate principal stress (σ_2) The principal stress which is intermediate between σ_1 and σ_3.

Deviator stress $(\sigma_1 - \sigma_3)$ The difference between the major and minor principal stresses; in a triaxial compression test the stress due to the axial load which is applied in excess of the all-round confining pressure.

Mohr circle The graphical representation of the state of stress on any plane, in terms of normal stress and shear stress.

Mohr failure envelope The line or curve which is tangential to Mohr circles representing the state of stress at failure of several specimens of soil tested under different confining pressures.

Failure plane The plane on which the maximum strength of the soil has been mobilised when failure occurs. Theoretically it is inclined at an angle of $(45° + \phi/2)$ to the horizontal in a typical triaxial test specimen.

Shear plane or surface The plane or surface on which slip of one portion of a test specimen relative to another occurs. It may or may not be inclined at the same angle as the failure plane.

Normal stress at failure (σ_f) The normal stress acting on the failure plane at failure.

Shear stress at failure (τ_f) The shear stress acting on the failure plane at failure.

Mohr–Coulomb failure criterion The shear strength of a soil obtained from triaxial compression tests expressed by the equation, $\tau_f = c_u$ when $\phi = 0$.

Undrained shear strength (c_u) The shear strength of a soil measured under undrained conditions, before drainage of water due to application of stress can take place.

Sensitivity (S_t) The ratio of the undrained shear strength of an undisturbed clay specimen to that of the same specimen after remoulding at the same moisture content.

13.3 Theory

13.3.1 Principal stresses

The concepts of force, normal stress and shear stress were discussed in Sections 12.3.1 and 12.3.2, with regard to shear along a predetermined place. In a cylindrical compression test, failure also occurs due to shear but there is no constraint to induce failure along a particular surface. It is therefore necessary to consider the relationship between shear stress (τ) and normal stress (σ) acting on any plane within the compression test specimen.

In a compression test a soil specimen is subjected to compressive forces acting in three directions at right angles to each other, one in the longitudinal direction, the other two laterally. (In the special case of uniaxial or unconfined compression the lateral stresses are zero.) The three perpendicular planes on which these stresses act are known as the principal planes, and the stresses are known as principal stresses. The shear stresses on the principal planes are equal to zero. On planes inclined at angles other than 90° to the principal planes the shear stresses are not zero, and normal stresses are different from the principal stresses but do not exceed the largest of them.

In descending order of magnitude the principal stresses are referred to as follows:

Major principal stress (σ_1)

Intermediate principal stress (σ_2)

Minor principal stress (σ_3)

The principal stresses acting on the faces of a typical elemental cube are shown in Figure 13.4, the faces on which they act being principal planes.

In many instances the major principal stress acts in the vertical direction, and the intermediate and minor principal stresses act horizontally at right angles to each other. Many soil problems are considered in two dimensions and only the major and minor principal stresses (σ_1 and σ_3) are used, the influence of σ_2 being neglected. In the special case of axial symmetry, such as cylindrical compression, σ_2 and σ_3 are equal.

13.3.2 Uniaxial compression

A cylindrical specimen of soil subjected to an axial compressive force P, and no other forces, is shown in Figure 13.5(a). If the area of cross-section of the specimen is denoted by A, on any horizontal plane XX (see Figure 13.5(b)) the normal stress is equal to P/A and the shear stress τ is zero because there are no forces tending to produce a sliding movement. In this simple case the normal stress on XX is the major principal stress and can be denoted by σ_1. On a vertical section YY (see Figure 13.5(c)), the normal stress (which would be the minor principal stress) is zero because there are no horizontal forces; and the shear stress is

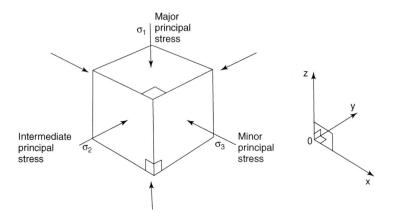

Figure 13.4 Principal stresses and principal planes

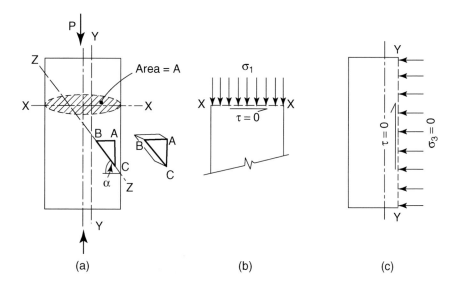

Figure 13.5 Stresses on cylindrical specimen subjected to uniaxial compression: (a) external forces and planes of section; (b) stresses on section XX: normal stress, $\sigma_1 = P/A$ shear stress $\tau = 0$; (c) stresses on section YY are zero

zero because the uniformly distributed vertical stress σ_1 induces no tendency for one part of the cylinder to move vertically with respect to another.

These conditions will be used to investigate the stress prevailing on a plane surface such as ZZ (see Figure 13.5(a)) which is inclined at any angle α to the horizontal. The small wedge-shaped element ABC of the specimen, bounded by vertical and horizontal planes and a plane parallel to ZZ, is shown enlarged in Figure 13.6(a). The inclined face BC is of unit length and the thickness of the wedge in the plane perpendicular to the paper is also of unit length.

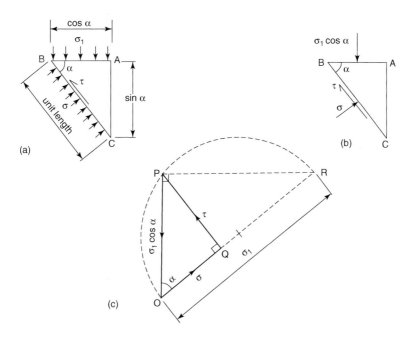

Figure 13.6 Stresses and forces on small element of uniaxial test specimen: (a) stresses on wedge element; (b) forces on element; (c) triangle of forces

The stresses on the three faces represented by the triangle ABC are as follows:

	Normal stress	Shear stress
AB	σ_1	0
AC	0	0
BC (to be determined)	s	t

In this two-dimensional analysis, stresses normal to the triangle ABC are not considered. The forces acting on the three faces of the wedge are obtained by multiplying the stress by the area of each. Since the wedge thickness is unity, the areas on which the stresses act are numerically equal to the lengths of the sides.

$$\begin{aligned}
\text{On AB: normal force} &= \sigma_1 \, (\cos \alpha) = \sigma_1 \cos \alpha \\
\text{tangential force} &= 0 \times (\cos \alpha) = 0 \\
\text{On AC: normal force} &= 0 \times (\sin \alpha) = 0 \\
\text{tangential force} &= 0 \times (\cos \alpha) = 0 \\
\text{On BC: normal force} &= \sigma \times 1 = \sigma \\
\text{tangential force} &= \tau \times 1 = \tau
\end{aligned}$$

These forces are shown in Figure 13.6(b).

Since the wedge ABC is in equilibrium the three forces acting upon it, if plotted as vectors, must form a triangle, as shown by the triangle of forces OPQ in Figure 13.6(c).

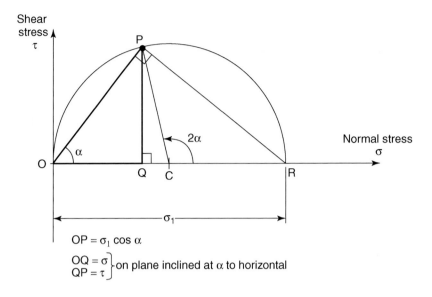

$OP = \sigma_1 \cos \alpha$

$\left. \begin{array}{l} OQ = \sigma \\ QP = \tau \end{array} \right\}$ on plane inclined at α to horizontal

Figure 13.7 Mohr diagram for uniaxial compression

This provides a relationship between σ, τ, σ_1 and α which can be derived mathematically. An alternative analysis is based on the geometry of Figure 13.6(c), as follows.

The line OQ is produced to R to make the length OR equal to σ_1. In the triangle OPR, since OP is equal to $\sigma_1 \cos \alpha$ and therefore equal to OR cos α, the angle OPR must be a right angle, whatever the value of α. Because the angle in a semicircle is a right angle, the point P must always lie on the semicircle of diameter OR, for any value of α. The semicircle OPR is also shown in Figure 13.7, in which the force diagram in Figure 13.6(c) has been rotated so that OR lies in the horizontal direction. The line OR forms a graphical axis (the abscissa) representing normal stresses, and vertical distances (ordinates) represent shear stresses.

Thus the stresses on a plane inclined at any angle α to the horizontal are defined by the point P, which is established by drawing the line OP at the angle α to the horizontal axis to meet the semicircle at P. The stress acting normal to the plane is given by the abscissa OQ, and the shear stress acting along the plane is given by the ordinate QP, both to the same scale as used when drawing OR to represent σ_1, the applied normal stress.

The significance of the graphical representation of stresses in the form of a circle will be discussed in Section 13.3.4.

13.3.3 Triaxial compression

A cylindrical specimen subjected to an axial major principal stress σ_1 and a radial minor principal stress σ_3, is shown in Figure 13.8(a). The stresses on the three sides of a wedge-shaped element ABC bounded by horizontal and vertical planes and a plane inclined at any

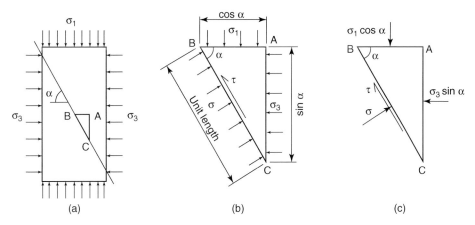

Figure 13.8 Stresses on cylindrical specimen subjected to triaxial compression: (a) external stresses and plane of section; (b) stresses on wedge element; (c) forces on element

angle α to the horizontal are shown in Figure 13.8(b), and are summarised as follows. As in Section 13.3.2, the following analysis is two-dimensional.

	Normal stress	Shear stress
AB	σ_1	0
AC	σ_3	0
BC (to be determined)	s	t

The forces acting on the three faces of the wedge are as follows (see Figure 13.8(c)):

$$\text{On AB: normal force} = \sigma_1 \cos \alpha$$
$$\text{tangential force} = 0$$
$$\text{On AC: normal force} = \sigma_3 \sin \alpha$$
$$\text{tangential force} = 0$$
$$\text{On BC: normal force} = \sigma$$
$$\text{tangential force} = \tau$$

There are now four forces, and because the wedge is in equilibrium they must form a closed polygon when plotted as vectors. The force diagram is shown as PVOQP in Figure 13.9, from which the relationship between σ, τ, σ_1, and σ_3 can be derived mathematically, but the following geometrical analysis is probably easier to understand.

The line OSQ is produced to R to make the length SR equal to $(\sigma_1 - \sigma_3)$. In the triangle OVS, since the angle OVS is a right angle it is evident that SV is equal to $\sigma_3 \cos\alpha$ and that OS is equal to σ_3.

Also

$$PS = PV - SV$$
$$= \sigma_1 \cos \alpha - \sigma_3 \cos \alpha$$
$$= (\sigma_1 - \sigma_3) \cos\alpha$$

$$= SR \cos \alpha$$

i.e. $$\frac{PS}{SR} = \cos \alpha$$

Therefore the angle RPS must be a right angle, whatever the value of α. Hence the point P must always lie on a semicircle of diameter SR, for any value of α. The semicircle SPR is also shown in Figure 13.10, in which the force diagram in Figure 13.9 has been rotated so that OSR lies in the horizontal direction (compare with Figure 13.7). The line OR forms a graphical axis (the abscissa) representing normal stresses, and the vertical distances (ordinates) represent shear stresses. However, this circle does not pass through the origin O, but intersects the horizontal axis at the two points $\sigma = \sigma_3$ and $\sigma = \sigma_1$. The centre of the circle is denoted by C, and is at a distance of $(\sigma_1 + \sigma_3)/2$ from O.

Hence in a specimen subjected to unequal compressive stresses σ_1 and σ_3 acting vertically and horizontally respectively, the stresses on a plane inclined at any angle α to the horizontal can be determined graphically by using a simple geometric construction. The stresses σ_1 and σ_3 are marked out along a horizontal axis, and represented to a suitable scale by OR and OS (see Figure 13.10). The semicircle of diameter RS (i.e. $(\sigma_1 - \sigma_3)$) is drawn, and the point P is established by drawing the line SP at the angle α to the horizontal axis to meet the semicircle at P. The stress acting normal to the plane is given by the abscissa OQ, and the shear stress acting along the plane is given by the ordinate QP, both to the same scale as used when plotting OR and OS to represent the applied stresses.

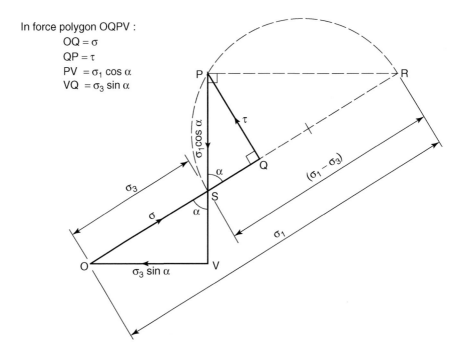

In force polygon OQPV :
$$OQ = \sigma$$
$$QP = \tau$$
$$PV = \sigma_1 \cos \alpha$$
$$VQ = \sigma_3 \sin \alpha$$

Figure 13.9 Polygon of forces for triaxial compression

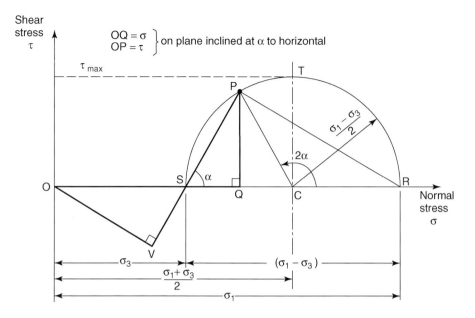

Figure 13.10 Mohr diagram for triaxial compression

13.3.4 Mohr circle of stress

The graphical construction for defining the locus of the point P by means of the circles referred to in Sections 13.3.2 and 13.3.3 (see Figures 13.7 and 13.10) is of great significance in soil mechanics. These circles are known as the Mohr circles of stress, after Otto Mohr (1871), the mathematician who introduced this method of analysis. It is described in greater detail in text books on strength of materials (e.g. Case and Chilver, 1971; Whitlow, 1973). Since we are normally concerned only with positive stresses, only the upper half of the circles are drawn. The Mohr circle for unconfined compression is a special case of triaxial compression in which $\sigma_3 = 0$.

In the Mohr circle shown in Figure 13.10, the following points should be noted:

1. The horizontal axis represents total normal (principal) stresses, and the vertical axis represents shear stresses, all drawn to the same scale.
2. The ends of the diameter of the circle are defined by the values of σ_3 and σ_1, measured from the origin.
3. The point P, whose coordinates are the normal and shear stresses on a plane inclined at an angle α to the horizontal, is determined by drawing a line from S inclined at α to the horizontal. Alternatively, P can be found by drawing a radius from the centre C at an angle 2α to the horizontal axis. On the plane inclined at α, the normal stress is equal to OQ and the shear stress is equal to PQ.
4. The diameter of the circle is equal to $(\sigma_1 - \sigma_3)$, the principal stress difference which is also known as the 'deviator stress' (see Section 13.3.5).
5. The maximum shear stress is represented by the point T (the topmost point of the circle) and is equal to the radius, i.e. $(\sigma_1 - \sigma_3)/2$.

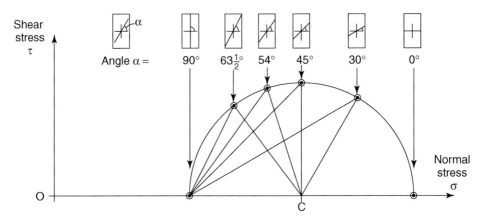

Figure 13.11 Representation of stresses on various planes in triaxial test specimen as points on Mohr circle of stress

6. A plane on which the maximum shear stress occurs is inclined at 90°/2 or 45° to the horizontal.
7. The centre of the circle C is at a distance OC = $(\sigma_1 - \sigma_3)/2$ from the origin.

The Mohr circle in Figure 13.11 shows points representing stresses σ and τ on several inclined planes (including those at 0° and 90° to the horizontal) within a cylindrical test specimen.

13.3.5 Deviator stress

When a cylindrical soil specimen of length L and diameter D (see Figure 13.12(a)) is subjected to a triaxial compression test, it is loaded in two stages as shown in Figure 13.12(b).

1. An all-round pressure (the cell pressure) denoted by σ_3 is applied. This acts equally in all directions, so the axial stress and radial stresses are all equal to σ_3 and no shear stress is induced in the specimen.
2. An axial load P is applied from outside the cell and is progressively increased. The additional stress caused by P is in the axial direction only and is equal to P/A.
3. The total axial stress, denoted by σ_1, is therefore equal to $(\sigma_3 + P/A)$, i.e.

$$\sigma_1 = \sigma_3 + \frac{P}{A} \tag{13.1}$$

This equation may be rewritten as

$$(\sigma_1 - \sigma_3) = \frac{P}{A} \tag{13.2}$$

The principal stress difference $(\sigma_1 - \sigma_3)$, which is the axial stress in excess of the all-round pressure σ_3, is known as the 'deviator stress' and is the stress which is applied to the specimen by means of a force from outside the cell.

In a test the cell pressure σ_3 is maintained constant at a given value, while the deviator stress is gradually increased. The increasing axial stress causes the specimen to compress in the axial direction, and from the measured change in length the corresponding strain is

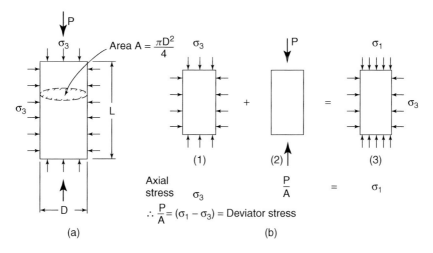

Figure 13.12 Explanation of deviator stress: (a) cylindrical specimen subjected to triaxial compression; (b) applied loading separated into two components

calculated (see Section 12.3.3). The deviator stress P/A can be calculated at any stage from measurements of load and strain, as described in Section 13.6.3. Deviator stress is plotted against strain and the strength of the soil is usually taken to be the stress at failure, that is the maximum or 'peak' deviator stress on the stress–strain plot. The major principal stress at failure is calculated from the peak deviator stress by using Equation (13.1), enabling the Mohr circle of stress representing failure under a particular confining pressure σ_3 to be drawn.

13.3.6 Shear strength parameters

Mohr envelope

From a series of compression tests on identical specimens of a soil carried out at different confining pressures a set of Mohr circles representing failure (as defined by the maximum deviator stress) can be drawn. A line (which may be straight or curved) drawn tangential to these circles is called the Mohr envelope, which represents the Mohr–Coulomb failure concept (see Figure 13.13).

If a Mohr circle for a particular state of stress in a soil lies entirely below the Mohr envelope, the soil is in a stable condition. If the Mohr circle touches the envelope, the maximum strength of the soil has been reached (i.e. failure has occurred) on some plane through the specimen.

A circle which intersects the Mohr envelope and rises above it has no physical meaning because once the envelope is reached, failure occurs and the soil can offer no further shear resistance.

The Mohr envelope is normally used for the derivation of shear strength parameters c', ϕ', when effective stresses are considered, and this requires knowledge of the pore water pressure. The widely used practice of plotting a set of Mohr circles of total stress and deriving values of so-called total stress parameters c and ϕ from a linear envelope is not, and never has been, part of the BS 1377 procedure. However, in the special case of

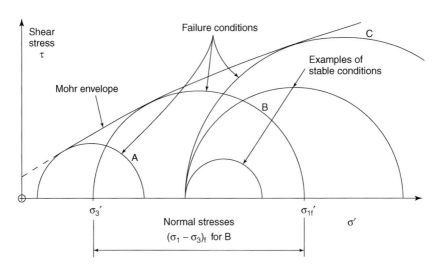

Figure 13.13 Mohr circles and Mohr envelope representing failure

saturated soils in undrained conditions (i.e. where no change in water content can take place), measurement of pore pressure might not be necessary and total stresses can be used in certain applications.

Undrained saturated soils ($\phi = 0$ concept)

In a saturated soil where no change in water content can occur, any increase in the mean normal stress results in a virtually equal increase in pore pressure. The mean normal effective stress therefore remains unchanged, and so the shear strength of the soil is not affected. If a set of similar specimens is prepared from a sample of saturated soil and each is tested under undrained conditions at a different confining pressure, the deviator stress measured at failure will be the same for each test. The Mohr circles of total stress will all be of the same diameter, and the set of circles will give a Mohr envelope which is horizontal, as shown in Figure 13.14. The value of ϕ is zero (the $\phi = 0$ condition) and the shear strength, c_u, is the same for all specimens. Thus when $\phi = 0$, Coulomb's equation for failure becomes

$$\tau_f = c_u \tag{13.3}$$

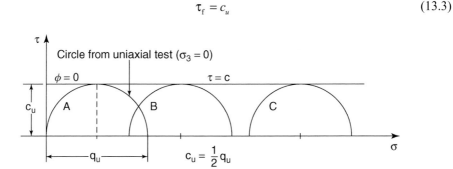

Figure 13.14 Mohr circles for saturated clay ($\phi = 0$)

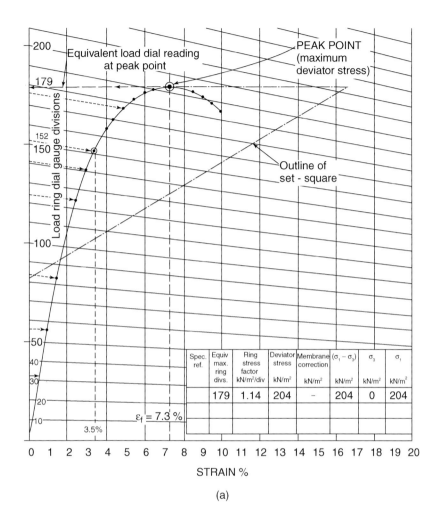

The table within the figure:

Spec. ref.	Equiv max. ring divs.	Ring stress factor kN/m²/div	Deviator stress kN/m²	Membrane correction kN/m²	$(\sigma_1 - \sigma_3)$ kN/m²	σ_3 kN/m²	σ_1 kN/m²
	179	1.14	204	–	204	0	204

(a)

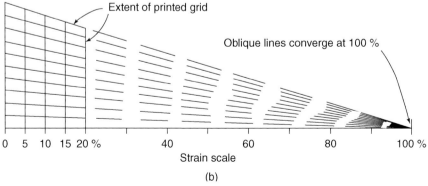

(b)

Figure 13.16 Area-correction graph sheet: (a) method of use; (b) construction of grid

13.3.8 Membrane correction

The restraining effect of the rubber membrane enclosing a triaxial specimen makes a small contribution to the resistance offered against compression. A correction has to be subtracted from the measured stress at failure to allow for this. The correction depends upon the manner in which the specimen deforms under axial load, which for this purpose is assumed to be either plastically or as a brittle material.

For soils of high strength, such as stiff clays, the effect of the membrane restraint is insignificant and is usually neglected. The same applies to large diameter specimens (100 mm diameter and upwards) except for low strength soils. For soft and very soft clays the membrane effect can form an appreciable proportion of the measured strength and omission of the correction could lead to errors on the unsafe side.

Plastic deformation

For a barrelling type of failure which occurs in a plastic soil, the magnitude of the membrane correction (denoted here by c_M kPa) depends upon the axial strain in the specimen ($\varepsilon\%$), the compression modulus of the membrane material (M N/mm width) and the initial diameter of the specimen (D mm). Henkel and Gilbert (1952) showed that the membrane correction is equal to

$$\frac{4M\varepsilon(1-\varepsilon)}{D}$$

or in terms of the units defined above

$$c_M = \frac{0.4M\varepsilon(100-\varepsilon)}{D} \text{kN/m}^2 \tag{13.8}$$

The compression modulus of the membrane material, M, is assumed to be equal to its extension modulus, which can be determined by the method given in Section 13.7.4.

The membrane correction to be applied is specified in BS 1377:Part 7:1990 by means of a graph (Figure 11), which is reproduced in Figure 13.17. This relationship applies to a specimen 38 mm diameter fitted with a rubber membrane 0.2 mm thick. For specimens of any other diameter (D mm) with any other membrane thickness (t mm), the correction obtained from this graph is multiplied by

$$\left(\frac{38}{D} \times \frac{t}{0.2}\right)$$

If more than one membrane is used, then t refers to the total thickness of rubber.

Brittle failure

In a brittle soil a plane of slip often develops before failure is reached and one portion of the specimen slides along that plane relative to the other portion. The effective area of cross-section then begins to decrease, instead of increasing as described in Section 13.3.7, and at the same time the membrane distorts in a manner completely different from that referred to above. These effects require opposing corrections to be made to the measured compressive strength. They are not taken into account in BS 1377, and are not usually allowed for in undrained tests. Corrections due to slip-plane effects will be discussed in Volume 3.

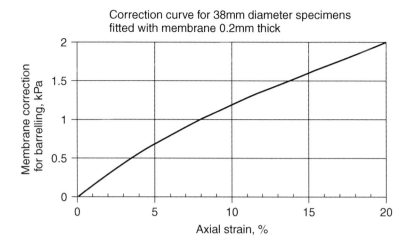

Figure 13.17 Membrane correction graph for barrelling (reproduced from Figure 11 of BS 1377:Part 7:1990)

13.3.9 Sensitivity

When a clay is remoulded by kneading and working without change in moisture content, it usually becomes softer than it was when undisturbed due to the breakdown of its structure. An unconfined compressive strength test on a remoulded specimen will give a compressive strength q_r which is lower than the compressive strength q_u of the undisturbed clay. The ratio of the undisturbed to the remoulded strength is known as the 'sensitivity' of the clay, and is denoted by S_t (Terzaghi and Peck, 1967, Art. 8):

$$S_t = \frac{q_u}{q_r} \tag{13.9}$$

This ratio can be quite large for some clays (see Section 13.4.3).

13.4 Applications

13.4.1 General applications of undrained shear strength

The relevance of the shear strength of soils to geotechnical analysis was outlined in Section 12.4.1. For some applications, where the undrained or short-term shear strength of saturated soil is relevant, the undrained shear strength parameter c_u, obtained from the tests described in this chapter, is sufficient for a stability analysis. Some examples were listed in Section 12.4.1. However, the value of c_u is much more dependent on the type of test and on the rate of testing than are the effective stress shear strength parameters (see Section 12.3.8).

The most usual application of the results of these tests is to foundation analysis, on the lines discussed in Section 13.4.2.

Analysis of long-term stability, and analysis of cases where drainage and pore water pressures are significant, require the use of effective stress parameters which will be dealt with in Volume 3.

13.4.2 Application to foundations

For foundations placed on clay soil, the condition immediately after completion of construction is nearly always the most critical. This is because the full load is then applied to the soil, but the soil has not had time to gain additional strength by consolidation. For saturated soils of low permeability the undrained shear strength, as determined by unconfined or triaxial compression tests, is relevant, where loading takes place under essentially undrained conditions.

For many small projects on which expenditure on an extensive testing programme is not justified, the bearing capacity of virtually saturated intact clay soils can be evaluated quickly and easily from unconfined compression tests. However, this can be done only if the ground conditions have been investigated and classified to a sufficient depth, and if settlement criteria are satisfied. In fissured clays, unconfined tests can be misleading and triaxial tests on large diameter specimens are needed.

13.4.3 Application to sensitive clays

One of the classification criteria of soft clays is the sensitivity to remoulding (see Section 13.3.9). This applies especially to soft alluvial and marine clays and gives an indication of the reduction in shear strength, which can be very large, which may occur if the clay is subjected to disturbance. A practical example is the remoulding effect of driving piles or sheet piling through a soft clay layer to reach a firmer stratum.

Many clays, including most found in the UK, are relatively insensitive to remoulding and their ratio of undisturbed to remoulded strength, i.e. the sensitivity, S_t, is normally less than 4. 'Sensitive' clays are those having sensitivities between 4 and 8; clays with a sensitivity exceeding 8 are referred to as 'quick' or 'extra-sensitive' clays. Some highly sensitive quick clays can have sensitivities up to 500. These are often clays which were originally laid down as marine deposits and from which the salt in the pore spaces has been removed by leaching with non-saline ground water. These clays require extreme care in sampling and handling to avoid disturbance.

The remoulded shear strength of many clays can be related to the liquidity index, I_L (see Volume 1 (third edition), Section 2.3.3), i.e. the value of the moisture content with respect to the liquid and plastic limits. (Skempton and Bishop, 1954.) For a clay at the liquid limit ($I_L = 1$) the shear strength is about 1.7 kPa, and at the plastic limit ($I_L = 0$) it is about 100–150 times greater (170–250 kPa) (Wood and Wroth, 1976, 1978).

13.4.4 Advantages and limitations of compression tests

While the direct shear procedures using the shearbox, described in Chapter 12, provide a relatively simple means of measuring undrained shear strength, they have a number of limitations, which were outlined in Section 12.4.5. On the other hand compression tests are suitable for most soils from which it is possible to prepare undisturbed specimens. Recompacted soils of all types, and remoulded clays, can also be tested. Specimens of cohesionless soils such as sands can be difficult to prepare and these are perhaps more conveniently tested in the shearbox apparatus. Soils containing gravel-size particles require large diameter specimens, 100 or 150 mm diameter, and the same applies to soils such as stiff fissured clays which contain discontinuities or other surfaces of potential weakness; and to non-homogeneous soils.

Some of the advantages of compression tests on cylindrical specimens over direct shear tests, and a limitation which should be borne in mind, are summarised below.

Advantages

1. The specimen is not constrained to induce failure on a predetermined surface, but failure can occur on any surface.
2. Consequently, a compression test may reveal a surface of weakness relating to some natural feature of the soil structure.
3. Specimens can be orientated if necessary to encourage failure to occur along a particular feature, but without imposing restraint.
4. The stresses which are applied in a triaxial compression test are a closer approximation to those which occur in-situ than are the conditions imposed in a direct shear test.
5. The applied stresses are principal stresses and close control is possible over stresses and rates of deformation.
6. Drainage conditions during test can be controlled and a variety of test conditions is possible. In the QU test described here, the specimen is completely enclosed and sealed so that the drainage is positively prevented for all types of soil.
7. Results from triaxial tests on good quality samples of intact clays have been found to show good correlation with values of shear strength found in the field (this is not true for fissured clays, see below).

Limitations

In highly fissured clays (e.g. London clay) the size of specimen tested can have an important effect on the measured strength. Tests on small (38 mm) diameter specimens give unrealistically high strengths, and tests on 100 mm diameter samples may give strengths higher than those measured by in-situ plate tests in boreholes. To obtain realistic results, test specimens must be large enough to enable the structure of the soil, particularly fissures and other discontinuities, to be adequately represented (Skempton and La Rochelle, 1965).

13.4.5 Typical values of shear strength

Values of the shear strength of non-cohesive soils, in terms of the angle of shearing resistance, were given in Table 12.3 (Chapter 12).

Typical undrained shear strengths of saturated clays found in Britain, for which $\phi = 0$, classified according to their description of consistency, are given in Table 13.1. In a particular clay stratum below the upper horizon which is subjected to seasonal wetting and drying, the undrained shear strength (taken as half the unconfined compressive strength) often increases uniformly with depth.

Table 13.1 Shear strength of clays

Consistency description	Undrained shear strength (kPa)
Very soft	< 20
Soft	20–40
Firm	40–75
Stiff	75–150
Very stiff	150–300
Hard	>300

13.5 Unconfined compression tests

13.5.1 Load frame method
(BS 1377:Part 7:1990:7.2 and ASTM D 2166)

The test described here is the definitive method given in BS 1377 for the determination of unconfined compressive strength of cylindrical specimens of soil. Axial compression is applied to the specimen at a constant rate of deformation (the strain-controlled procedure). The ASTM controlled strain procedure is similar, and that standard also includes a procedure using controlled stress, in which the axial force is increased incrementally at regular intervals of time.

Apparatus

1. Mechanical load frame, either hand-operated or machine drive, capable of providing platen speeds in the range 0.5–4 mm/min. A machine of 10 kN capacity is suitable, but a larger machine may be used if it has the required range of speeds and can accommodate the test specimen.

2. Calibrated force measuring device (usually a load ring (see Section 8.2.1, subsection on Conventional and electronic measuring instruments)), having a capacity and sensitivity appropriate to the strength of the test specimen. A 2 kN load ring reading to the nearest 1 N is suitable for specimens of most soils up to 100 mm diameter.

3. Platen with strain dial gauge mounting designed for unconfined compression testing. The assembly (shown in Figure 13.18) consists of:
   Lower platen, fitting onto load frame platen
   Post and bracket for dial gauge
   Upper platen, fitting on to load ring spigot
   Specimens up to 76 mm diameter can be accommodated. A specimen of 100 mm diameter would require a larger top platen, of about 120 mm diameter. Alternatively a standard triaxial cell of the appropriate size can be used, without cell fluid.

4. Dial gauge or linear transducer (see Section 8.2.1, subsection on Conventional and electronic measuring instruments), 25 mm travel reading to 0.01 mm, for measuring the axial compression of the specimen.

5. Supporting pillar for dial gauge or linear transducer, and seating for its stem.

6. Apparatus for extruding and trimming undisturbed soil specimens (see Section 9.1.2).

7. Vernier callipers, reading from 150 mm to 0.1 mm.

8. Steel rule, try-square.

9. Balance, reading to 0.1 g.

10. Drying oven and other standard moisture content apparatus.

11. Clinometer or protractor.

12. Stop-clock or seconds timer.

The apparatus set up with a test specimen in a 10 kN load frame is shown in Figure 13.18.

Procedural stages

1. Prepare apparatus
2. Prepare specimen
3. Measure specimen
4. Set up specimen

Figure 13.18 10 kN load frame with unconfined compression test apparatus

5. Record zero readings
6. Apply compression
7. Take readings
8. Unload
9. Sketch mode of failure
10. Remove specimen
11. Remould and re-test (if required)
12. Measure moisture content
13. Plot graphs
14. Calculate
15. Report results

Test procedure

1. *Preparation of apparatus*

   Ensure that the load frame stands firmly on a solid level bench top or support.

   Attach the load ring to the cross-head of the frame and fit any necessary extension pieces, and the upper platen, securely to the lower end of the ring. Check that the load dial gauge is securely held and that the end of the stem makes contact with the adjustable stop on the ring.

   Locate the lower platen centrally on the machine platen and set the dial gauge post vertically upright.

   Adjust the level of the lower platen to allow enough clearance to insert the test specimen.

   If a motorised unit is used, select the gear position which will give a platen speed of between 0.5% and 2% of the specimen length per minute. The time to failure should not exceed 15 min. With a height:diameter ratio of 2:1, the following speeds are usually appropriate.

Specimen diameter (mm)	*Approximate platen speed (mm/min)*
38	1.5
50	2
75	3
100	4

   If a hand drive is used, ascertain by trial the rate at which the handwheel should be turned to give the appropriate platen speed.

   Soft soils which require large deformations to failure will require somewhat higher rates of strain, whereas stiff or brittle materials which fail at small deformations will require lower rates of strain.

2. *Preparation of test specimen*

   The method of preparation depends upon the type of sample available, the most usual being as follows:

   1. Sample contained in a 38 mm diameter tube (or in a tube of any diameter which is the same as the test specimen diameter). This procedure is described in Section 9.2.4
   2. Sample contained in a U-100 tube or 100 mm diameter piston sampling tube (see Section 9.2.6)
   3. Block sample and the like (see Section 9.3.2)

   The height of the specimen should be about twice its diameter.

3. *Measurement of specimen*

   If the specimen has been prepared in a split mould of known dimension, no further measurement should be necessary other than weighing. Otherwise measure the diameter and height of the specimen to 0.1 mm using vernier callipers. Make two or three measurements of each dimension, and average the readings, as described in Volume 1 (third edition), Section 3.5.2. Weigh the specimen to the nearest 0.1 g, preferably on a small tray or moisture tin which has been weighed previously.

The specimen should be handled carefully, especially if soft, to avoid disturbance, distortion and loss of moisture. Thin plastic gloves should be worn to reduce loss of moisture due to handling. If possible, preparation and measurement should be carried out in a humidified atmosphere.

4. *Setting up specimen*
    Place the specimen centrally on the lower platen on the machine, and check that the specimen axis is vertical. Wind up the platen by hand until the specimen just makes contact with the top platen; this will be indicated by a fractional movement of the load dial gauge.

    Adjust the strain dial gauge on the pillar to read zero, or a convenient initial reading. Ensure that most of the travel of the dial gauge is available for recording compressive movement.

    If a triaxial cell is used, set up the specimen on the pedestal and assemble the cell as described in Section 13.6.3, stages 7 and 8. A rubber membrane need not be used because the cell is not filled with water.

5. *Recording zero readings*
    Record the readings of the strain dial, and the load dial, in the starting position under zero compressive load. It is convenient if both readings are zero, or an exact number of hundreds.

6. *Compression test*
    If a motorised unit is used, switch on the motor and start the clock at the same time. The clock can be used to verify that the correct rate of strain is being applied.

    If a hand-operated machine is used, an assistant is required to turn the handle at the correct speed, checking against the clock, while readings are being taken.

7. *Taking readings*
    Record the readings of the load dial at regular intervals, such as every 0.2 mm (or for larger specimens, 0.5% intervals), of the strain dial readings. If the machine speed is to be checked, record the time from the start also. When the rate of increase of the load dial reading becomes small, fewer readings need to be taken. If the load dial readings increase rapidly near the start of the test, take readings at regular intervals of load dial readings so that enough readings are taken to define the stress–strain curve before failure. At least 12 sets of readings should be obtained up to failure.

    Continue loading and taking readings until it is certain that failure has occurred, according to one of the following criteria:

    Three or more consecutive readings of the load dial show a decreasing or a constant load

    A strain of 20% (15 mm compression of a 38 mm diameter specimen) has been reached

    If the load dial reading is plotted against deformation, or strain percent, as the test proceeds (see stage 13), the point of failure can be identified easily.

8. *Unloading*
    When the specimen has failed, stop the machine, allow the motor to stop completely and put it into reverse; or wind down by hand until the load is taken off the specimen. Read the load dial gauge as a check on the initial reading under zero load. Lower the machine platen far enough to enable the specimen or triaxial cell to be removed.

9. *Sketching mode of failure*

    Make a sketch of the specimen as it appears after failure, to indicate the manner in which it has failed and especially to which of the three main types of failure it belongs (see Figure 13.3).

    > Plastic failure
    > Brittle failure
    > Semi-plastic failure

    If a shear surface is evident, measure the angle of inclination of the plane to the horizontal, if possible to the nearest 1°, using a clinometer or protractor (see Figure 13.40). This should be done as quickly as possible to avoid loss of moisture.

10. *Removing specimen*

    Carefully remove the specimen from the base platen, keeping it together in one piece. Place it on a small weighed tray or moisture container, together with any soil adhering to the upper and lower platens.

11. *Remoulded test*

    If the remoulded strength is to be measured for the determination of sensitivity, refer to Section 13.5.3 before proceeding with stage 12.

12. *Measurement of moisture content*

    Weigh the specimen and container to 0.1 g (m_2), place in a standard drying oven overnight, and weigh dry (m_3), as in the standard moisture content test. The container (mass m_1) should be large enough to retain any pieces of the specimen which might drop off.

    For a large specimen it is usually preferable to take selected portions of soil (e.g. from near the top, in the middle, and near the bottom) to determine the moisture content.

13. *Plotting graphs*

    The load–strain relationship may be plotted using a special area-correction graph sheet as shown in Figure 13.16, from data recorded on a test sheet of the type shown in Figure 13.39 (see Section 13.6.3) on which percentage strain values are pre-printed. (The latter is designed only for 38 mm diameter specimens about 80 mm long, and a sheet printed with different values is necessary for other sizes.) This method has the added advantage that a curve equivalent to the corrected form of the stress–strain curve can be plotted while the test is in progress.

    From each observed load dial reading on the vertical axis, project parallel to the nearest oblique line to the corresponding value of strain. For example, one of the marked points in Figure 13.16(a) represents a load dial reading of 152 divisions at a strain of 3.5%. Plot all the observed points in this way and draw a smooth curve through them.

    To obtain the corrected maximum value of compressive stress, place a ruler along the vertical axis and hold a set-square against it so that a horizontal line can be drawn to just touch the peak of the curve (see Figure 13.16(a)). Draw this line across to the vertical axis and record the equivalent load dial reading (R_c).

    Draw a vertical line from the peak of the curve to the horizontal axis, and record the strain corresponding to this point.

    If the area-correction graph sheet is not used, the compressive stress corresponding to each reading must be calculated, as described in stage 14, and plotted against percentage strain as an ordinary graph. The maximum value of the stress, and the corresponding strain, are read off and recorded. The peak point may not necessarily coincide with an observed point, but could be on the curve between two readings.

14. *Calculations*

The following symbols are used to calculate the stress and strain values:

Initial length of specimen $= L_0$ mm

Diameter of specimen $= D$ mm

Initial area of cross-section $= A_0$ mm², where $A_0 = \pi D^2/4$

Amount of compression at any stage $= x$ mm

Strain $\varepsilon = x/L_0 \times 100\%$

Mean calibration of load ring $= C_R$ N/division

Load ring reading at strain $\varepsilon = R$ divisions

Area of cross-section at strain $\varepsilon = A$ mm₂

Load on specimen at strain $\varepsilon = R \times C_R$ N

Compressive stress at strain $\varepsilon = \sigma$

$$\text{where } \sigma = \frac{R \times C_R}{A} \times 1000 \text{ kPa}$$

$$\text{But } A = \frac{100\, A_0}{100 - \varepsilon\%} \quad \text{(see Section 13.3.7)} \tag{13.10}$$

$$\therefore \sigma = \frac{R C_R (100 - \varepsilon\%)}{100\, A_0} \times 1000 \text{ kPa}$$

If the ring calibration C_R is assumed to be constant, the value of $1000\, C_R/A_0$ need be calculated once only as the 'stress constant' in terms of kPa per division. The calculation on each compressive stress is then simplified to

$$\sigma = \frac{10 C_R}{A_0} (100 - \varepsilon\%) R \text{ kPa} \tag{13.11}$$

Values of σ are plotted against $\varepsilon\%$ as described in stage 13.

If the area-correction graph is used, this calculation need be done for the peak point value only, i.e. at failure

$$\sigma_f = \frac{C_R}{1000} \times R_f \text{ kPa} = q_u$$

where R_f is the load ring reading at failure.

Other calculations are as follows:

$$\text{Initial volume of specimen} = V = \frac{\pi D^2 L_0}{400} \text{ cm}^3$$

$$\text{If the initial mass of specimen} = m_0 \text{ g,}$$

$$\text{its bulk density} \qquad \rho = \frac{m_0}{V} \text{ Mg/m}^3$$

$$\text{Initial moisture content} = w = \frac{m_2 - m_3}{m_2 - m_1} \times 100\%$$

$$\text{Dry density } \rho_D = \frac{100 \rho}{100 + w} \text{ Mg/m}^3$$

15. *Reporting results*

The following data are reported:

Specimen dimensions (to 0.1 mm)

Moisture content (to 0.1%)

Bulk density and dry density (to 0.01 Mg/m³)

Sketch showing mode of failure

Stress–strain curve (either stress against strain or represented by the area-correction grid plot)

Maximum compressive stress at failure (the unconfined compressive strength, q_u) to nearest 1 kPa

Strain (%) at failure, to nearest 0.2%

Rate of strain applied

Description of soil specimen

Type of sample from which prepared

Method of preparation

The test method, i.e. the load frame method for determination of unconfined compressive strength in accordance with Clause 7.2 of BS 1377:Part 7:1990.

13.5.2 Autographic unconfined compression test (BS 1377:Part 7:1990:7.3)

Apparatus

1. Portable self-contained hand-operated autographic apparatus in carrying case shown in Figure 13.19. The principle of operation is explained below

Figure 13.19 Autographic unconfined compression test apparatus

2. Accessories normally supplied with the apparatus:
   (a) Specimen end platens, flat and polished, up to 50 mm diameter
   (b) Set of calibrated springs of different stiffnesses (usually 4)
   (c) Printed charts on which the graph is recorded (see Figure 13.26 later in this section)
   (d) Transparent mask for use with the charts (see Figure 13.20)
3. Hand extruder for 38 mm diameter specimens (see Section 9.1.2)
4. Split former, 38 mm diameter
5. End trimming tool
6. Trimming knife, wire saw
7. Steel straight edge
8. Steel rule, try-square
9. Vernier callipers
10. Graduated specimen extruder dolly
11. Clinometer or protractor
12. Balance, reading to 0.1 g
13. Drying oven and other standard moisture content apparatus

Principle of apparatus

The main features of the apparatus are shown in Figure 13.21. The drawing plate (shown in dotted outline) which holds the paper chart is fixed to the upper moving plate C. The pivot K of the L-shaped arm G which holds the pencil D is mounted on the lower moving plate J. The lower end of arm G carries a knife-edge M which rests on the adjustable stop L, providing a fulcrum at a fixed level. The upper end of the spring is fixed to C and the lower end to

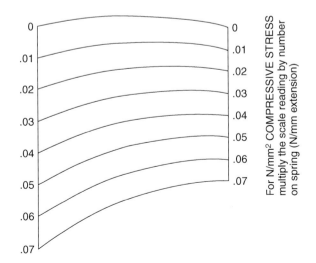

Figure 13.20 Transparent mask for use with autographic compression test apparatus

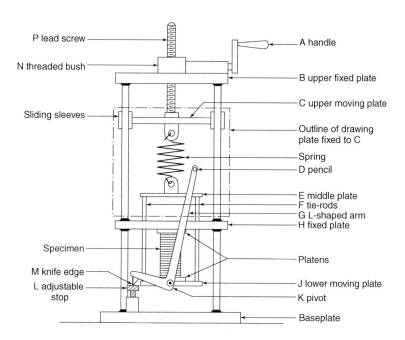

Figure 13.21 Main features of autographic unconfined compression test apparatus

the middle plate E, which is attached to J by the tie-rods F. The specimen is held between the lower platen, supported by the plate J, and the upper platen attached to the underside of the fixed plate H. The upper fixed plate B carries a threaded bush N which is rotated by the handle A and which engages with the lead screw P. The lower end of the lead screw is attached to the moving plate C.

Clockwise rotation of the handle A rotates the bush N, thereby raising the lead screw P because it cannot rotate. This lifts the plate C, say by a distance x (see Figure 13.22). If the spring extends, due to the resistance offered by the specimen, the plate J rises by a smaller distance y, equal to the amount by which the specimen is compressed. The relative vertical movement $(x - y)$ is equal to the extension of the spring, and is equal to the vertical movement of the pencil D relative to the chart, denoted by q. The vertical movement of the pencil in the chart is therefore proportional to the applied load, if it is assumed that the spring has an elastic calibration, i.e. force proportional to amount of extension.

If the lengths of the two limbs of the pencil arm G are denoted by a and b (see Figure 13.23), it can be seen from the geometry of the arrangement that the horizontal movement p of the pencil D, for a vertical movement y of the lower plate, is determined by the relationship $p = (a/b)y$. That is, the horizontal movement of the pencil is directly proportional to the amount of compression of the specimen, i.e. to the strain.

The following limiting cases illustrate the principle of operation:

If the specimen does not compress, the pencil moves along a vertical line on the chart by a distance equal to the extension of the spring (loading producing zero strain)

If the spring does not extend, the pencil moves along the arc of a circle centred at K (compressive strain under constant load)

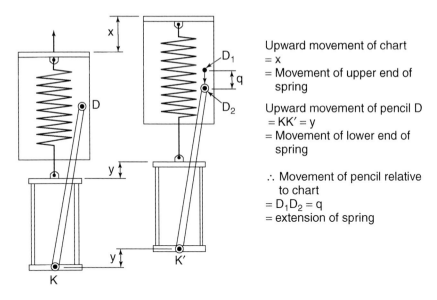

Upward movement of chart
= x
= Movement of upper end of spring

Upward movement of pencil D
= KK′ = y
= Movement of lower end of spring

∴ Movement of pencil relative to chart
= D_1D_2 = q
= extension of spring

Figure 13.22 Vertical movement of pencil relative to chart

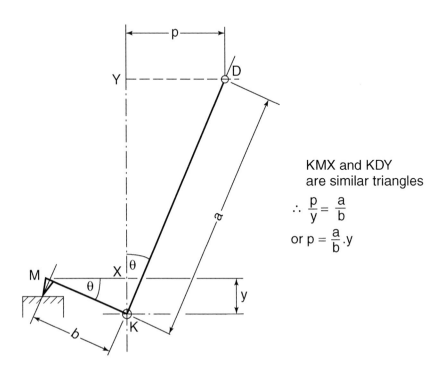

KMX and KDY
are similar triangles

∴ $\dfrac{p}{y} = \dfrac{a}{b}$

or $p = \dfrac{a}{b} \cdot y$

Figure 13.23 Horizontal movement of pencil relative to chart

From the above it can be seen that the load axis is a vertical straight line, and the ordinates are parallel straight lines; also that the deformation (strain) axis is a circular arc, and the abscissae are non-concentric circular arcs all of equal radius.

The curves on the transparent mask (see Figure 13.20) are drawn to make allowance for the area correction (see Section 13.3.7) and are not necessarily circular or parallel.

Procedural stages

1. Prepare apparatus
2. Prepare specimen
3. Measure specimen
4. Place in apparatus
5. Adjust apparatus
6. Apply compression
7. Remove specimen and chart
8. Sketch mode of failure
9. Remould and re-test (if required)
10. Measure moisture content
11. Derive result from chart
12. Report results

Test procedure

1. *Preparation of apparatus*

   Fit the platens to the apparatus by screwing them securely into the threaded connections. Select the spring which is best suited to the shear strength of the soil to be tested. This can be estimated from experience, but a general guide, based on the descriptive terms given in Table 13.1 is as follows.

Descriptive terms for shear strength of specimen	Suggested spring calibration (N/mm)
Very soft	2
Soft	4
Firm (lower range)	8
Firm (upper range)	16

   The spring used should be stiff enough to ensure that failure of the specimen occurs within the range of deformation permitted by the apparatus, but no stiffer than is necessary for providing reasonable sensitivity.

   Fit the spring by hooking it into the upper and lower sockets, then adjust the height of the lower moving platen by rotating the handle, so that there is enough clearance to insert the specimen.

   Record the specimen details, and the number and calibration of the selected springs, on a printed chart and attach it, top side uppermost, to the drawing plate with the spring clips. Align the edge of the chart with the edge of the plate.

   Fit a sharpened pencil into the holder, and clamp it so that it is just clear of the chart.

   Attach the split mould to the sample extruder and lightly oil the inside faces.

2. *Preparation of specimen*
   The procedure is the same as for stage 2 of Section 13.5.1 depending upon the type of specimen.

3. *Measurement of specimen*
   If the specimen has been trimmed in the split mould, it needs only to be weighed (to 0.1 g), provided that the dimensions of the mould have been verified.

   For a soft specimen which may be difficult to handle without causing disturbance, the tube measurements can be accepted. Otherwise the specimen can be measured with vernier callipers in several places, and the mean dimensions used. It is weighed to 0.1 g.

   Soil that has been trimmed from immediately adjacent to the test specimen may be used for the determination of moisture content.

4. *Setting up*
   Place the specimen centrally on the lower platen. Wind the handle to raise the lower platen until the specimen just makes contact with the upper (fixed) platen, and is in correct alignment.

5. *Adjustments*
   Adjust the pencil arm horizontally by turning the knurled screw L (see Figure 13.21) so that the pencil is over the vertical zero line on the chart (see Figure 13.24). Move the chart vertically so that the pencil is also over the curved line representing zero

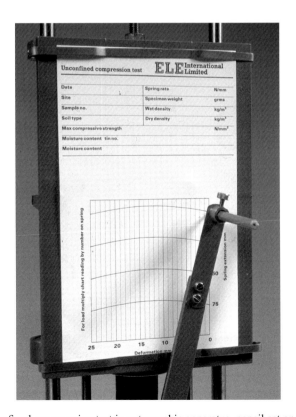

Figure 13.24 Unconfined compression test in autographic apparatus, pencil set on zero line

Figure 13.25 Unconfined compression test specimen after failure

load, i.e. at the origin. Push the pencil forward so that it bears on the chart with enough pressure to make a clear mark. Make final adjustments if necessary so that the pencil point is exactly at the origin of the printed graph axes.

6. *Compression of specimen*

Compress the specimen by rotating the handle steadily at the correct rate, usually one turn every 2 s. According to the BS the rate of deformation should be approximately 8 mm/min, giving a test time of about 2 min to achieve 20% strain.

The test is completed either when the specimen has failed (see Figure 13.25) and a definite peak on the graph plot is shown, as in Figure 13.26, curve (a); or when a strain of 20% is reached (seec Figure 13.26, curve (b)) whichever occurs first.

7. *Removal of specimen and chart*

Rotate the handle in the opposite direction so as to wind down the platen until the load is removed and there is enough clearance to take the specimen out. Place the specimen on a small weighed tray or moisture container, together with any soil adhering to the platens.

Remove the chart from the drawing plate. Check that the curve is clearly drawn, and that the identification details of the specimen, the date of test, and the number of calibration of the spring used, are written on the chart.

Autographic Unconfined Compression Test

Specimen diameter *38* mm		length *80* mm		
Date	*12-3-79*	Spring rate	*2*	N/mm
Site	*Halesby*	Specimen weight	*168*	g
Sample No.	*T-16*	Wet density	*1.85*	Mg/m³
Soil type	*Brown silty clay*	Dry density	*1.38*	Mg/m³

Max compressive strength *(a) 43 × 2 = 86 kN/m²*   *(b) 26 × 2 = 52 kN/m²*

Moisture content: Tin No.   *T.27*

Moisture content   *34 %*

(Superimposed mask contours __ kN/m² _ _ _ _ _)

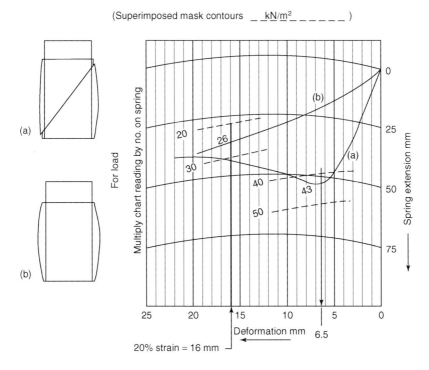

Figure 13.26 Typical results from unconfined compression test in autographic apparatus, illustrating two types of failure

8. *Sketching mode of failure*
    Make a sketch on the chart or on the back of it, to indicate the mode of failure of the specimen. If a surface of failure is visible, measure its inclination to the horizontal to the nearest 1° if possible, using a clinometer or protractor. Record any other features which are visible.

    If the specimen is likely to fall apart after testing, it should be sketched before removing it from the machine platen.

9. *Remoulded test*

   If the remoulded strength is to be measured for the determination of sensitivity, refer to Section 13.5.3 before proceeding with stage 10.

10. *Measurement of moisture content*

    Weigh the specimen on the tray, place in the oven, weigh when dry and cool and calculate the moisture content, as in the standard moisture content test.

11. *Derivation of result*

    Place the transparent mask over the test chart so that the vertical axis coincides with the vertical axis on the chart and the zero position on the mask coincides with the starting point (normally the origin) of the test curve (see Figure 13.27). The mask is applicable only to the specimen diameter for which it is designed (in this instance 38 mm).

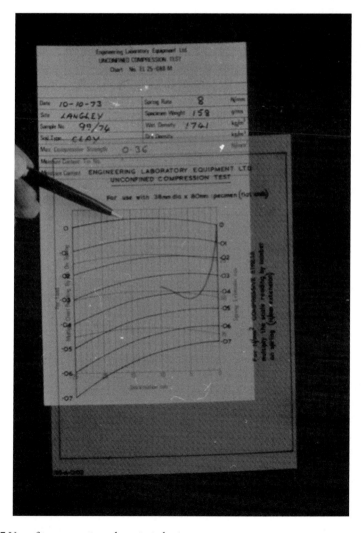

Figure 13.27 Use of transparent mask on test chart

Read off the number on the curved line on the mask corresponding to the lowest point (the 'peak') on the test graph, interpolating between the curved lines if necessary, as shown in Figure 13.26, curve (a). Record the reading and multiply it by the spring calibration (N/mm) to obtain the compressive stress (N/mm^2) on the specimen at failure. The unconfined compressive strength q_u (kPa) is equal to that value multiplied by 1000.

If a peak value was not obtained, a similar calculation is made using the mask reading corresponding to a strain of 20% (curve (b) in Figure 13.26).

With some machines the chart may be calibrated to give a direct reading of compressive strength in N/mm^2 or kPa on a 38 mm diameter specimen. Older machines may be calibrated to give lb/ft^2, in which case multiply by 0.0479 to obtain kPa. If the specimen diameter D is not 38 mm, the result should be multiplied by $(38/D)_2$ to obtain the corrected strength.

Read off the deflection (mm) from the chart corresponding to the peak stress. Divide it by the specimen length (mm) and multiply by 100, to obtain the strain (%) at failure.

Calculate the bulk density and dry density of the specimen, as in Section 13.5.1.

12. *Reporting results*

The compressive stress at failure (either peak stress, or at 20% strain) is reported as the unconfined compressive strength of the specimen to two significant figures. The autographic plot should be included as part of the test report. Report the strain at failure to the nearest 1%.

Also report:

> Mode of failure (plastic, brittle, or intermediate)
> Any other visible features
> Specimen identification details, dimensions and type of soil
> Moisture content
> Bulk density, dry density

The test method, i.e. the autographic method for determination of unconfined compressive strength in accordance with Clause 7.3 of BS 1377:Part 7:1990.

13.5.3 Remoulded test and sensitivity

The remoulded strength of a saturated clay, which is used for calculating sensitivity, is best determined immediately after measuring the undisturbed strength. The procedure described below is used as stage 11 of Section 13.5.1 or as stage 9 of Section 13.5.2. It is not included in BS 1377:1990.

1. Enclose the specimen in a small polythene bag, together with a little more of the soil at the same moisture content, and remould it thoroughly by squeezing and kneading it with the fingers for a few minutes.
2. Take the soil out of the bag and work it into a 38 mm diameter tube or split mould, without entrapping any air, as quickly as possible to avoid moisture loss, using a tamping rod.
3. Trim the ends flush and square.
4. Extrude or remove the specimen from the tube or split mould.
5. Measure and weigh the specimen.
6. Place it in the machine and carry out a repeat compression test in the same way as for the undisturbed specimen (see Section 13.5.1, stages 4–10; or Section 13.5.2, stages 4–8).

7. After sketching the mode of failure, measure the moisture content as described in Section 13.5.1, stage 12 or Section 13.5.2, stage 10.
8. Calculate the remoulded compressive strength q_R, in the same way as described in Section 13.5.1, stages 13 and 14 or Section 13.5.2, stage 11, for the undisturbed strength q_u.
9. Calculate the sensitivity, S_t, of the clay from the equation $S_t = q_u/q_R$.
10. Report the remoulded strength in the same way as the undisturbed strength and report the sensitivity to the first decimal place.

13.6 Triaxial compression tests

13.6.1 General

Scope

The principle of the triaxial compression test is outlined in Section 13.1.3. The test described in Section 13.6.3 is the 'definitive' method given in Part 7 of BS 1377:1990. A very similar test is given in ASTM D 2850.

The definitive test applies to specimen diameters in the range 38–110 mm as described in Section 13.6.3 and sometimes three individual specimens are tested at different cell pressures. Tests on specimens of large diameter, i.e. 100 mm and upwards, are described in Section 13.6.4. Variations from the definitive test, and methods of preparation of test specimens, are given in subsequent sections.

Test procedures are based on the use of conventional measuring instruments requiring manual observation and recording, although electronic logging and data processing is being used increasingly in commercial work (see Section 8.2.6).

Specimen sizes

To meet the requirements of Eurocode 7 for triaxial testing, rotary core samples or thin-walled tube samples should be used, which typically fall into the range 70–100 mm diameter. Nominal measurements for height:diameter ratios of 2:1 for commonly used samples are summarised in Table 13.2.

Types of soil

The definitive triaxial test (see Section 13.6.3) is intended mainly for use with fine-grained homogenous cohesive soils. The largest size of particles should not exceed one-fifth of the specimen diameter. If particles larger than this are found in the specimen after test, their size and mass should be reported. If larger particles are present, larger diameter specimens should be used. Table 13.3 indicates the suggested maximum particle size for the most widely used specimen sizes.

Soils which contain fissures or other discontinuities should be tested by using a specimen of as large a diameter as practicable; for instance a U-100 tube sample of fissured clay should be extruded and tested as a 100 mm diameter specimen, 200 mm long. Smaller specimens trimmed from this type of soil are likely to be representative of the intact material only (see Figure 13.28). If a small specimen includes fissures it is likely to break up and be rejected in favour of a 'better' specimen of intact soil. The discontinuities have an important

Table 13.2 Typical measurements of triaxial specimens

Diameter (in)	Specimen size diameter × height (mm)	Area (mm²)	Volume (cm³)	Approximate mass (g)**
	35 × 70	962.1	67.35	140 g
1.5	38 × 76	1134	86.19	180 g
	50 × 100	1963	196.3	410 g
	70 × 140	3848	538.8	1130 g
2.8	71.12 × 142.24*	3973	565.1	1190 g
	75 × 150	4418	662.7	1392 g
	100 × 200	7854	1571	3.3 kg
4	101.6 × 203.2*	8107	1647	3.5 kg
	150 × 300	17670	5301	11.1 kg
6	152.4 × 304.8*	18240	5560	11.7 kg

* Exact conversion of inch size
** Based on bulk density of 2.1 Mg/m³

Table 13.3 Maximum particle sizes for triaxial specimens

Nominal specimen size diameter × length (mm)	Suggested maximum size of particles (mm)
38 × 76	6.3
50 × 100	10
70 × 140	14
100 × 200	20
150 × 300	28

influence on the shear strength of the soil in-situ and a test specimen should be large enough to represent these features (Skempton and Henkel, 1957).

Confining pressure

The cell confining pressure to use for a test should be related to the in-situ conditions at the point from which the sample was taken; there are no 'standard' confining pressures. As a general guide, if three specimens from one sample are to be tested as a set, and the total vertical in-situ stress at the sample location is σ_v, confining pressures of 0.5 σ_v, σ_v and 2 σ_v might be appropriate. The range of confining pressures should cover the range of vertical stress likely to be experienced by the soil in-situ. For overconsolidated clays the lowest cell pressure should not normally be less than σ_v.

For compacted specimens the confining pressures should likewise be related to the total stresses likely to occur in the field.

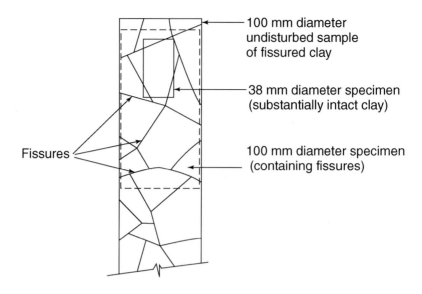

100 mm diameter
undisturbed sample
of fissured clay

38 mm diameter specimen
(substantially intact clay)

Fissures

100 mm diameter specimen
(containing fissures)

Figure 13.28 Representation of tube sample of fissured clay

13.6.2 Apparatus for definitive test

Preparation and measurement of specimens

Items 1–5 listed below are described in Section 9.1.2 and the preparation of test specimens is described in Sections 9.2.3 and 9.2.4.

1. Extruder for vertical extrusion of samples from sample tubes, hand-operated or motorized.
2. Adaptor for holding three 38 mm diameter sample tubes (where required).
3. Sample tubes 38 mm internal diameter and about 230 mm long (where required for preparation from larger diameter samples), with sharp cutting edges and end caps, with or without relieved bore (three required).
4. Extruder for 38 mm diameter tubes.
5. Split mould for forming a specimen of the diameter specified for the test.
6. Trimming knife, wire saw, spatula.
7. Straight edge.
8. Steel rule, try-square.
9. Vernier callipers.
10. Balance reading to 0.1 g.
11. Oven and other standard moisture content apparatus.

Setting up and testing

Many of the items listed below are described in Chapter 8, in the sections indicated. Triaxial cells and ancillary equipment are described in Section 13.7.2 and 13.7.3, including notes on their proper use.

12. Load frame, 10 kN capacity or larger, preferably motorised (see Section 8.2.3). For 38 mm diameter specimens a range of platen speeds in the range 0.05–4 mm/min is required, in order to induce failure within a period of 5–15 min as specified by the BS. For larger specimen diameters, for example, 100 mm, a load frame of 50 kN capacity should be used, with a range of platen speeds of 2–5 mm/min.

    The available travel should be capable of compressing a 38 mm diameter specimen by about 25 mm and a 100 mm diameter specimen by about 65 mm.

13. Calibrated force measuring device, normally a load ring (see Section 8.2.1). A load ring of 2 kN capacity and a sensitivity of 1–1.5 kN per division is suitable for smaller diameter specimens of most cohesive soils. A ring of higher capacity (e.g. 4.5 kN, reading to 3 N per division) may be needed for non-cohesive soils when tested under a high confining pressure. Higher capacity rings will be needed for 100 mm diameter specimens, up to 5 kN for cohesive soils, 20 kN for granular soils and 50 KN for weak rocks.

    Calibration of load rings, based on BS EN ISO 7500-1:1998, is described in Section 8.4.3. Significant readings taken during a test (i.e. those used for determining soil parameters) must lie within the range of calibration.

14. Triaxial cell, capable of sustaining an internal water pressure up to 1000 kPa. The main features of a typical triaxial cell are shown diagrammatically in Figure 13.31, and are summarised as follows:

    - Corrosion-resistant cell top with air bleed plug, piston and close-fitting guide bushing, post and support for stem of axial deformation dial gauge. The piston should be properly cleaned and lightly oiled.
    - Cylindrical cell body, of Perspex or similar transparent material, which is removable and can be secured and sealed to the top and base.
    - Corrosion-resistant base, incorporating a connection port (for applying pressurised water to the cell) fitted with a valve, and the base pedestal. If the cell base is fitted with other connection ports these should be blanked off with a plug or a valve which is kept closed.

    Further details of triaxial cells are given in Section 13.7.2.

15. Upper and lower solid specimen end caps of suitable diameter to match the specimen, of non-corrodible metal or plastic. The upper end cap (pressure pad) has a spherical seating for a ball bearing of about 12 mm diameter for 38 mm diameter triaxial cells, or is fitted with an integral hemispherical dome. This will vary according to the size of cell used (see Section 13.6.4 for guidance on 100 mm diameter specimens). A lower cap (base cap), if needed, fits on the cell pedestal.

16. Dial gauge or linear transducer (see Section 8.2.1), 25 – 50 mm travel reading to 0.01 mm, for strain measurement (see Sections 8.2.1 and 8.3.2).

17. Mounting bracket for attaching strain dial gauge or linear transducer to lower end of load ring.

18. Constant pressure system for maintaining cell pressure up to 1000 kPa at a constant level to within 5 kPa. Pressures up to about 700 kPa may be sufficient for many tests. Five types of constant pressure system are described in Section 8.2.4, of which the second (a motorised air system) is referred to in Section 13.6.3.

19. Nylon tubing and appropriate couplings for connecting the pressure system to the cell.

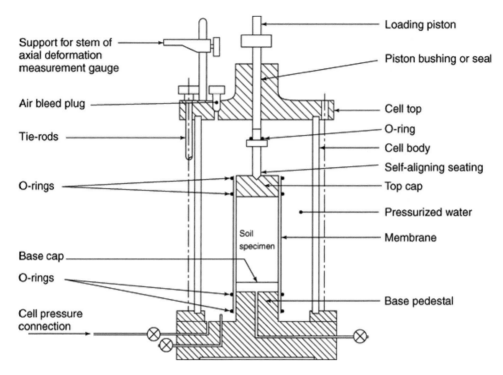

Support for stem of
axial deformation
measurement gauge

Air bleed plug

Tie-rods

O-rings

Base cap

O-rings

Cell pressure
connection

Loading piston

Piston bushing or seal

Cell top

O-ring

Cell body

Self-aligning seating

Top cap

Pressurized water

Membrane

Soil
specimen

Base pedestal

Figure 13.29 Unconfined compression test on 100 mm diameter specimen in 5 tonne compression frame

20. Pressure gauge of 'test' grade covering the range 0–1000 kPa, reading to 10 kPa (see Section 8.2.1). The gauge should be calibrated regularly (see Section 8.4.4) and the calibration data should be displayed with the gauge.
21. Latex rubber membranes, of dimensions matching the specimen (see Table 13.3), typically 0.2 mm thick, in the form of an open tube; three required, one for each specimen. (Details specified in BS 1377 are given in Section 13.7.4.)
22. Rubber O-ring sealing ring to fit tightly on each end cap. (Details specified in BS 1377 are given in Section 13.7.4.)
23. Suction membrane stretcher, fitted with a short length of rubber tube and pinch clip.
24. Small metal tray.
25. Clinometer or protractor.
26. Wiping cloths, sponge.

13.6.3 Triaxial test procedure (38–110 mm diameter specimens) (BS 1377:Part 7:1990:8, and ASTM D 2850)

The definitive test specified in BS 1377: Part 7:1990, Clause 8, relates to the determination of the undrained shear strength of a single test specimen. A set of three similar specimens can

each subjected to a different confining pressure as described below for a set of three 38 mm diameter specimens.

Procedural stages

1. Prepare test apparatus
2. Prepare sample extrusion apparatus
3. Extrude sample
4. Prepare specimens
5. Measure specimen
6. Fit membrane and end caps
7. Set up in triaxial cell
8. Assemble cell
9. Pressurise cell
10. Select machine speed
11. Adjust gauges
12. Compress specimen
13. Unload
14. Dismantle cell
15. Remove specimen
16. Sketch mode of failure
17. Measure moisture content
18. Clean equipment
19. Repeat stages 4–18 on other specimens
20. Plot graphs
21. Calculate
22. Plot Mohr circles
23. Report results

Test procedure

1. *Preparation of apparatus*

   Attach the appropriate load ring securely to the cross-head of the load frame. Check that the dial gauge is secure and that its foot makes contact with the anvil on the ring. A small deflection should be indicated when the ring is compressed slightly by hand.

   Wind down the machine base or raise the cross-head if necessary, to provide adequate clearance for inserting the triaxial cell. Ensure that the rubber seal on the base of the cell body is in good condition and is properly seated (see Section 13.7.2). Wipe the piston with a clean dry cloth and see that it moves freely in the bush. Place the cell base with fitted 38 mm base adaptor on the machine platen, and check that the pedestal is clean.

   If an air–water pressure system using a bladder is used, ensure that the bladder is initially almost deflated. Check that the air regulator valve is operating correctly and maintains a constant pressure when set.

2. *Preparation of extrusion apparatus*

   See Sections 9.2.2, 9.2.4, and 9.2.5 stage 1.

3. *Extrusion of sample*

   See Sections 9.2.4, and 9.2.5 stages 2–13.

4. *Preparation of test specimens*

   Remove the end caps from one of the 38 mm tubes (referred to as specimen A) and mount it on the hand extruder, fitted with the split mould (see Figure 9.22). Extrude and trim the specimen as described in Section 9.2.3.

5. *Measurement of specimen*

   The dimensions of the specimen are normally assumed to be the same as the internal measurements of the split mould, which should have been verified. Weigh the specimen on a small weighed container or tray to the nearest 0.1 g.

6. *Fitting end caps and membranes*

   Place the specimen on the lower end cap and the top cap on the upper end of the specimen. Fit two rubber O-rings over the membrane stretcher and roll them to near the middle of its length (see Figure 13.30(a)). Fit a rubber membrane inside the suction membrane stretcher and fold back the ends over the outside of the tube (see Figure 13.31) so that the membrane is not wrinkled or twisted (see Figure 13.30(b)).

   Apply suction through the rubber tube attached to the membrane stretcher by sucking with the mouth, so as to draw the membrane tightly against the inside wall of the stretcher tube. Maintain the suction while lowering the device carefully over the specimen (see Figure 13.30(c)). When the stretcher is positioned centrally over the specimen the suction can be released so that the membrane clings to the specimen (see Figure 13.30(d)).

   Roll the membrane ends gently onto the end caps, making a smooth fit with no entrapped air (see Figure 13.32). Hold the membrane stretcher steady with one hand, with the lower end about level with the middle of the lower end cap, and roll off one of the O-rings to seal the membrane onto the cap. Raise the membrane stretcher and repeat the operation to seal the top cap (see Figure 13.30(e)), then remove the stretcher.

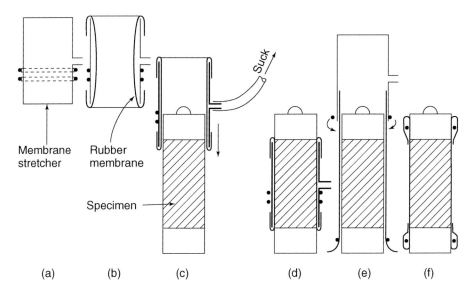

Figure 13.30 Stages (a)–(f) of fitting rubber membrane to triaxial test specimen

Finally, fold back the ends of the membrane neatly over the sealing rings at each end (see Figures 13.33 and 13.30(f)).

With soft or friable specimens it is preferable to fit the lower end cap in place on the triaxial cell pedestal first, before placing the specimen, to minimise the amount of handling.

7. *Setting up in triaxial cell*

Place the specimen on the triaxial cell base with the lower end cap correctly seated or screwed into the cell base (see Figure 13.34). Check that the specimen is aligned vertically. Place the ball bearing (if needed) in the recess in the top loading cap. Ensure that the surface of the cell base and the sealing ring are clean.

8. *Assembling cell*

Withdraw the cell piston to its maximum extent. Ensure that the cell sealing ring is in position. Lower the cell body into position over the specimen, taking care not to knock the specimen either with the cell walls or with the end of the piston. Position the cell so that the index marks (if there are any) on the cell and base coincide. Fit the tie-rods into their slots or locate them into their respective threaded holes in the base, depending on the design of the cell. Ensure that the tie-rods are vertical, and the clamping nuts properly seated, before tightening down lightly. Final tightening should be done systematically, first two opposite pairs moderately tight, then two other opposite pairs, and so on, and gradually increase the tightness in similar sequence. This procedure should ensure that the cell is tightened down evenly and is in true vertical alignment with the axis of the

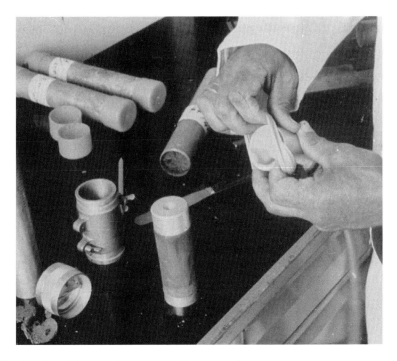

Figure 13.31 Fitting rubber membrane to membrane stretcher

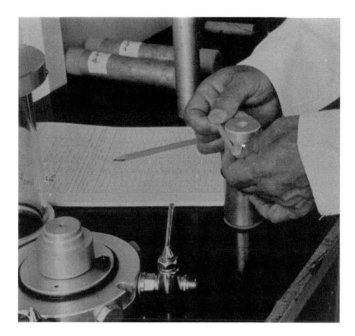

Figure 13.32 Rolling membrane on to end caps

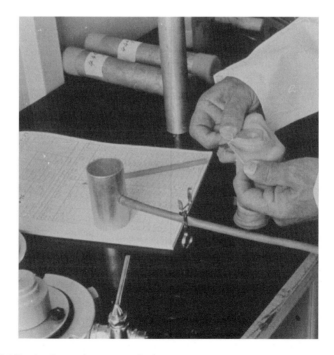

Figure 13.33 Folding back membrane over O-rings

Figure 13.34 Placing specimen on triaxial cell pedestal

load frame. It should not be necessary to use a tool on wing nuts or knurled nuts that are designed for hand-tightening only.

Allow the piston to fall so that it makes contact with the ball bearing or hemispherical dome on the top cap. A proper fit in the hollow at the end of the piston confirms correct alignment. Wind up the base pedestal by hand until the piston has 2 or 3 cm of free movement upwards before being restrained from further movement by the load ring.

Connect the water supply line to the connection on the base of the cell which leads into the cell chamber, and tighten the joint.

9   *Pressurising the cell*
The procedure described below relates to the use of a motorised compressed air system, of the type described in Section 8.2.4, item (2). The arrangement is shown in Figure 13.35. The principle is the same, although detailed procedures will differ, if some other type of pressure system is used.

Open the air bleed valve (valve e in Figure 13.35) on the triaxial cell. Open valve d on the cell base, and allow water into the cell from the supply line by opening valve a. Slow down the rate of filling as the water level nears the top, and as soon as water emerges from valve e close it and shut off the water supply.

Open the connection from the constant pressure line to the cell (valve c) (or disconnect the water supply line and connect the pressure system to the cell if they have to be connected separately). Open valves f and g, and increase the pressure in the cell gradually by turning the air regulator valve R steadily clockwise, up to the required value as indicated by the pressure gauge, making the appropriate calibration correction. The piston will be pushed upwards by the pressure until it is stopped by the load ring. Check after a minute or two that the pressure remains constant, and adjust the regulator

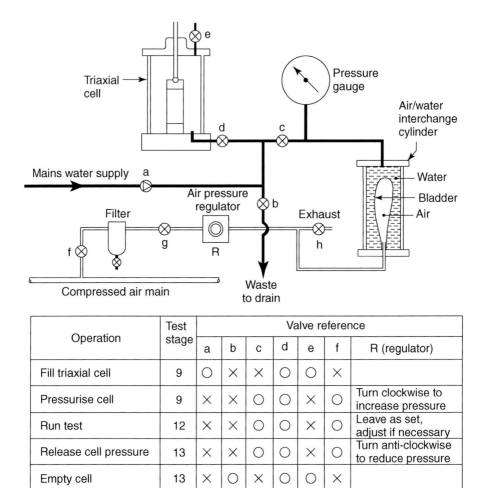

Figure 13.35 Connections to triaxial cell from air–water pressure system

if necessary. Throughout the test the connection to the constant pressure system *must* remain open, and the pressure regulator should be adjusted only if the pressure gauge indicates a change in pressure.

Operation of the valves of the system referred to above is summarised in the table in Figure 13.35.

10. *Selecting machine speed*

Select the machine speed for the test so that the maximum deviator stress is achieved within a period of 5–15 min. For soils that are plastic a rate of strain of up to 2% per minute is suitable (e.g. 1.5 mm/min for a specimen 76 mm high). For brittle soils which reach a maximum stress at strains of 10% or less the machine speed should be correspondingly lower.

The gear settings on older machines for given speeds are usually tabulated on the machine. With some machines the direction of movement of the platen is reversed when the gear setting is changed from one position to the next, so check that the reversing switch is in the correct position for upward movement. The manufacturer's instructions should be followed carefully.

11. *Adjustments*

Switch on the motor to raise the base pedestal, and as the piston is pushed into the cell towards the specimen either record the reading of the force gauge when it becomes steady (the initial reading, R_0) or re-set the load ring gauge so that it reads zero. This will eliminate the effects of both cell pressure and piston friction on the load dial reading, which will then indicate the axial force applied to the specimen in excess of that exerted by the confining pressure.

Switch off the motor and continue winding up the base pedestal by hand until the piston just makes contact with the top cap. This will be indicated by a small movement of the load ring dial gauge.

Secure the strain dial gauge so that its stem is vertical. Adjust and secure the foot attached to the post on the cell so that the dial reads zero or a convenient initial reading, and the stem has a clear travel of at least 25 mm. The clearance between the load ring with its attachments and the top of the cell, and the projection of the piston, should allow for at least 25 mm relative downward movement of the piston.

A specimen ready for a triaxial test in a 10 kN load frame is shown in Figure 13.36.

12. *Compression test*

Switch on the motor and record the readings of the load dial gauge at regular intervals of the strain dial gauge. Suitable intervals are indicated on the printed form shown in Figure 13.37, which shows the corresponding percentages for a specimen 80 mm long. These percentages can be considered to be valid for specimen lengths of about 70–85 mm. At least 15 sets of readings should be taken up to the point of failure.

For a very stiff soil, readings should be taken at frequent intervals of the force gauge reading rather than of deformation, in order to obtain the required number of sets of readings.

The compression test is continued until failure of the specimen has occurred, that is when the maximum value of the compressive stress has been passed or a strain of 20% (a compression of about 16 mm) reached. Beyond this point the specimen becomes severely distorted and further readings have little meaning.

With experience it is possible to plot the load–strain curve as the test proceeds. If the area-correction grid sheet referred to in Sections 13.5.1 and 13.5.1 (see Figure 13.16) is used, the correct form of stress–strain curve is produced and the failure condition can be easily identified. The test should be continued until three or four consecutive readings show a decreasing stress, so as to guard against a small momentary drop in the load dial reading (which can occur when coarse particles are present) being prematurely accepted as indicating that a maximum reading has been reached.

13. *Unloading*

When it is clear that failure of the specimen has occurred, stop the motor and allow it to come to rest before switching on in the reverse direction. Allow the specimen to unload,

Figure 13.36 Specimen in triaxial cell ready for test (operator is adjusting axial deformation gauge)

or wind the machine platen down by hand, until the platen returns to its starting position. There will then be a gap between the top cap and the piston, and the load dial should indicate the zero reading.

Reduce the pressure in the cell to zero by opening the regulator valve R (see Figure 13.35). The piston should then fall slowly under its own weight to rest on the top cap. Open valve b to the waste line, or connect the cell outlet to waste, and open the air bleed valve e so that the water drains out of the cell.

14. *Dismantling cell*

When the cell is empty, close valves b and d. Slacken the cell body securing screws or clamps progressively and pull up the piston before easing the cell off the base. Take care not to knock the cell against the specimen. Any water remaining on the cell base can be mopped up with a sponge.

15. *Removing specimen*

Take the specimen off the cell pedestal and stand it on a small tray. Pull up the folded-back length of rubber membrane at the top and carefully roll the O-ring off the top cap. The membrane can then be pulled down on to the lower end cap and the top cap removed.

TRIAXIAL COMPRESSION TEST
QUICK UNDRAINED

Location   *Halesby*

Location No. *3419*

Sample No. *T-31 A*

Date   *13.3.79*

Undisturbed

	Whole Sample 38 mm dia. × 80 mm	Part Sample in Tin no.
Length *78.2* mm	Wet weight g *185.6*	(+ tin)
Dia. *38.0* mm	Dry weight g *157.7*	(+ tin)
Area *1134* mm²	Tin weight g *21.5*	(tin)
Volume *88.7* ml	Dry weight g *136.2*	
	Moisture loss g *27.9*	
Machine No. *T.2* Cell *3*	MOISTURE CONTENT % *20.5*	
P.R.No. *118-13-54*	BULK DENSITY Mg/m³ *1.85*	
P.R. Calibration. *1.71* N/div.	DRY DENSITY Mg/m³ *1.54*	

Cell pressure *100* kN/m²   Rate of Strain   *1*   % per minute

Strain dial mm	Stress dial div.	Strain %	Strain dial mm	Stress dial div.	Strain %	Strain dial mm	Stress dial div.	Strain %
0	*0*	0	5.20		6.5	11.20		14
0.20	*28*	0.25	5.60		7	11.60		14.5
0.40	*46*	0.5	6.00		7.5	12.00		15
0.60	*63*	0.75	6.40		8	12.40		15.5
0.80	*81*	1.0	6.80		8.5	12.80		16
1.20	*107*	1.5	7.20		9	13.20		16.5
1.60	*152*	2	7.60		9.5	13.60		17
2.00	*187*	2.5	8.00		10	14.00		17.5
2.40	*216*	3	8.40		10.5	14.40		18
2.80	*245*	3.5	8.80		11	14.80		18.5
3.20	*272*	4	9.20		11.5	15.20		19
3.60	*291*	4.5	9.60		12	15.60		19.5
4.00	*297*	5	10.00		12.5	16.00		20
4.40	*252*	5.5	10.40		13			
4.80		6	10.80		13.5			

Mode of Failure
Inclination of shear plane to axis
53°

Description   *Stiff light brown*
Notes   *silty & sandy clay*
Extruded from U -100 tube

Figure 13.37 Typical data from triaxial compression test on 38 mm diameter specimen recorded on printed form

Alternatively, these operations could be done while the specimen is standing on the base pedestal, depending on the design of the end caps.

16. *Sketching mode of failure*

    Sketch the mode of failure of the specimen, from two directions at right angles (front and side) if one view does not fully convey the details, and record any other features observed. If a surface failure is visible, measure its inclination to the horizontal using a clinometer or protractor (see Figure 13.38).

17. *Measurement of moisture content*

    Slide the specimen off the lower end cap (or cell pedestal), and place it together with any loose fragments of soil adhering to the end caps or membrane on a weighed moisture container or small tray. Weigh, dry overnight, cool and weigh dry, as in the standard moisture content test. The wet mass provides a check against the initial mass before testing.

    It is sometimes desirable to measure the moisture content of a particular part of the specimen, for instance the zone adjacent to a failure slip surface. The soil from the zone should be carefully cut out and measured separately from the remainder of the specimen, and this procedure should be recorded with the help of a sketch.

Figure 13.38 Specimen in triaxial cell ready for test. Photo courtesy of Newton Technology Geomechanics Laboratory

18. *Cleaning equipment*

Remove the membrane and O-ring from the lower end cap. Wash the end caps, rings and membrane. The membrane can be used again if careful inspection reveals no flaws, but two uses should be a maximum; a new membrane is less costly than a spoiled specimen (see Section 13.7.4).

Clean the cell body and base, and wipe them dry. Ensure that the screw threads and the rubber sealing ring are clean.

The piston should be kept clean and dry, and not oiled. Before storing, assemble the cell and place a dust cover such as a small polythene bag over the top of the cell to protect the piston and bush.

Specimen tubes should be cleaned and wiped dry after use, and given a very thin coat of light oil inside. If the cutting edge is not sharp or is burred, it should be made true by filing or grinding and wiped free of swarf.

19. *Tests on other specimens*

Repeat stages 4–18 for the other two specimens (designated B and C) of the set, using different cell pressure, in ascending order of magnitude.

20. *Plotting graphs*

For each specimen plot the stress–strain relationship as described in Section 13.5.1, stage 13, either in terms of load dial reading against strain on the area-correction grid sheet, or by using values of compressive stress calculated as described in step 21 below. Curves for all three specimens may be plotted on one sheet, as shown on the area-correction graph in Figure 13.39. Curve A is drawn from the observed data tabulated in Figure 13.37.

The corrected peak load dial readings, and the corresponding strains at failure, are read off and tabulated as shown in the bottom right-hand corner of Figure 13.39.

21. *Calculations*

From each set of readings calculate the difference (R) between the observed load ring reading and the initial reading R_0 determined in step 11. (If the gauge was set to zero, $R_0 = 0$.) Calculate the strain (%) from the axial deformation gauge readings, and then calculate the measured deviator stress $(\sigma_1 - \sigma_3)_m$, as described in Section 13.5.1, stage 14.

Equation (13.10) becomes

$$(\sigma_1 - \sigma_3)_m = \frac{RC_R(100 - \varepsilon\%)}{100A_0} \times 1000 \quad \text{kPa}$$

The measured deviator stress must be corrected to allow for the effect of the rubber membrane, which depends upon strain as shown in Figure 13.17. The membrane correction, adjusted if necessary for specimen diameter and membrane thickness as explained in Section 13.3.8, is subtracted from $(\sigma_1 - \sigma_3)_m$ to give the corrected deviator stress $(\sigma_1 - \sigma_3)$. These values are plotted against strain as described in 20 above. The corrected stress representing failure (usually the maximum or 'peak' deviator stress) is denoted by $(\sigma_1 - \sigma_3)_f$.

If the axial force or load ring reading is plotted on an area-correction graph sheet, the above calculations need be carried out only on the peak value taken from the graph as shown in Figure 13.39.

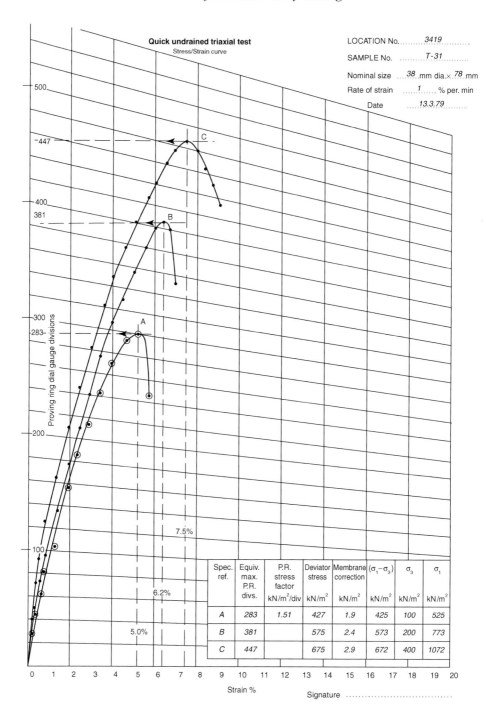

Figure 13.39 Set of three triaxial tests plotted on area-correction graph sheet

The value of the major principal stress at failure, σ_{1f}, is calculated from the equation

$$\sigma_{1f} = (\sigma_1 - \sigma_3)_f + \sigma_3$$

where σ_3 is the cell confining pressure which remains constant throughout the test. These calculations are shown in tabular form in Figure 13.39.

For each specimen calculate the moisture content, bulk density and dry density, as described in Section 13.5.1, stage 14.

Calculate the undrained shear strength, c_u (kPa) for each specimen tested, and tabulate with the cell confining pressure. The value of c_u is given by

$$c_u = \frac{1}{2}(\sigma_1 - \sigma_3)_f$$

22. *Plotting Mohr circles*

Using the values of σ_3 and σ_{1f} tabulated in Figure 13.39, the Mohr circle at failure for each specimen can be drawn, if required. The scale intervals on the vertical (shear stress axis) must be made the same as those used on the horizontal (principal stress) axis.

23. *Reporting results*

The following data are reported:

The test method, i.e. determination of undrained shear strength in triaxial compression, is carried out in accordance with Clause 8 of BS 1377:Part 7:1990
Initial specimen dimensions (to 0.1 mm)
Moisture content of each specimen and of sample trimmings (to 0.1%)
Bulk density and dry density of each specimen (to 0.01 Mg/m³)
Set of stress–strain curves for the three specimens, either as load dial readings against strain on the area-correction graph, or as calculated stress against strain %
Tabulated values of σ_3, σ_1 and $(\sigma_1 - \sigma_3)$ at failure, with membrane corrections
Value of shear strength (c_u) at failure, equal to $0.5(\sigma_1 - \sigma_3)_f$ (kPa), for each confining pressure
Strain at failure for each specimen (to 0.2%)
Rate of strain (% per min)
Sketch of each specimen at failure
Description of soil
Type of sample from which obtained
Method of preparation of each specimen
A suitable form for reporting the numerical data is shown in Figure 13.40.

13.6.4 Large diameter triaxial tests

Triaxial tests on large diameter specimens (100 mm diameter or larger) are similar in principle to those described above for 38 mm diameter specimens, but there are differences in some of the details. The special features relating to tests on 100 mm diameter undisturbed specimens obtained from U-100 sample tubes, which is a standard size in the UK, are dealt with below. Tests on specimens up to 110 mm diameter are included in the definitive method specified in Clause 8 of BS 1377:Part 7:1990. Additional comments relating to 150 mm diameter specimens, which are less frequently used, are given at the end of this section.

Undrained triaxial compression test

Location	Halesby							Job reference		3419	

Soil description	Stiff light brown silty & sandy clay	Borehole/Pit no. T

Sample no. 31

4.65

Undisturbed / Specimens			Location and orientation of test specimens	Top
Method of preparation	Extruded from U –100 tube into 38 mm dia. tubes			65 mm

Nominal dimensions	Diameter 38 mm	Date	13.3.79
	Height 76 mm		

Rate of strain	1.0 mm per minute	Membrane thickness 0.2 mm

Specimen ref.	Bulk density	Moisture content	Dry density	Cell pressure	At failure		Membrane correction	Shear strength	Mode of failure
					Compressive Stress	Strain			
	Mg/m³	%	Mg/m³	kPa	kPa	%	kPa	kPa	
A	1.85	21	1.53	100	427	5.0	0.7	213	
B	1.83	20	1.52	200	575	6.2	0.8	287	
C	1.86	21	1.54	400	675	7.0	0.9	337	

Figure 13.40 Summary of results for a set of triaxial compression tests

Apparatus

The following items listed in Section 13.6.2 are required: items 1, 6–9, 11, 17–20, 25. In addition the following items, specifically intended for 100 mm diameter specimens, are required in place of the equivalent items in Section 13.6.2.

27. Split mould for 100 mm diameter specimen 200 mm long.
28. Balance, 7 kg capacity reading to 1 g.
29. Load frame, 50 kN capacity. The platen speed for a rate of strain of 2% per min is 4 mm/min.
30. Load measuring ring or load cell (see Section 8.2.1). Suggested capacities are given in Table 13.4.
31. Triaxial cell, with fittings, for 100 mm diameter specimens.
32. Dial gauge or linear transducer (see Section 8.2.1), 50 mm travel reading to 0.01 mm.
33. Base adaptor and upper end cap (pressure pad) 100 mm diameter, solid or fitted with blanking plug and seal.
34. Rubber membranes, 100 mm diameter and 330 mm long, nominal thickness 0.5 mm.
35. Rubber O-ring sealing rings to fit tightly on 100 mm diameter end caps; four required.
36. Suction membrane device for 100 mm diameter specimens, with rubber tube and pinch clip.

Table 13.4 Suggested load ring capacities

Type of soil	Capacity (kN)	Approximate sensitivity (N/division)
Clays	4.5	3
Granular materials	20	18
Weak rocks, granular soils at high cell pressures	50	45

37. Metal tray to hold 100 mm diameter specimen.
38. Half-round plastic rainwater guttering of about 100 mm diameter; two lengths about 450 mm long and two lengths 200 mm long.

Procedure

The procedure is generally similar to that described in Section 13.6.3. Where modifications are necessary they are detailed below against the relevant stage numbers shown in brackets.

Specimens 100 mm diameter are usually tested singly, direct from the sampling tube if undisturbed, although samples from several tubes may be grouped together as a set. Samples of this size require very careful handling and may need two people when extruding, transporting, and setting up.

Sample preparation (stage 4)

See Chapter 9, Section 9.2.5.

Measurement of sample (stage 5)

Measure the diameter of the specimen with vernier callipers at three or four positions along its length, and average the measurements. Measure the length to the nearest 0.5 mm, unless the split former of known length has been used for trimming.

Weigh the sample on a weighed tray or length of guttering, to the nearest 1 g. Support it between pieces of guttering when moving it to and from the balance.

Setting up (stages 6 and 7)

Fit the 100 mm diameter base adaptor, if one is necessary, on the cell pedestal. A slightly oversize sample (say 106 mm diameter) may require a special base or an additional plate to support it, but this is not essential except for a soft clay which may be liable to local plastic deformation. If a separate plate is used, the corners should be well rounded, with no exposed sharp edges, so as not to puncture the rubber membrane. A fillet of plasticine is also desirable (see Figure 13.41). A similar arrangement should be made next to the top cap.

Fit the membrane and the O-ring seals, two at each end, over the sample and end caps using the 100 mm membrane stretcher, in the same way as for a 38 mm specimen. Ensure that the specimen is aligned vertically. Place the ball bearing in the recess in the top loading cap if needed.

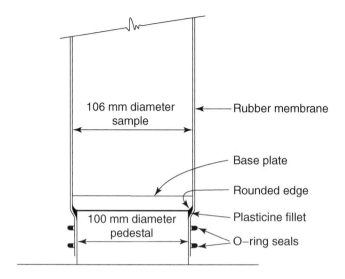

Figure 13.41 End plates for 106 mm diameter sample

Assembling cell and adjustments (stages 8–11)

The 100 mm cell body is heavier and more difficult to handle than the 38 mm cell and needs extra care, and perhaps an assistant. Particular care should be taken not to knock the sample.

The procedures for tightening down, connecting up, filling the cell, pressurising, and making final adjustments are similar to stages 8–11 of Section 13.6.3. A much larger volume of water, and therefore a longer time, is required to fill the cell than for a 38 mm cell. When the strain dial gauge is set in place allowance should be made for a clear travel of 50 mm. A specimen 100 mm diameter is shown ready for test in a 50 kN load frame in Figure 13.42.

Compression test (stage (12))

The compression test is similar to that for 38 mm specimens, but requires a machine platen speed of about 4 mm/min. Suitable intervals for reading the load dial, to give convenient intervals of strain, are shown on the printed test form in Figure 13.43. The strain percentages are valid for sample lengths in the range 185–210 mm.

The test is continued until the maximum compressive stress has been passed, or until a strain of 20% has been reached, as for 38 mm specimens. The same area-correction graph sheet as shown in Figure 13.39 may be used for plotting the load–strain curve if the horizontal axis represents percentage strain.

After testing (stages (13)–(18))

Procedures are the same, in principle, as those used for 38 mm diameter specimens. For the measurement of moisture content the sample can be cut down the middle so that two or three representative portions can be taken and the results averaged. Other portions can be used for

Figure 13.42 100 mm diameter specimen in triaxial cell mounted in a 50 kN load frame

index tests. If the whole sample is weighed and dried for measuring moisture content, several check weighings should be made after drying to ensure that constant mass has been reached.

For a detailed inspection of the specimen, cut about one-third to halfway through along its axis and then break it open. The broken surface will reveal details of soil fabric more clearly than a cut surface, especially if it is left overnight to partially air dry, when it will be in an ideal condition for photographing. Breaking open will also reveal if the sample contains a large particle, or particles, which could invalidate the test results.

Plotting, calculation and reporting (stages 20–23)
Results are plotted and calculated in the same way as for 38 mm diameter specimens. If two or more samples are to be grouped together as a set their Mohr circles may be drawn on one diagram. The test data should be tabulated, together with cell pressure and shear strength at failure for each specimen.

It should be stated that 100 mm diameter 'whole core' samples were used.

TRIAXIAL COMPRESSION TEST
QUICK UNDRAINED

100 mm

Cell pressure   kN/m^2 Rate of Strain   % per minute

Strain dial mm	Stress dial div.	Strain %	Strain dial mm	Stress dial div.	Strain %	Strain dial mm	Stress dial div.	Strain %
0		0	13.00		6.5	28.00		14
0.50		0.25	14.00		7	29.00		14.5
1.00		0.5	15.00		7.5	30.00		15
1.50		0.75	16.00		8	31.00		15.5
2.00		1.0	17.00		8.5	32.00		16
3.00		1.5	18.00		9	33.00		16.5
4.00		2	19.00		9.5	34.00		17
5.00		2.5	20.00		10	35.00		17.5
6.00		3	21.00		10.5	36.00		18
7.00		3.5	22.00		11	37.00		18.5
8.00		4	23.00		11.5	38.00		19
9.00		4.5	24.00		12	39.00		19.5
10.00		5	25.00		12.5	40.00		20
11.00		5.5	26.00		13			
12.00		6	27.00		13.5			

Figure 13.43 Test form for 100 mm diameter triaxial test

Tests on larger diameter samples

Triaxial tests on samples 150 mm diameter and larger require a large testing machine, probably of 100 kN capacity or more, with sufficient horizontal and vertical clearances to accommodate the large cell. A load ring of up to 100 kN capacity may be required for some soils. A typical apparatus is shown in Figure 13.44.

The cell and accessories are similar in principle to those for 100 mm samples, but their size and construction means that physical handling is more difficult. For instance the cell itself weighs about 50 kg, about 3.5 times the weight of a 100 mm cell. A typical soil sample would weigh about 12 kg.

Undisturbed samples of 150 mm diameter are not often taken from boreholes, but when they are the provision of a specially constructed extruder would be justified. Alternatively an extrusion device could be fitted to a large load frame, using the drive unit to provide the extruding force while the frame provides the reaction.

Figure 13.44 Triaxial apparatus for 150 mm diameter sample (100 kN load frame)

This size of sample is more likely to be required for testing recompacted soils which contain particles up to 37.5 mm. For this purpose a split mould and ancillary items are used (see Figure 9.7). Preparation of compacted samples for triaxial tests is discussed in Section 13.6.9. Testing procedures are similar to those already described, but special attention must be given to the handling of samples and equipment.

Triaxial compression tests on samples 254 mm (10 in) diameter using 'free' ends (see Section 13.6.6) have been reported by Rowe (1972).

13.6.5 Multi-stage tests (BS 1377:Part 7:1990:9)

When it is not practicable to obtain three small specimens from a U-100 sample for triaxial tests at three different cell pressures, the whole sample is tested but this gives only one value of shear strength. A method which enables three sets of data to be obtained from a single specimen is the 'multi-stage' triaxial test. This procedure is usually associated with effective stress tests (Kenney and Watson, 1961), but its application to QU tests has been described by Lumb (1964) and Anderson (1974). It is given here not as a recommended procedure, but as an expedient for economising on soil samples when whole core samples have to be tested. The test is satisfactory for plastic soils which require large strains for failure, but it should not be used for brittle soils or for soils that are sensitive to re-moulding. It is therefore particularly useful for stony clay soils such as those referred to as 'boulder clay', from which it may be impracticable to obtain small specimens.

In this test the axial load is applied to the sample under the first cell pressure (stage A) until the load–strain curve indicates that failure is imminent, i.e. it is reaching a maximum value. The cell pressure is then raised to the second value (stage B), and compression of the sample is resumed. The process is repeated and the third cell pressure is applied (stage C), under which the sample is allowed to reach failure. The three stages provide data from which a set of three Mohr circles can be plotted if required. Selection of cell pressures should be related to in-situ conditions (see Section 13.6.1).

Procedure

The multi-stage test is usually carried out on a 100 mm diameter sample 200 mm long, using the apparatus and procedure given in Section 13.6.4. Since it is necessary to plot the load–strain curve as the test proceeds, it is convenient to run the test a little slower than the standard rate; about 1% per min is suitable.

After applying the first cell pressure, the compression test is started and readings of the load dial gauge are observed at the usual intervals of strain, and plotted immediately on an area-correction grid sheet (stage A in Figure 13.45(a)). It may be necessary to take readings at strain intervals which are closer than usual, in order to see clearly when the curve bends over and approaches a peak value. When the curve shows that a peak value of deviator stress is imminent, the next stage is started by increasing the cell pressure immediately (the BS 1377 method). If a pronounced peak deviator stress is not evident before 20% strain is reached, stop the test and treat it as a single-stage test.

Increase the cell pressure to the second value without stopping the machine and continue taking readings. Record the point at which the cell pressure is increased.

When the next maximum deviator stress is indicated, repeat the above procedure using the third value of cell pressure. Continue this stage until a peak deviator stress is clearly defined and the stress is decreasing, if possible; otherwise terminate the test at 20% strain.

When the stage is completed, remove the axial force on the sample and lower the machine platen so that the top cap is clear of the piston. Start the machine again in the upward direction, at the same speed as during the test, and record the reading of the load ring when it becomes steady. Repeat this operation after reducing the cell pressure to each of the other pressures used for the test. These readings give the initial load reading R_0 (taking account of cell pressure on the piston, and piston friction) for each stage.

A multi-stage test usually comprises three stages, as described above, but in some instances only two stages may be practicable, and in others it may be possible to extend the test to four stages.

If the soil behaves in a plastic manner and a clearly defined maximum stress is not indicated, an arbitrary procedure (not given in BS 1377) is to terminate each stage at the following values of strain:

Stage A: 16%
Stage B: 18%
Stage C: 20%

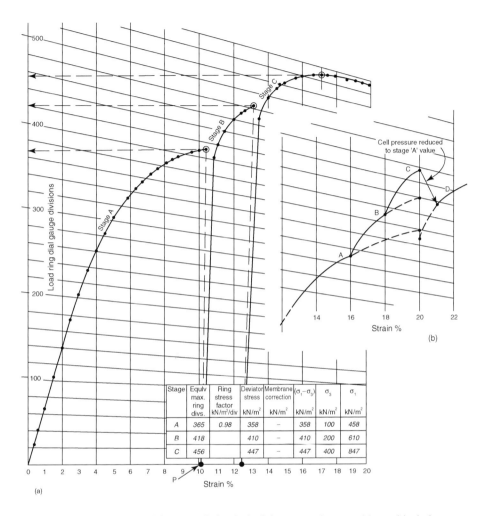

The table within the figure:

Stage	Equiv max. ring divs.	Ring stress factor kN/m²/div	Deviator stress kN/m²	Membrane correction kN/m²	$(\sigma_1-\sigma_3)$ kN/m²	σ_3 kN/m²	σ_1 kN/m²
A	365	0.98	358	–	358	100	458
B	418		410	–	410	200	610
C	456		447	–	447	400	847

Figure 13.45 Results from a multistage undrained triaxial compression test: (a) graphical plot on area correction grid; (b) extrapolation of curves to 20% strain for plastic deformation

Anderson (1974) recommends that the deviator stress–strain curves for stages A and B should be extrapolated to a strain of 20%, as shown in Figure 13.45(b). By carrying out a fourth stage (D) after reducing the cell pressure to the initial (stage A) value, he found that the deviator stress at 20% strain obtained by producing curve D backwards (Figure 13.45(b)), was in good agreement with that obtained from the first stage, for many 'boulder clay' samples.

13.6.6 Tests using 'free' ends (lubricated Ends)

In the conventional method of mounting triaxial specimens described in Sections 13.6.3 and 13.6.4, frictional or adhesion forces between the loading caps and the specimen inevitably restricts free lateral movement of the specimen ends (Bishop and Green, 1965). This results in the formation of 'dead zones' adjacent to the platens (see Figure 13.46(a)), and the familiar barrelling effect (see Figure 13.46(b)) which occurs in plastic soils. The soil is unrestrained only within the middle third of a specimen of 2:1 height:diameter ratio, which is why a smaller ratio is not normally used.

The restraining effects at the ends can be reduced considerably by a simple method described by Rowe and Barden (1964). Special end caps are used, of a diameter slightly larger than the specimen diameter, and made of stainless steel with a highly polished surface. Between each end cap and the specimen, two discs of rubber membrane material of the same diameter as the specimen are inserted, separated from each other and from the end caps by layers of silicone grease (see Figure 13.46(c)). This arrangement is not claimed to eliminate friction altogether, but the end friction is so small that they are referred to as 'free' ends, or lubricated ends.

When subjected to compression, specimens with 'free' ends maintain an approximate cylindrical shape (see Figure 13.46(d)), instead of barrelling, resulting in a more uniform stress distribution. This applies to specimens of height:diameter ratios of less than 2:1, as well as to conventional specimens, making it practicable to test specimens of a ratio of 1:1. Testing of 100 mm diameter specimens of 100 mm length is practicable, so that three separate specimens can be obtained from a U-100 undisturbed sample. Triaxial tests on specimens 254 mm diameter and 254 mm height using 'free' ends were referred to by Rowe (1972). The use of 'free' ends is also advantageous for multi-stage tests (Section 13.6.5) whether or not the height:diameter ratio is less than 2. The area correction is still valid.

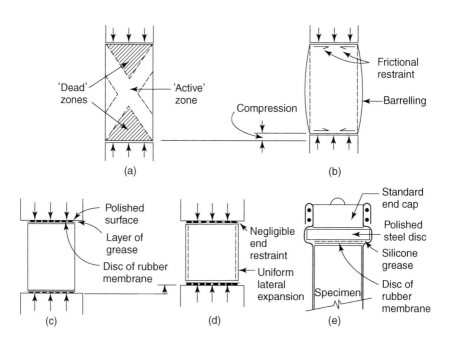

Figure 13.46 Lubricated ends for triaxial test specimens: (a) 'dead' zones in conventional test; (b) frictional restraint resulting in barreling; (c) provision of end cap lubrication; (d) resulting specimen deformation; (e) assembly using polished discs with standard end caps

It is not necessary to obtain special end platens if two discs of polished stainless steel are used, one at each end of the specimen, as shown in Figure 13.46(e). These are about 6 mm thick, and about 108 mm diameter for 100 mm specimens and 41 mm diameter for 38 mm specimens. The edges should be well rounded so as not to cut into the rubber membrane.

Apart from economy of soil samples, the main advantage of using 'free' ends is that the results obtained are more consistent than those obtained from conventional tests. However, to take advantage of the more uniform stress distribution, and to ensure equalisation of pore water pressures within the specimen, tests should be run more slowly than usual. For 100 mm diameter specimens of low permeability (clay) soils, a rate of strain of 2% per hour might be appropriate.

13.6.7 High pressure tests

Scope

Triaxial tests on specimens of relatively strong materials that are likely to be subjected to high stresses require cell pressures higher than those normally used for soils. Typical materials in this category are soft rocks, such as friable or weakly cemented sandstones, for which cell pressures up to 3.5 MPa or 7 MPa would be appropriate. Stronger materials would be treated as rocks for which special equipment using pressures up to 70 MPa are required, but tests of this kind are beyond the scope of this volume.

Equipment

To obtain and maintain constant pressures in the medium pressure range (i.e. 0.1–7 MPa) a hand-operated or motorised hydraulic pressure system designed for the purpose is needed.

Specially designed steel-bodied triaxial cells are required to withstand pressures of this magnitude. Cells with acrylic bodies should never be pressurised beyond the manufacturer's stated working capacity (typically 1 MPa for unbanded cells or 1.7 MPa for cells reinforced with nylon banding). Large cells for high pressures may incorporate a porthole for illuminating and observing the test specimen. Connecting hoses and pipework must be capable of withstanding the applied pressures.

A high capacity load frame and load measuring device are required. Correct alignment of the specimen with the piston, and on the axis of the load frame, is essential.

Test specimens

Test specimens should be carefully prepared with flat ends, which are at right angles to the axis. Specimens containing coarse grains or surface irregularities may cause piercing of the normal rubber membrane. This can be avoided by applying a thin coat of paraffin wax to the curved surface of the specimen followed by a layer of aluminium foil, before placing two rubber membranes separated by a layer of silicone grease.

Test procedure

Undrained tests on relatively hard materials should be run at a much slower rate of strain than that normally used for softer materials, because failure usually occurs at quite small strains. Instead of recording the load ring reading at regular intervals of strain, the strain dial reading should be recorded at suitable increments of the load ring reading, in order to provide the specified minimum of 15 sets of readings up to the point of failure. It is important to observe and record the highest reading reached by the load ring, together with the corresponding strain if possible. After that the load is likely to fall dramatically.

13.6.8 Specially orientated specimens

The standard method for obtaining a set of three triaxial specimens from a U-100 tube sample, described in Section 9.2.4, produces cylindrical specimens with their axes vertical, as in Figure 13.47(a), assuming that the U-100 sample was taken vertically. If the soil contains discontinuities or lithological features which are horizontal, or nearly horizontal, they are not likely to influence the measured strength because a surface of shear failure will cut across them at an angle of 45° or more to the horizontal (see Figure 13.47(b)). These discontinuities or other features are referred to below, for brevity, as laminations.

It may sometimes be desirable to induce failure to occur along one of these surfaces, to ascertain whether or not they represent planes of weakness. To do this a set of specimens of the form shown in Figure 13.48(b) is required, which need to be orientated in the U-100 sample in the manner shown in Figure 13.48(a). Clearly, it is impracticable to obtain such specimens by jacking directly into 38 mm tubes. It would be possible to extrude the whole sample and cut the specimens by hand or to push 38 mm tubes in by hand using a clinometer as a guide to the orientation, but the orientation cannot be controlled to any accuracy by either process.

The essential requirement is to obtain a set of specimens whose axes are inclined at a certain angle, θ to the axis of the tube. The value of θ depends upon the inclination δ of the laminations to the horizontal, and on the angle, α at which the laminations are to lie relative to the specimen axis, as shown in Figure 13.49. From the geometry of the arrangement it can be seen that

$$\theta + \alpha = (90° - \delta)$$
$$\text{i.e.} \quad \theta = 90° - (\alpha + \delta)$$

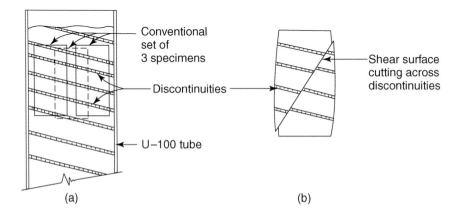

Figure 13.47 Triaxial specimens from soil containing 'laminations': (a) sample in U-100 tube from which conventional specimens are prepared; (b) test specimen after failure

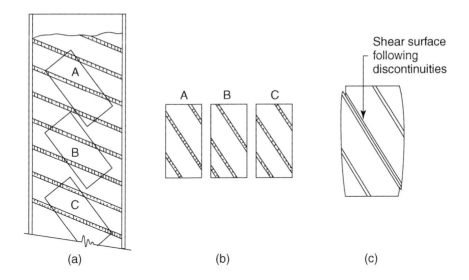

Figure 13.48 Triaxial specimens prepared to a specified orientation: (a) orientation in U-100 tube; (b) set of three specimens of required form; (c) failure on a discontinuity surface

Theoretically the angle α should be equal to $(45° - \phi/2)$, where ϕ is the angle of shear resistance of the soil. However, the value of α is not critical, and may be assumed to be between 27° and 30°. If it is less than $(\tan^{-1} 0.5)$, about 26.5°, the plane of shear surface would intersect one or both end caps. The angle δ can be measured to the nearest 1°, so the specimen orientation θ needs to be accurate to within 1° at least. An apparatus which was designed by the original author to achieve this is shown in use in Figure 13.50. The sample is extruded from the U-100 tube and laid with the laminations in the vertical plane in an alloy support of semicircular section mounted on a turntable. The mounting can slide horizontally in vee-groove bearings, and can be locked into position by thumbscrews. The turntable carries a scale graduated in degrees, which can be read to the nearest 0.5°. The turntable is mounted on a rigid base on which a rack-and-pinion operated specimen extruder is also mounted. Thin-walled 38 mm diameter tubes are rigidly fixed to the back end of the extruder ram.

The turntable carrying the sample can be rotated to any angle between 0° and 90°, and is locked in the position corresponding to the angle θ. The tube is then pushed steadily into the sample, and is withdrawn. The sample carrier is moved horizontally by the required distance to allow a second tube to be inserted, and a second specimen to be taken, maintaining the same orientation. Several specimens inclined at exactly the same angle to the axis of the sampling tube can thus be obtained, to an accuracy of within 1°. The influence of specimen orientation has been discussed by Bishop and Little (1967).

13.6.9 Reconstituted specimens

Compaction

Specimens of recompacted soil for compression tests may be prepared by applying the standard compaction procedures described in Volume 1 (third edition), Chapter 6 (Section 6.5).

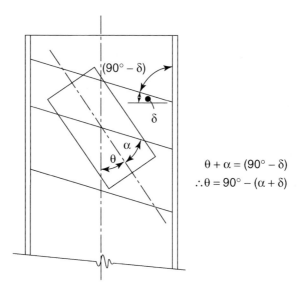

$$\theta + \alpha = (90° - \delta)$$
$$\therefore \theta = 90° - (\alpha + \delta)$$

Figure 13.49 Geometry of orientated specimen in tube

Figure 13.50 Apparatus designed by author for preparing specimens to a required orientation (photograph courtesy of Soil Mechanics Ltd)

Usually it is required to prepare specimens either at a specified dry density or by applying a specified compactive effort. Methods of specimen preparation are given in Sections 9.5.2, 9.5.4 and 9.5.5 for 38 mm diameter specimens, and Section 9.5.6 for larger specimens.

Procedures for testing recompacted specimens, including calculation, plotting and presentation of results, are the same as for similar tests on undisturbed specimens. The reported test data should include details of the procedure used for preparing the specimens, and the relationship of their moisture content and dry density to the optimum conditions defined by the relevant compaction curve.

Compaction procedures referred to above apply mainly to cohesive soils and to partially saturated cohesionless soil (such as 'damp' sand). Preparation of specimens of dry and fully saturated cohesionless soils requires special procedures which are described below. For convenience the soil is referred to as sand, although it may also contain silt or gravel-size particles.

Saturated sand specimens

The following procedure for the preparation of a triaxial specimen 38 mm diameter of saturated sand is based on that given by Bishop and Henkel (1962).

A triaxial cell with a pore water pressure connection fitted to the base pedestal is required. The outlet on the cell base is connected to a burette by about 1.2 m length of flexible tubing. The burette and tubing are filled with de-aired water, without entrapping any air, and by raising the level of the burette the connection to the pedestal is also filled with de-aired water, displacing all the air. The hole in the pedestal is covered by a porous stone which has been saturated by boiling in water, and the burette is adjusted to the equilibrium position as shown in Figure 13.51.

A rubber membrane is sealed to the cell pedestal by two O-rings, and a split former incorporating a recess for the rings is assembled around it and clamped in position after first applying a thin film of grease to the mating surfaces. The top end of the membrane is fitted around a metal ring, secured there by two O-rings, and supported by a clamp held by a burette stand. A rubber bung and funnel are fitted into the top of the ring, as shown in Figure 13.52. Water is poured in slowly through the funnel, thereby pressing the membrane against the side of the former. Entrapped pockets of air must be avoided. Water is added until the funnel is about half-filled, and the rubber stopper on the end of a glass rod is inserted in the mouth of the funnel. If the cell pedestal connection is fitted with a valve, it should be closed.

A quantity of dry sand slightly in excess of that required to fill the specimen former is weighed out and mixed in a beaker with enough water to just cover the sand, and the mixture is boiled to remove air. The sand is transferred to the funnel with a spoon.

The specimen is formed by displacing the stopper so that the sand flows steadily, but fairly rapidly, into the former. A steady rate of flow should minimise segregation, and the rate should be standardised if several identical specimens are to be prepared. This procedure will result in a specimen of low density (high porosity), which will not be maintained if the specimen is subjected to disturbance such as by vibration or knocking against it, therefore the subsequent setting-up operations must be performed with extreme care. A higher density (lower porosity) can be obtained by tamping or by light vibration using an electric engraving tool referred to in Section 9.1.2 (see Figure 9.14).

Excess water is siphoned from the funnel, enabling the funnel, bung and upper ring to be removed. Surplus sand not used in the specimen is dried and weighed, so that the actual mass of sand in the specimen can be determined. The top of the specimen is carefully levelled, the loading cap is placed in position and the rubber membrane is sealed on to it with two O-rings. The cap should be slightly smaller in diameter than the former, so that downward movement is not restricted if the specimen should consolidate.

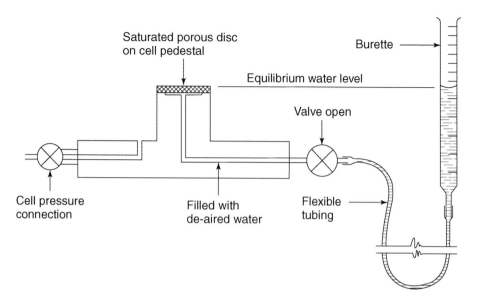

Figure 13.51 Triaxial cell base fitted with burette for preparing specimen of saturated sand

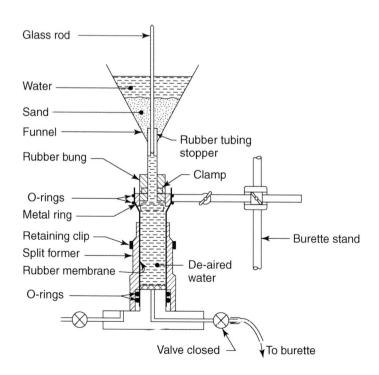

Figure 13.52 Preparation of triaxial specimen of saturated sand (after Bishop and Henkel, 1962)

Saturated sand possesses no cohesion, and to enable the specimen to support itself a small negative pore pressure has to be applied. This is done by lowering the burette, so that the free water level is below the base of the specimen (see Figure 13.53), with the valve on the cell pedestal open. The suction required depends upon the size and density of the specimen. The difference in level, d, (see Figure 13.53) may need to be only about 200 mm for a 38 mm diameter specimen, but 500 mm or more for a 100 mm diameter and 200 mm high specimen. When the valve on the cell base is opened, consolidation of the specimen occurs almost at once and is indicated by a small rise in water level in the burette.

The split mould can then be removed and the consolidated height and diameter of the specimen are measured, using vernier callipers and steel rule, allowing for the thickness of the rubber membrane. This requires extreme care if the specimen is in a loose state. The consolidated density and porosity are calculated. If a number of specimens are to be prepared at the same density, the suction procedure described above should be standardised.

The cell body is carefully fitted into place and tightened down, and the cell is filled with water and pressurised. Any additional volume change resulting from consolidation under the confining pressure can be measured by the change in water level in the burette. The burette is then placed alongside the cell with the water level at the mid-height of the specimen.

For this type of specimen a drained test would normally be carried out by applying the axial load slowly so that failure takes place after about one hour. During the test the drainage valve remains open and the burette reading is recorded along with load and strain readings. The burette should be adjusted if necessary to maintain the water level always at about the mid-height of the specimen. (Drained tests will be covered in greater detail in Volume 3.)

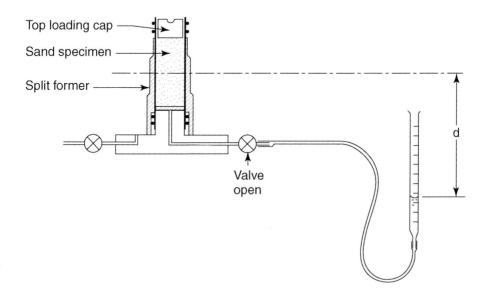

Figure 13.53 Application of suction to saturated sand specimen

By closing the drainage valve before applying the cell pressure, an undrained test can be carried out at the 'quick' rate of strain. This can be used to demonstrate that it is possible to obtain a value of ϕ close to zero in sands, under undrained conditions.

Dry sand specimens

The following procedure is based on that described by Bishop and Henkel (1962) for the preparation of triaxial specimens of dry sand and of other dry materials such as grain and sugar.

A split former fitted with a vacuum connection, enclosing a rubber membrane, is clamped to the pedestal of the triaxial cell as described above, except that a dry porous stone is used. Two vacuum lines are required, one for connection to the split former for holding the membrane in contact with its inner wall, the other (which is capable of being controlled to give a very low suction) for connection to the pore pressure outlet on the triaxial cell. Vacuum for the specimen can be obtained from a water filter pump, or if a vacuum line is used by incorporating an air bleed which can be easily regulated. A vacuum gauge, or a water or mercury manometer, should be incorporated so that the degree of vacuum can be closely controlled.

The specimen is formed by pouring a weighed quantity of sand into the mould from a funnel, fitted with a length of rubber tubing (see Figure 13.54), while applying a vacuum to the split former. To obtain a 'loose' specimen of low density (high porosity), continuous rapid pouring from a small drop, which should be kept constant by steadily raising the funnel, should be used. A loose specimen should not be subjected to shock or vibration. A higher density (lower porosity) may be obtained by pouring at a slower rate from a higher drop (Kolbuszewski, 1948). Alternatively, the specimen may be vibrated, using the tool referred to above, or tamped in layers, taking care not to damage the rubber membrane.

The top surface of the specimen is carefully levelled, the top loading cap is placed in position and the membrane is sealed on to it using two O-rings. A small suction (only about 2–5 kPa below atmospheric pressure, say 200–500 mm of water, or 15–40 mm of mercury) is applied to the base of the specimen to give it sufficient strength to stand while the split former is removed. The specimen is carefully measured and the cell body is fitted, filled with water, and pressurised as described above.

The vacuum line is removed and the pressure in the specimen is restored to atmospheric before proceeding with a quick compression test, keeping the drainage valve open.

13.7 Triaxial test equipment

13.7.1 General items

Triaxial tests require the use of numerous items of equipment which also have a more general application. These items were described in Chapters 8 and 9, in the following sections:

Load frames (see Section 8.2.3)
Constant pressure systems (see Section 8.2.4)
Load rings (see Sections 8.2.1 and 8.3.3)
Pressure gauges (see Sections 8.2.1 and 8.3.4)
Dial gauges (see Sections 8.2.1 and 8.3.2)
Specimen preparation equipment (see Section 9.1.2)

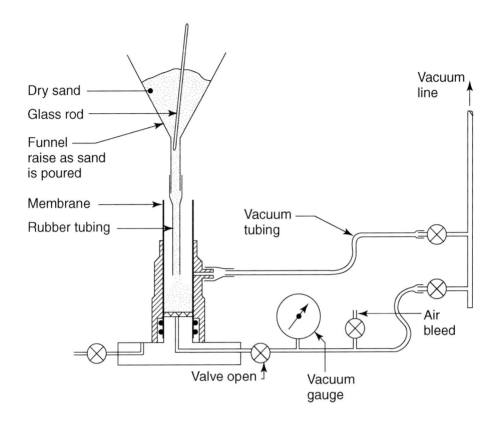

Figure 13.54 Preparation of triaxial specimen of dry sand

The sections which follow give details of equipment required specifically for triaxial tests, and include the care of triaxial cells and the calibration of rubber membranes.

13.7.2 Triaxial cells

Triaxial cells are available in several sizes, each of which can accommodate several different specimen diameters by means of interchangeable base and top cap fittings. Typical cell sizes are given in Table 13.5, and a corresponding range of cells is shown in Figure 13.55. A cell for specimens up to 50 mm diameter is shown in Figure 13.56.

Cells are made of a corrosion-resistant metal, with an acrylic plastic transparent cylindrical body. In addition to the standard cells which are designed for pressures up to 1000 kPa, cells reinforced with bonded fibreglass, or with thicker walls, are available for withstanding pressures up to 1700 kPa. Steel cells are also manufactured for tests at higher pressures (up to 7 MN/m²), and for testing rocks (up to 70 MN/m²). The stated working pressures of a cell must never be exceeded and only water should be used as the pressurising fluid. It is dangerous to pressurise triaxial cells with air or other gases. The maximum piston load as stated by the manufacturer should never be exceeded, otherwise the cell is likely to become distorted. Incorrect alignment can also cause distortion, even under a moderate load.

Cells should be used under stable ambient conditions at normal temperatures. Excessively high or low operating temperatures may cause leakage past the piston, or increased friction between the piston and its bush.

Table 13.5 Typical sizes of triaxial cells

Type of cell	Specimen diameters		Typical maximum piston load (kN)
	(mm)	*(in)*	
Small	35, 38, 50	1.5, 2	13.5
Intermediate	35, 38, 50, 70	2.8	29
100 mm	100	4	45
Large	150	6	82

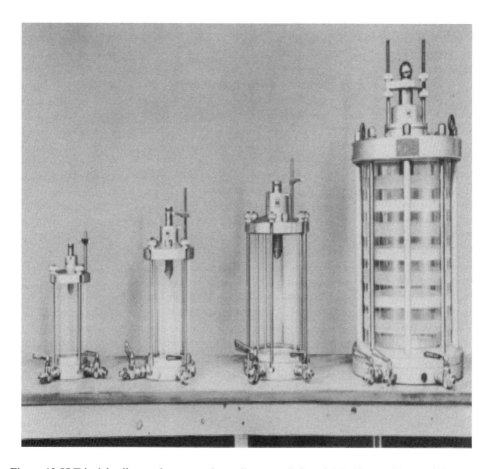

Figure 13.55 Triaxial cells; maximum specimen diameters (left to right): 50 mm, 70 mm, 100 mm, 150 mm (see Table 13.6)

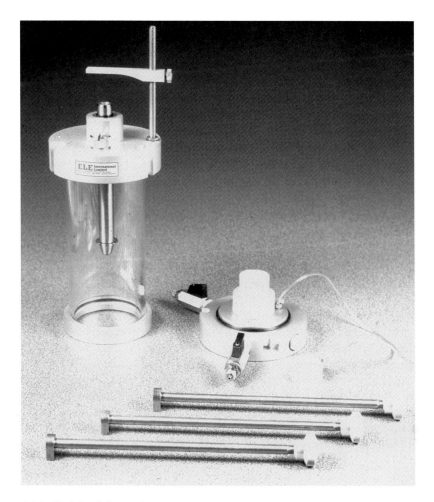

Figure 13.56 Triaxial cell for specimens up to 50 mm diameter

The piston and bush are manufactured to match each other and are ground, honed and lapped to a very close tolerance (within 0.002 mm). This provides the necessary watertight yet almost frictionless fit. Pistons should be kept dry and free from grease and dust, but the manufacturer's recommended silicone grease lubricant should be applied occasionally and sparingly to the nipple on the cell bush, using a grease gun. A well-fitting piston should allow no water to escape when the cell is pressurised, yet should fall slowly under its own weight when the cell is empty. When not in use the piston and bush should be protected from dust by covering with a small polythene bag, and the piston should be wiped dry with a clean cloth. Dust and dirt will rapidly cause scoring of the piston and bush, causing leakages or binding or both.

Some older cells have an annular recess around the bush, for collecting any slight leakages which may occur during a test. The water can be led away to a beaker by fitting a length of rubber tubing to the outlet if provided. For tests of long duration a layer of oil can be inserted to float on top of the water in the cell, and this both reduces leakage and acts as a piston lubricant.

A sticking piston must *never* be gripped with pliers or a similar tool. If a piston becomes seized in the bush, gentle heat (hot water) applied to the cell top while keeping the piston cool may free it. If the piston cannot be moved the whole cell top should be sent to a precision workshop or returned to the manufacturer for attention.

The cell base and base adaptors should be carefully cleaned before assembly. In particular, the O-ring seals must be free from dust and dirt, undamaged, and correctly seated. They should be given a light smear of silicone grease immediately before fitting into place, but they must not be allowed to pick up any dirt. When placing a cell on its base, ensure that any index marks correctly coincide. Screws or tie-rods should be tightened progressively, lightly at first, then by gradually tightening opposite pairs a little at a time. Ensure that the tie-rods are vertical and nuts properly seated before final tightening. Wing nuts or knurled nuts should be tightened only by hand, without the use of a spanner or other tool.

13.7.3 Triaxial cell fittings and accessories

A base adaptor and top cap of the appropriate diameter are required for each size of triaxial specimen, which are listed in Table 13.2. Base adaptors fit on the cell pedestal and are generally suitable only for the type of cell for which they are designed. Adaptors are either solid (as used for quick tests) or perforated with a small diameter central hole (for drained tests or pore water measurements). A solid plug can be fitted over a perforated adaptor when a drainage connection is not required.

Top caps, otherwise known as loading caps or pressure pads, are also either solid, or perforated with an eccentric small diameter hole to which a drainage connection can be fitted. Top caps for smaller specimen diameters are usually of acrylic plastic and are bonded to a hemispherical stainless steel ball which fits the recess in the end of the cell loading piston. Larger diameter caps are usually of aluminium alloy and may be formed with a central recess into which a steel ball bearing is placed to transmit the load from the piston.

Other accessories which are required for QU triaxial tests and which must conform to the specimen diameter, are

Split formers (see Sections 9.1.2 and 13.6.9)
Suction membrane stretchers (see Section 13.6.3, stage 6)
Rubber membranes (see Section 13.7.4)
O-rings (see Section 13.7.4)
O-ring placing tool

Base adaptors, top caps and other accessories for four different specimen diameters are shown in Figure 13.57.

13.7.4 Membranes and sealing rings

Membranes

Membranes of latex rubber for triaxial test specimens are supplied in the standard sizes shown in Table 13.6. The unstretched internal diameter should be not less than 90% of the specimen diameter and not greater than the specimen diameter. Membranes should be long enough to cover the specimen and end caps, and their thickness should not exceed 1% of the specimen diameter. For specimens up to 50 mm diameter, membranes of 0.2 mm

Figure 13.57 Triaxial test accessories for specimens of (left to right) 38, 50, 100 and 150 mm diameter: suction membrane stretchers, rubber membranes (150 mm membrane is fitted on stretcher), O-rings, base adaptors, top caps

thickness are suitable. Two or more membranes, separated by silicone grease, may be fitted to specimens containing angular particles.

Ideally, a fresh membrane should be used for each test specimen. However for quick triaxial tests a membrane may be used a second time provided that it is carefully examined before re-use. Flaws or pinholes show up if the membrane is held against the light while stretching it first lengthways and then widthways. If defects are visible the membrane should be thrown away immediately, or cut along its length if the material can be used for other purposes such as for 'free ends'.

Membranes suitable for re-use should be washed carefully in clean water and hung up to dry on a rack made of lengths of wood dowelling projecting from a vertical board. When completely dry they should be lightly dusted inside and out with French chalk and stored in a cool dark place.

Table 13.6 Rubber membrane sizes

Specimen diameter		Membrane length
(mm)	*(in)*	*(approximate)* × *thickness (mm)*
35		
38	1.5	150 × 0.3
50	2	200 × 0.4
70	2.8	250 × 0.4
100	4	330 × 0.5
150	6	510 × 0.5

Calibration of membranes

The extension modulus of the rubber membrane material, which is required for calculating the membrane correction to be applied to the measured compressive strength of triaxial specimens (see Section 13.3.8), can be obtained as follows (Bishop and Henkel, 1962).

Cut a length of 25 mm from a rubber membrane of the size to be used. Set up the apparatus shown diagrammatically in Figure 13.58, after dusting the inside of the membrane and the glass rods with French chalk. The original author's apparatus in use is shown in Figure 13.59. Add weights to the suspended pan in increments and measure the resulting extended lengths of the membrane, denoted by x in Figure 13.58. This can be done to the nearest 0.5 mm by using callipers and a steel ruler to measure the distance from the upper and lower edges of the glass rods at both sides. A graph of pan mass against distance x can be plotted, as in Figure 13.60, from which the load giving say 15% strain can be read off. This enables the modulus M of the rubber membrane to be obtained, from which a membrane correction curve of the type shown in Figure 13.17 can be derived as described in Section 13.3.8.

Sealing rings

Rubber O-rings are required for making a watertight seal between the rubber membrane and the top cap and base adaptor. The unstretched diameter of the O-rings should be between 80%

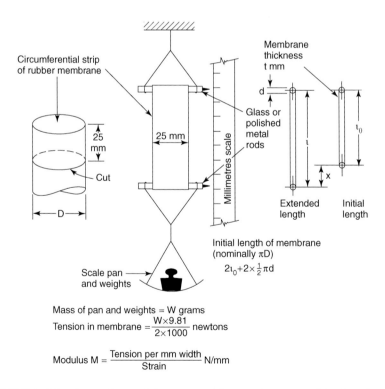

Mass of pan and weights = W grams

$$\text{Tension in membrane} = \frac{W \times 9.81}{2 \times 1000} \text{ newtons}$$

$$\text{Modulus } M = \frac{\text{Tension per mm width}}{\text{Strain}} \text{ N/mm}$$

Figure 13.58 Principle of extension modulus test on rubber membrane material (after Bishop and Henkel, 1962)

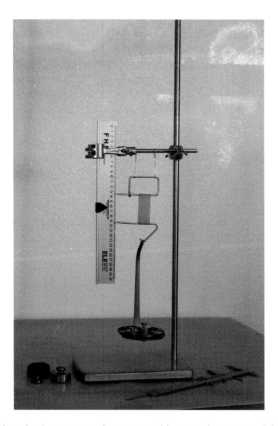

Figure 13.59 Original author's apparatus for test on rubber membrane material

and 90% of the specimen diameter, and they should be free from flaws and should show no necking when stretched. One ring is normally used at each end of the specimen, but for tests of long duration, two at each end are preferable. The rings must be of the correct size and should be examined before use to ensure that they are free from defects such as necks or cuts. Fitting O-rings in place is made easier if they are first fitted around the membrane stretcher, and then rolled off into place. A thin film of silicone grease between cap and membrane helps to ensure a watertight seal.

References

ASTM D 2166-06 Standard Test method for unconfined compressive strength of cohesive soil. American Society for Testing and Materials, Philadelphia, PA, USA

ASTM D 2850-AR07 Standard Test method for unconsolidated, undrained compressive strength of cohesive soils in triaxial compression. ASTM, Philadelphia, PA, USA

Anderson, W. F. (1974) The use of multi-stage triaxial tests to find the undrained strength parameters of stony boulder clay. *Proc. Inst. Civ. Eng.*, Technical Note No. TN89

Bishop, A. W. and Henkel, D. J. (1962) *The Measurement of Soil Properties in the Triaxial Test* (second edition). Edward Arnold, London

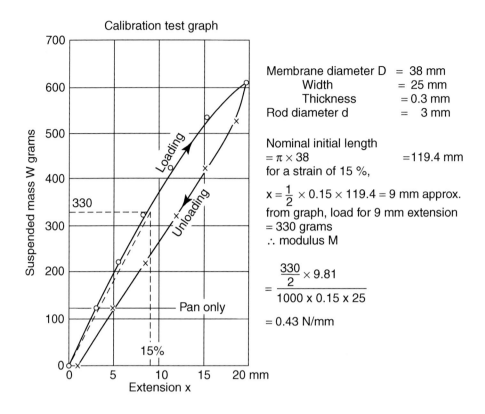

Figure 13.60 Typical data from extension modulus test on rubber membrane

Bishop, A. W. and Green, G. E. (1965) The influence of end restraint on the compression strength of a cohesionless soil. *Géotechnique*, Vol. 15, No. 3

Bishop, A. W. and Little, A. L. (1967) The influence of the size and orientation of the sample on the apparent strength of the London clay at Maldon, Essex. *Proc. Geotech. Conf.*, Oslo, Vol. 2, pp. 89–96

BS EN ISO 7500-1:1998, Metallic materials verification of static uniaxial testing machines. Tension/compression testing machines. Verification of the force measuring system. British Standards Institution, London

Case, J. and Chilver, A. H. (1971) *Strength of Materials and Structures*. Edward Arnold, London

Cooling, L. F. and Golder, H. Q. (1940) Portable apparatus for compression tests in clay soils. *Engineering*, Vol., 149 (3862), pp. 57–58

Cooling, L. F. and Smith, D. B. (1936) The shearing resistance of soils. *Proc. 1st Int. Conf. Soil Mech. and Found. Eng.*, Vol. 1. Harvard, MA, USA

Henkel, D. J. and Gilbert, G. D. (1952) The effect of the rubber membrane on the measured triaxial compression strength of clay samples. *Géotechnique*, Vol. 3, No. 1

Kenney, T. C. and Watson, G. H. (1961) Multiple-stage triaxial tests for determining *c'*and *φ'* of saturated soils'. *Proc. 5th Int. Conf. Soil Mech.*, Paris, Vol. 1

Kolbuszewski, J. (1948) An experimental study of the maximum and minimum porosities of sands. *Proc. 2nd Int. Conf. Soil Mech. and Found. Eng.*, Rotterdam, Vol. 1

Lumb, P. (1964) Multi-stage triaxial tests on undisturbed soils. *Civ. Eng. and Public Works Review*, May 1964

Marschi, N. D., Chan, C. K. and Seed, H. B. (1972) Evaluation of properties of rockfill materials. *J. Soil Mech. Found. Div. ASCE*, Vol. 98, Paper No. 8672

Mohr, O. (1871) Beiträge zur Theorie des Erddruckes *Z. Arch. u. Inng.* ver. Hannover, Vols. 17 and 18

Rowe, P. W. (1972) The relevance of soil fabric to site investigation practice'. 12th Rankine Lecture, *Géotechnique*, Vol. 22, No. 2

Rowe, P. W. and Barden, L. (1964) Importance of free ends in triaxial testing. *J. Soil. Mech. Found. Div. ASCE*, Vol. 90, SMI, January, 1964

Skempton, A. W. and Bishop, A. W. (1954) Soils. Chapter X of *Building Materials — their Elasticity and Inelasticity* (Reiner, M. and Ward, A. G. eds.). North Holland Publishing Company, Amsterdam

Skempton, A. W. and Henkel, D. J. (1957) Tests on London Clay from deep boring at Paddington, Victoria and the South Bank. *Proc. 4th Inst. Conf. Soil Mech. and Found. Eng.*, Vol. 1, pp. 100–106. London

Skempton, A. W. and La Rochelle, P. (1965) The Bradwell slip: a short-term failure in London clay. *Géotechnique*, Vol. 15, No. 3

Terzaghi, K. and Peck, R. B. (1967) *Soil Mechanics in Engineering Practice*. Wiley, New York (currently available as Terzaghi, K., Peck, R. B. and Mesri, G. 1996, *Soil Mechanics in Engineering Practice* (third edition). Wiley, New York)

Transport and Road Research Laboratory (1952) *Soil Mechanics for Road Engineers*. Chapters 19, 22, HMSO, London

Whitlow, R. (1973). *Materials and Structures*. Longmans, London

Wood, D. M. and Wroth, C. P. (1976) The correlation of some basic engineering properties of soils. *Proc. Int. Conf. on Behaviour of Offshore Structures*, Trondheim, Vol. 2

Wood, D. M. and Wroth, C. P. (1978) The use of the cone penetrometer to determine the plastic limit of soils. *Ground Engineering*, Vol. 11, No. 3

Chapter 14

Oedometer consolidation tests

14.1 Introduction

14.1.1 Scope

The standard oedometer consolidation test for saturated clays is the main feature of this chapter. Characteristics of both normally consolidated and overconsolidated clays are described. Analysis of data from tests on clays is presented as a standard conventional procedure, variations on which are described for application to tests on silty soils. Special test procedures are described separately for soils having a swelling potential, for partially saturated soils and for peats. Tests for the direct measurement of permeability in the consolidation cell are included.

As with other load testing devices, calibration of the consolidation loading frame is important and this is covered together with other practical aspects of the apparatus and testing procedure.

14.1.2 Purpose

The oedometer consolidation test is used for the determination of the consolidation characteristics of soils of low permeability. The two parameters normally required are

- The compressibility of the soil (expressed in terms of the coefficient of volume compressibility; also known as modulus of volume change), which is a measure of the *amount* by which the soil will compress when loaded and allowed to consolidate.
- The time related parameter (expressed in terms of the coefficient of consolidation) which indicates the *rate* of compression and hence the *time period* over which consolidation settlement will take place.

1. *Compressibility*

Whenever a load, such as that due to a structural foundation, is placed on the ground, some degree of settlement will occur even if the applied pressure is well within the safe bearing capacity of the soil. The limitation of settlements to within tolerable limits is sometimes of greater significance in foundation design than limitations imposed by bearing capacity requirements derived from shear strength.

2. *Time effects*

Settlements in sands and gravels take place in a short time, usually as construction proceeds, and these rarely cause major problems. But in clay soils, because of their low permeability, settlements can take place over much longer periods, which may be months, years, decades, even centuries, after completion of construction. Estimates of the rate of settlement, and of the time within which settlement will be virtually complete, are therefore important factors in foundation design.

14.1.3 Principle of test

The test is carried out by applying a sequence of some four to eight vertical loads to a laterally confined specimen having a height of about one-quarter of its diameter. The vertical compression under each load is observed over a period of time, usually up to 24 hours. Since no lateral deformation is allowed it is a one-dimensional test, from which the one-dimensional consolidation parameters are derived.

The consolidation cell consists essentially of a mould for containing and rigidly supporting the test specimen; an upper and lower drainage surface; a loading cap; and an outer casing containing water in which the whole can be immersed. Details of the cell, and the loading frame in which it is mounted, are given in Section 14.5.3.

14.1.4 Historical development

Attention was first drawn to the problem of the long-term consolidation of clays by Terzaghi (1925), with the publication in Vienna of *Erdbaumechanik*. Terzaghi proposed a theoretical approach to the consolidation process, and he had already designed the first consolidation apparatus which he named an 'oedometer' (from the Greek *oidema*, swelling). In the early 1930s, consolidation tests on specimens of various sizes were carried out in the USA and were reported by Casagrande (1932), Gilboy (1936), and Rutledge (1935). The mathematical theory of consolidation was published by Terzaghi and Fröhlich in 1936.

In 1938, Skempton at Imperial College, London, developed an oedometer for a 1 in thick specimen based on the Casagrande principle, using a bicycle wheel to support the beam counterbalance weight. A more compact oedometer, for a specimen 3 in diameter and 0.75 in high, was designed by Nixon in 1945, and four of these were mounted on one bench. Other machines, based on the same principle, were developed by the leading manufacturers of testing equipment, and many are still in use today.

When oedometer consolidation testing became recognised as a standard laboratory procedure after 1945, two types of oedometer cell were developed, known as the fixed-ring cell and the floating-ring cell. In the fixed-ring cell, the mould into which the specimen was transferred from the cutting ring was clamped in the cell on top of a lower porous disc having a diameter larger than that of the specimen (see Figure 14.1). The upper porous disc, immediately beneath the loading cap, was fractionally smaller so that it could enter the ring as the specimen consolidated. Only the top surface of the specimen was displaced during consolidation.

In the floating-ring cell the ring into which the specimen was initially trimmed was used to hold the specimen during the test, but it was supported only by friction from the specimen itself (see Figure 14.2). Both the upper and lower porous discs were slightly smaller than the inside diameter of the ring, so that the specimen was compressed about equally from top and bottom, and it was claimed that the amount of side friction was half that of a fixed-ring cell (Lambe, 1951). The floating-ring cell was cheaper than the fixed-ring type, and another advantage was that specimen disturbance was less because transfer from ring to mould was eliminated. Its disadvantages were that only a light ring could be used to hold the specimen,

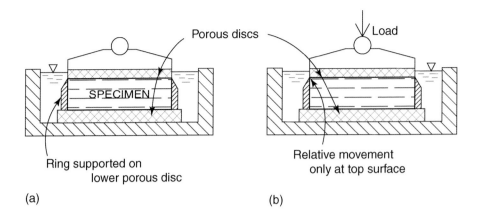

Figure 14.1 Principle of fixed ring oedometer consolidation cell: (a) initially; (b) after consolidation

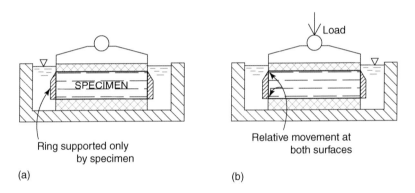

Figure 14.2 Principle of floating ring oedometer consolidation cell: (a) initially; (b) after consolidation

and therefore it was likely to undergo some lateral deformation under high pressure; and that the weight of the ring could cause some disturbance to soft clays. In addition, this type of cell could not be adapted to make direct measurements of permeability.

The fixed-ring cell is the type now most often used and is standard in the UK. The specimen is held in the cutting ring, which is accurately located and rigidly restrained by a retainer or the cell body, which avoids damaging the cutting edge (see Figure 14.3). Setting up and dismantling are simple operations which entail little risk of disturbing the specimen. Provision of 'O' ring seals enables direct permeability measurements to be made while the specimen is under load during the test (see Figure 14.46, Section 14.6.6). The suggestion by Lambe (1951), that applied loads should be increased by 10% to allow for side friction, is not taken into account in British practice, but friction is minimised by using a smooth polished ring with light lubrication.

Oedometer presses designed for working in SI units were introduced in 1971 and the normal British oedometer uses a specimen 75 mm diameter and 20 mm high. A suggested

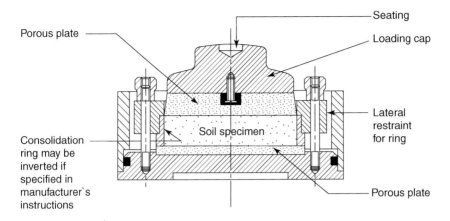

Porous plate

Seating

Loading cap

Consolidation ring may be inverted if specified in manufacturer's instructions

Soil specimen

Lateral restraint for ring

Porous plate

Figure 14.3 Details of a typical oedometer consolidation cell (reproduced from Figure 1(a) of BS 1377:Part 5:1990)

way of using earlier oedometers, which were designed for traditional (imperial) units, is given in Section 14.8.4.

The major advances in recent years have been based on the use of displacement transducers for the measurement of vertical compression of specimens, so that automatic recording (day and night), data processing, and automatic printing and graphical plotting systems are now commercially available (see Section 8.2.6). However, the procedures given in this chapter relate to manual operation and recording and it is good practice to provide a manual backup to electronic instrumentation.

14.2 Definitions

Consolidation The process whereby soil particles are packed more closely together over a *period* of time under the application of continued pressure. It is accompanied by drainage of water from the pore spaces (voids) between solid particles.

Voids ratio (e) The ratio of the volume of voids (water and air) to the volume of solid particles in a mass of soil.

Degree of saturation (S_r) The volume of water contained in the void space between soil particles expressed as a percentage of the total voids

$$S_r = \frac{w\rho_s}{e}\%$$

where w is the soil moisture content (%); ρ_s is the particle density (Mg/m³); and e is the voids ratio.

Pore water pressure (u) The hydrostatic pressure of the water in the voids, or pores, between solid particles. Also referred to as pore pressure, or the neutral stress; abbreviated as p.w.p.

Total stress (σ) The actual stress in a soil mass due to the application of an applied external pressure or force.

Effective stress (σ') The difference between the total stress and the pore water pressure

$$\sigma' = \sigma - u$$

Effective stress approximates to the stress carried by the solid soil structure.

Excess pore pressure The increase in pore water pressure due to the sudden application of an external pressure or stress. Also called the excess hydrostatic pressure.

Degree of consolidation (U) The ratio of the excess pore pressure lost after a certain time due to drainage, to the initial excess pore pressure, at any instant during the consolidation process. It is usually expressed as a percentage and is sometimes referred to as the percentage pore pressure dissipation.

$$U = \frac{u_1 - u_w}{u_1 - u_o} \times 100\%$$

where u_w is the pore pressure at the time considered; u_1 is the initial pore pressure; and u_o is the final equilibrium pore pressure when consolidation is complete.

Initial compression The amount of compression which takes place in a laboratory test between the instant of application of the load and the beginning of the primary consolidation phase.

Primary consolidation The part of the total compression under load to which the Terzaghi theory of consolidation applies. It is the phase during which drainage and pore pressure dissipation occur.

Secondary compression The compression which continues after primary consolidation has virtually finished and which is time-dependent.

Curve fitting The determination of the coefficient of consolidation by comparing a laboratory test curve with the characteristics of the theoretical curve.

Coefficient of compressibility (a_v) The change in voids ratio per unit pressure change as a result of consolidation due to that pressure change

$$a_v = -\frac{\delta e}{\delta p}$$

Coefficient of volume compressibilty (m_v) Sometimes known as modulus of volume change. The change in volume per unit volume, per unit pressure change, as a result of consolidation due to that pressure change

$$m_v = \frac{a_v}{1+e} = -\left(\frac{1}{1+e}\right)\frac{\delta e}{\delta p}$$

Coefficient of consolidation (c_v) The parameter which relates the change in excess pore pressure with respect to time, to the amount of water draining out of the voids of a clay prism during the same time, due to consolidation

$$c_v = \frac{k}{m_v \rho_w g}$$

Time factor (T_v) The dimensionless parameter which is related to time, t, the coefficient of consolidation, c_v, and the length of drainage path (as described in Section 14.3.4); used for defining the theoretical rate of consolidation curve.

$$T_v = \frac{c_v t}{H^2}$$

Coefficient of secondary compression (C_{sec}) The ratio of the change in height to the initial height of a consolidation specimen over one decade (one log cycle) of time during the secondary compression phase

$$C_{sec} = \frac{(\delta H)_s}{H_o} \text{ over one log cycle of time}$$
$$= \frac{1}{H_o} \times \frac{(\delta H)_s}{\delta (\log_{10} t)}$$

Virgin compression curve Also known as the field compression curve. The relationship between voids ratio and effective pressure for the soil in-situ.

Normally consolidated clay A clay which has never been subjected to a greater effective pressure than the present effective overburden pressure.

Overconsolidated clay A clay which in past geological times has been consolidated under an effective pressure greater than the present effective pressure, usually by overlying deposits which have since been eroded away.

Preconsolidation pressure The maximum pressure to which an overconsolidated clay was subjected.

Overconsolidation ratio (*OCR*) The ratio of the preconsolidation pressure to the present effective overburden pressure.

Swelling The process opposite to consolidation, i.e. the expansion of a soil on reduction of pressure due to water being drawn into the voids between solid particles.

Swelling pressure Also known as equilibrium load. The pressure required to maintain constant volume, i.e. to prevent swelling, when a soil has access to water.

Compression index (C_c) The numerical value of the slope of the curve relating effective normal stress (plotted to a logarithmic scale) against void ratio for primary consolidation

$$C_c = \frac{-\delta e}{\delta (\log_{10} \sigma')}$$

Swell index (C_s) The slope of the log effective normal stress–void ratio curve for swelling.

14.3 Consolidation theory

14.3.1 Principle of consolidation

Soils consist of solid particles between which are spaces (voids) which may be filled with a gas (usually air), a liquid (usually water), or a combination of both (see Volume 1 (third edition), Section 3.3.2.). The theory of consolidation applies to fully saturated soils, in which the voids contain water only.

When a soil is subjected to a compressive stress its volume tends to decrease, which for a saturated soil can take place by three means.

1. Compression of the solid grains
2. Compression of the water within the voids between grains
3. Escape of water from the voids

In most inorganic soils the effect of item 1 is extremely small, and is neglected in consolidation theory. For organic soils, especially peat, the compressibility of the solid matter can be considerable. (Peats are dealt with separately in Section 14.7.)

The compressibility of water is negligible in comparison with other effects, so item 2 can be ignored. Most sedimentary clay deposits are fully saturated or very nearly so, and it is in these soils that the process of consolidation is most significant. The presence of air in the voids is discounted; to allow for partial saturation would make the analysis much too complicated for practical use.

The theory of consolidation is therefore based on item 3, the escape or 'squeezing out' of water from the voids between the skeleton of the solid grains.

In a free-draining soil such as saturated sand the escape of water can take place rapidly. But in a clay, for which the permeability may range from tens of thousands to millions of times less than that of sand, the movement of water occurs very much more slowly, and therefore considerable time may be required for excess water to be squeezed out to permeable boundaries.

The volume change associated with consolidation occurs equally slowly, and the resulting settlement under load therefore takes place over a long time period. This process can be visualised by means of the mechanical model described below.

14.3.2 Spring and piston analogy

This simplified version of the model analogy described by Terzaghi and Peck (1948) is due to Taylor (1948).

Consider a cylindrical container fitted with a watertight but frictionless piston of negligible mass, of area A mm², and provided with a drainage valve connected to a small-bore outlet tube. The container is filled with water, and between the piston and the base is an elastic compression spring (see Figure 14.4(a)). Initially the system is in equilibrium with the valve closed and no load on the piston. The spring is not compressed and there is no excess pressure in the water.

A weight of 200 N is now applied to the piston (see Figure 14.4(b)). Water is not allowed to escape, so the piston cannot move down and the spring is not compressed. The downward force is therefore supported by an upward force on the piston due to an additional pressure in the water. This pressure, called the excess hydrostatic pressure, is equal to $200/A$ N/mm². At a certain instant (time = 0), the drainage valve is opened and the timer clock is started. Water can now begin to escape from the cylinder (see Figure 14.4(c)), but only slowly because of the small bore of the outlet tube. The piston sinks slowly, resulting in progressively more load being carried by the spring and less by the pressure of the water (see Figure 14.4(d)–(f)). Finally the spring is fully compressed by the applied force and carries the whole of the load. There is now no excess pressure in the water, equilibrium is restored and movement has ceased (see Figure 14.4(g)).

The loads carried by the spring and by the water at various time intervals from the start are shown in Figure 14.4(c)–(g), together with the percentage of the final compression of the spring, which is the same as the percentage of the final total load which it carries at any instant. When equilibrium is reached, as in Figure 14.4(g), compression is 100% complete.

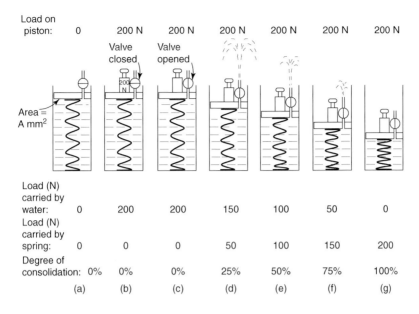

Figure 14.4 Spring and piston analogy illustrating the principle of consolidation (after Taylor, 1948)

In this model the time required to reach a given percentage compression depends on the following factors:

1.　Size of drainage outlet
2.　Viscosity of water (which depends on temperature)
3.　Compressibility of spring

Items 1 and 2 give the rate at which water can escape through the outlet. Item (3) is significant because a spring of greater compressibility would shorten more under the 200 N load and would require more water to escape, therefore a longer time would elapse before equilibrium was reached.

14.3.3 Consolidation of soils

The behaviour of the mechanical model described above is analogous to the behaviour of soils during the consolidation process. Those properties of the model and real soil which relate to each other are summarised in Table 14.1.

The stress induced by the externally applied load is known as the 'total stress' and is denoted by σ. The pressure in the water in the voids between solid particles in a soil is known as the 'pore water pressure' (p.w.p.), or pore pressure, and is denoted by u, or sometimes u_w. When an external load is applied to a saturated clay soil, the entire load is at first carried by the additional pore water pressure that is induced, referred to as the 'excess pore water pressure', which is equal to the total applied stress.

If the clay is bounded by surfaces from which water can escape (such as adjoining sand layers shown in Figure 14.5, Section 14.3.4) the excess pressure will cause water to flow out of the clay into the adjoining layers. This will occur slowly, because of the low permeability of the clay, but as water drains out an increasing proportion of the load is transferred to the

Table 14.1 Comparison of properties of mechanical model and soil

Item		Mechanical model	Soil
1		Rate at which water can escape depends on:	Rate of drainage depends on:
	(a)	Size of outlet	Size of pore spaces (i.e. permeability)
	(b)	Viscosity of water	Viscosity of pore water (depends on temperature)
	(c)	Length of outlet tube	Length of drainage path
2		Compressibility of spring controls:	Compressibility of soil structure determines:
	(a)	Amount of compression	Amount of consolidation settlement
	(b)	Time to reach equilibrium	Time to achieve 100% consolidation
3		Initial pressure in water	Initial excess pore water pressure u_o
4		Pressure in water at any time t	Average excess pore water pressure u at any time t
5		Load in spring	Stress carried by soil skeleton
6		Percentage of final compression	Percentage consolidation

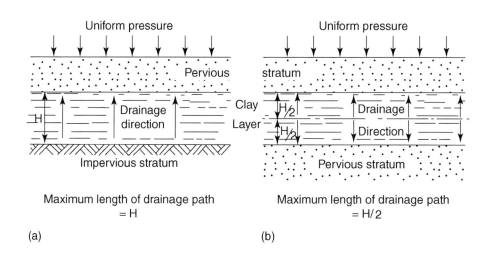

Figure 14.5 Clay layer under consolidation: (a) single drainage; (b) double drainage

grains forming the soil 'skeleton' and the pore pressure correspondingly falls. The difference between the total applied stress and the pore water pressure at any instant is known as the 'effective stress', and is approximately the same as the stress carried by the soil skeleton (Simons and Menzies, 1977). This is written in the form of an equation

$$\sigma' = \sigma - u \tag{14.1}$$

which is one of the most fundamental relationships in the field of soil mechanics (Terzaghi, 1926). In essence, the consolidation process consists of the gradual transfer of stress from the pore water to the soil skeleton. As the pore water pressure decreases, the effective stress increases.

At any stage during this process the extent to which this transfer of stress has progressed is known as the 'degree of consolidation', and is expressed as a percentage and denoted by U (which must not be confused with pore pressure u).

If u_1 = initial excess pore pressure and u_w = excess pore pressure at time t from start of consolidation, then the degree of consolidation at time t is given by

$$U = \frac{u_1 - u_w}{u_1 - u_o} \times 100\% \tag{14.2}$$

where u_o is the final equilibrium pore pressure. The value of U is sometimes referred to as the 'percentage pore pressure dissipation'. The solution to the consolidation equation (Equation (14.4) in Section 14.3.5) is expressed in terms of U.

The pore water pressure falls more rapidly near the drainage surfaces than at points remote from them. If the percentage U is related to the *average* pore pressure at time t, it can be assumed that the degree of consolidation is proportional to the amount of settlement which has taken place by time t. If

ΔH = settlement up to time t

ΔH_f = settlement which will ultimately take place (i.e. when $U = 100\%$) then

$$U = \frac{\Delta H}{\Delta H_f} \times 100\% \tag{14.3}$$

Since no measurements of pore water pressure are made in the oedometer test, the degree of consolidation has to be related to the change in height of the specimen. The beginning and end conditions are as follows. At the start of consolidation,

$$\text{time } t = 0, \quad u = u_1, \quad \Delta H = 0 \quad \text{and} \quad U = 0\%$$

At completion of consolidation

$$t = \infty \text{ (theoretically)}, \quad u = u_o, \quad \Delta H = \Delta H_f \quad \text{and} \quad U = 100\%$$

Since the test specimen is open to atmosphere, $u_o = 0$.

According to the theory, 100% consolidation is never quite reached. Nevertheless the magnitude of the ultimate settlement can be calculated, and the time required to reach any percentage of consolidation can be evaluated. Times for 50%, 90% and 95% consolidation are usually significant.

In practice the flow of water and displacements which take place during consolidation are nearly always three-dimensional. The analysis of three-dimensional effects is extremely complex and is rarely practicable (Davis and Poulos, 1965). For most applications Terzaghi's one-dimensional analysis provides a sound basis for the estimation of the magnitude of settlements, although the rate at which they develop has to be interpreted with caution. For cases in which a wide foundation rests on a relatively thin layer of clay between pervious layers, the drainage pattern is very similar to the one-dimensional assumptions.

14.3.4 Assumptions for consolidation theory

The assumptions on which the Terzaghi theory of consolidation is based, some of which have already been referred to, are summarised below.

1. The layer of soil being consolidated is horizontal, homogeneous, of uniform thickness, and is laterally confined
2. The soil is fully saturated, i.e. the voids are completely filled with water
3. Soil particles and water are incompressible
4. Darcy's law (see Section 10.3.2) for the flow of water through soil is valid
5. The coefficient of permeability and other soil properties remain constant during any one increment of applied stress
6. The applied pressure is uniform along a horizontal plane
7. Flow of water takes place only in a vertical direction, i.e. drainage and compression are one-dimensional
8. A change in effective stress in the soil causes a corresponding change in voids ratio and their relationship is linear during any one stress increment
9. The initial excess pore pressure due to the application of load is uniform throughout the depth of the clay layer
10. The extended duration of the consolidation period is due entirely to the low permeability of the soil
11. One or both of the strata adjacent to the clay layer are perfectly free-draining in comparison with the clay
12. The weight of the soil itself may be neglected

In this chapter the symbol H is used for the thickness of the clay layer or for the thickness of the soil specimen in the test.

When calculating or using the coefficient of consolidation, c_v, the significant measurement is not the thickness of the clay, H, but the length of the longest drainage path, denoted by h. They are the same where the clay drains only to one pervious surface and the other surface is impervious. Where the clay drains to both surfaces, the longest drainage path is equal to half the thickness of the layer. Thus, in Figure 14.5(a) (single drainage), $h = H$. In Figure 14.5(b) (double drainage) $h = 0.5H$.

The same theoretical equations apply to both cases, provided that they are expressed in terms of h. The symbol h can be subsequently replaced by H or $0.5H$ whichever is appropriate. In the standard oedometer consolidation test, double drainage conditions apply, as in Figure 14.5(b).

14.3.5 Theory of consolidation

Details of the mathematical theory of consolidation are not given here and are not necessary for an understanding of the consolidation test nor for the derivation of parameters from the test data. The mathematical analysis is given by Terzaghi (1943) and in other text books on soil mechanics (e.g. Scott (1974)).

The simple one-dimensional case of consolidation of a clay layer subjected to uniform loading, based on the assumptions given in Section 14.3.4, was shown by Terzaghi to lead to the following differential equation:

$$\frac{\partial u}{\partial t} = \frac{k}{\rho_w g m_v} \frac{\partial^2 u}{\partial z^2} \tag{14.4}$$

where u = excess pore water pressure at time t, at a given point; z = vertical height of that point; k = coefficient of permeability of the clay; m_v = coefficient of volume compressibilty of the clay; ρ_w = mass density of water; and g = acceleration due to gravity.

In Equation (14.4) the compound coefficient on the right-hand side is replaced by the coefficient c_v, called the coefficient of consolidation, where

$$c_v = \frac{k}{\rho_w g m_v} \tag{14.5}$$

so that Equation (14.4) becomes

$$\frac{\partial u}{\partial t} = c_v \frac{\partial^2 u}{\partial z^2} \tag{14.6}$$

The solution of Equation (14.6) expresses the percentage consolidation, U (defined in Section 14.3.3), as some function of c_v, h and time t, where h is the length of the longest drainage path, i.e.

$$\frac{U}{100} = f\left(\frac{c_v t}{h^2}\right) \tag{14.7}$$

The expression $(c_v t/h^2)$ is a dimensionless number, and can be replaced by a 'time factor', T_v, where

$$T_v = \frac{c_v t}{h^2} \tag{14.8}$$

thus Equation (14.7) can be written as

$$\frac{U}{100} = f(T_v) \tag{14.9}$$

The relationship between U and T_v, representing the solution of Equation (14.6), as derived mathematically by Terzaghi (1943), is shown graphically in Figure 14.6. The same relationship is given in Figure 14.7, where U is plotted against T_v drawn to a logarithmic scale, and in Figure 14.8, where U is plotted against the square root of T_v. The second and third types of curve are commonly used in the analysis of one-dimensional consolidation, rather than the first. In mathematical terms the curve approaches the asymptote $U = 100\%$ as time approaches infinity; in other words consolidation approaches 'completion' after a very long time, but is never fully achieved. Computed values of U, T_v and $\sqrt{T_v}$ are listed in Table 14.2.

For cases of non-uniform loading, three-dimensional consolidation, and other conditions which depart from the one-dimensional analysis, the mathematical equations lead to different curves relating U and T_v, but these are not discussed here. Examples are given by Terzaghi and Peck (1967, Figure 108), and Tschebotarioff (1951, Figure 6-11).

An important consequence of Equation (14.7) is that the degree of consolidation reached after a certain time is inversely proportional to the square of the length of the maximum

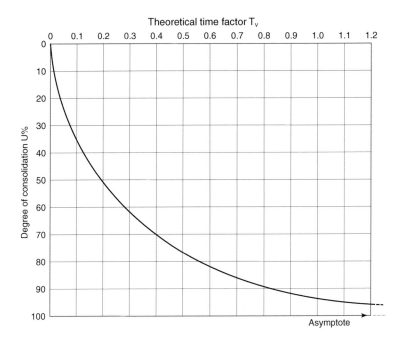

Figure 14.6 Time factor T_v related to degree of consolidation U%

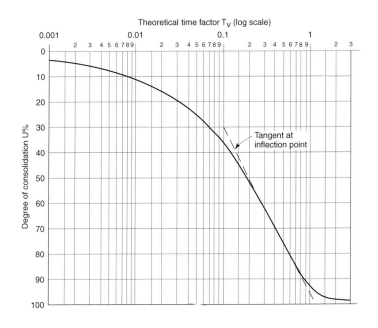

Figure 14.7 Time factor T_v (logarithmic scale) related to degree of consolidation U%

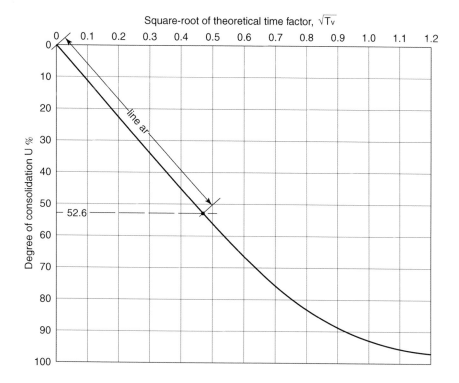

Figure 14.8 Square-root of time factor, $\sqrt{T_v}$ related to degree of consolidation U%

Table 14.2 Time factors for one-dimensional consolidation (from Leonards, 1962)

Degree of consolidation U%	Time factor	
	T_v	$\sqrt{T_v}$
0	0	0
10	0.0077	0.0877
20	0.031	0.176
30	0.071	0.266
40	0.126	0.355
50	0.196	0.443
60	0.286	0.535
70	0.403	0.635
80	0.567	0.753
90	0.848	0.921
95	1.129	1.063
100	∞	∞

drainage path. In practical terms this means that the consolidation time increases with the square of the thickness of the clay layer. For instance, loadings and other factors being equal, the time required to reach say 90% consolidation will be nine times greater for a clay layer 6 m thick than for a layer of the same clay 2 m thick.

14.3.6 Phases of consolidation

The relationship between degree of consolidation and time, derived from a typical laboratory test on a specimen of clay, is similar in general form to the theoretical relationship referred to in Section 14.3.5, but there are significant deviations which are referred to below. In this context the degree of consolidation is represented by the amount of compression (i.e. settlement) of the specimen at a particular time from the start.

Settlement is plotted against time drawn to a logarithmic scale (log-time–settlement curve) or against the square root of time in minutes (square-root-time–settlement curve). Typical laboratory curves of these types are shown in Figures 14.30 and 14.31 (see Section 14.5.5, stage 1), respectively, and their similarity to the theoretical curves in Figures 14.7 and 14.8 will be apparent.

For analytical purposes the compression of clays under load can be divided into three phases, known as

1. Initial compression
2. Primary consolidation
3. Secondary compression

In fact, these phases overlap and the time-dependent components 2 and 3 probably occur simultaneously. However, it is expedient to consider them separately (see Figure 14.9).

1. *Initial compression* takes place almost simultaneously with the application of a load increment in a laboratory test and before commencement of drainage. It is due partly to compression of small pockets of gas within the pore spaces and partly to bedding down of contact surfaces in the cell and in the load frame. A small proportion may be due to elastic compression, which is recoverable when the load is removed. This phase is responsible for a deviation from the theoretical curve near the beginning of a loading increment. In highly permeable relatively stiff soils the inclusion of some drainage, i.e. primary consolidation, is unavoidable during this phase.

2. *Primary consolidation* is the time-dependent compression due to the dissipation of the excess pore pressure under loading, and is accounted for by the Terzaghi consolidation theory. This phase relates closely to the theoretical curve for most clays. If the soil has access to water when the load is removed a small amount of recovery (swelling) may take place.

3. *Secondary compression* continues after the excess pore pressure of the primary phase has virtually dissipated. The mechanism is complex, but secondary compression is thought to be due to continued movement of particles as the soil structure adjusts itself to the increasing effective stress. Secondary compression is not usually recoverable on removal of the applied load, although secondary swelling has been observed, for instance, in peats.

In many applications only the primary consolidation phase is used for the estimation of settlements. For inorganic clays it is usually by far the most significant of the three phases, and the establishment of the magnitude of primary consolidation and of the primary consolidation–time curve, together with derived parameters, are the main objectives of a laboratory consolidation test. The primary phase is the only one which can properly be called 'consolidation' in the sense defined in Section 14.2. However, in peats and highly organic clays the secondary compression

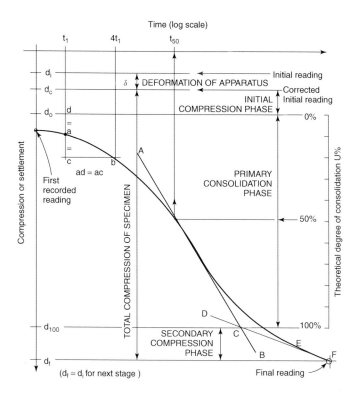

Figure 14.9 Phases of consolidation, and analysis of log-time–settlement curve

phase is of greater significance, and if taken over a sufficiently long time period can exceed the amount of primary consolidation.

The three phases are illustrated on the idealised log-time–settlement curve (see Figure 14.9) in terms of conventional boundaries. The derivation of the compression ratios, which indicate the relative magnitudes of the three phases, is explained in Section 14.3.12. However, the reporting of compression ratios is not a requirement of BS 1377:1990.

14.3.7 Primary consolidation: curve-fitting

The process of comparing a laboratory consolidation curve with the theoretical curve referred to in Section 14.3.5 is known as 'curve-fitting'. It relates only to the primary consolidation phase, and enables the coefficient of consolidation, c_v, in Equation (14.8) to be determined for each increment of loading.

Two curve-fitting procedures are used, one using the log-time–settlement curve (the log-time method), the other using the square-root-time–settlement curve (the square-root-time method).

1. Log-time method

This procedure was derived by Casagrande, hence it is also known as the Casagrande method. The principle of the method is illustrated in Figure 14.9, which is a representation of one stage of a log-time–settlement curve for a specimen under consolidation, and is explained below. An example of the application to an actual laboratory compression curve is shown in Figure 14.30 and is described in Section 14.5.6, item 1 (a).

In Figure 14.9, the settlement gauge reading at the instant of loading (time $t = 0$), i.e. the initial reading, is denoted by d_i, but the ordinate corresponding to zero time cannot be shown on a logarithmic scale. The displacement at the termination of the load increment (usually at $t = 24$ h, i.e. 1440 min), the final reading for the increment, is denoted by d_f. We need to establish the displacements representing the beginning and end of the primary consolidation phase, i.e. at theoretical degrees of consolidation of $U = 0\%$ (denoted by d_0) and $U = 100\%$ (denoted by d_{100}). These are not the same as the observed readings d_i and d_f, and are obtained as indicated below.

Theoretical 0%

The first half of the theoretical consolidation curve defined by Equation (14.9), from $U = 0\%$ to $U = 52.6\%$, can be represented to a very close approximation by

$$\frac{U}{100} = 2\sqrt{\left(\frac{T_v}{\pi}\right)} \tag{14.10}$$

which can be rewritten as

$$T_v = \frac{\pi}{4}\left(\frac{U}{100}\right)^2 \tag{14.11}$$

Equation (14.11) gives a parabolic curve and the properties of the parabola are used in the geometrical construction shown in Figure 14.9 to establish the position of the $U = 0$ abscissa. This construction is valid even though it is performed on the log-time graph, provided that the general form is similar to the theoretical curve in Figure 14.7.

Theoretical 100%

The point of inflexion of the log-time–settlement graph, which is the point at which the curvature changes, and is the steepest part of the curve, occurs at about 75% consolidation. The intersection of the tangent at this point with the backward extension of the secondary compression line defines the $U = 100\%$ abscissa, as shown in Figure 14.9.

2. *Square-root-time method*

This procedure was introduced by Taylor (1942), and is known as Taylor's method. The principle is illustrated in Figure 14.10 and is explained below. An example of its application to a laboratory square-root-time–settlement curve (derived from the same set of readings as was Figure 14.30) is shown in Figure 14.31 and is described in Section 14.5.6, item 1 (b).

Theoretical 0%

The first half of the laboratory curve has a straight-line relationship similar to the theoretical curve (see Figure 14.8) except for a deviation at the beginning, which is a consequence of

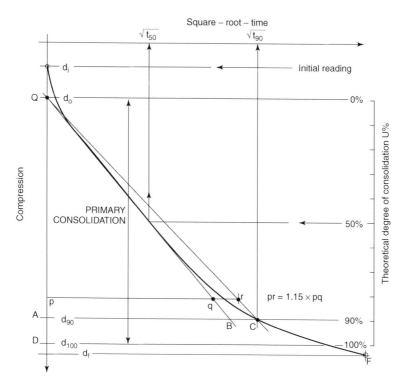

Figure 14.10 Analysis of square-root-time–settlement curve

the initial compression phase referred to in Section 14.3.6. This line extrapolated backwards gives the deformation representing a consolidation of $U = 0\%$, i.e. the value d_o, where it intersects the zero time axis (point Q in Figure 14.10).

Theoretical 100%

The equation to the theoretical straight-line portion, from Equation (14.10), is

$$\frac{U}{100} = 2\sqrt{\left(\frac{T_v}{\pi}\right)} = 1.128\sqrt{T_v}$$

This equation is represented by the line OB in Figure 14.11. At point B on this line, where $U = 90\%$

$$\sqrt{T_v} = \frac{0.90}{1.128} = 0.798$$

From Table 14.2, the value of $\sqrt{T_v}$ at $U = 90\%$ on the theoretical consolidation curve (point C, Figure 14.11) is 0.921. The ratio of these two values is $0.921/0.798 = 1.154$ (say 1.15).

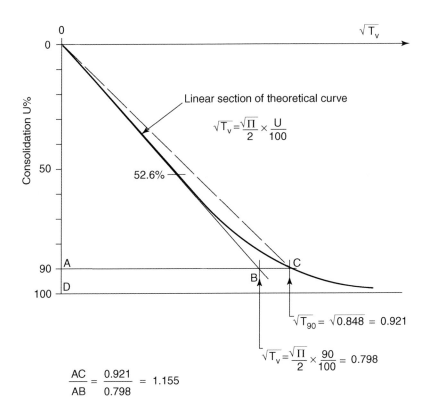

Figure 14.11 Derivation of square-root-time graphical analysis procedure

Using the above relationship, a line drawn from the theoretical origin (point Q in Figure 14.10) having abscissae 1.15 times greater than those of the line QB will intersect the laboratory curve at the point C where $U = 90\%$. The horizontal line through C intersects the vertical axis at A. The compression at which $U = 100\%$ is represented by D, where QD $= (100/90)$QA.

Relative merits of the two methods

Usually the log-time procedure is preferable to the square-root-time procedure if the settlement curves are of the conventional shape. When it is difficult to evaluate the d_0 point from the log-time plot, it may be possible to obtain d_0 from the square-root plot. It can then be transferred to the log-time plot if the latter provides a clear indication of the d_{100} point.

Evaluation of the d_{100} point from the square-root plot is not very satisfactory because the d_{90} point is defined by a line and a curve which intersect at a very small angle, and so the exact position is not easy to identify.

When starting to perform consolidation tests, or when testing an unknown soil for the first time, it is useful to plot both types of graph concurrently on separate sheets, using the same vertical scale of settlement readings for both. It can then be decided which type of graph is preferable or whether both should continue to be plotted. Generally, the square-root plot is better for determining d_0, and the log-time plot for d_{100}.

Tests on some soils may give curves which differ too much from the theoretical curves for the principles referred to above to be directly applicable. The analysis then departs somewhat from the conventional procedures given in Section 14.5.6, item 1. Suggested procedures for silty soils are given in that section under items 2 and 3, and for unsaturated clays under item 4. The analysis for peats is completely different, and is given in Section 14.7.

14.3.8 Coefficient of consolidation

When the actual time t for a given percentage of primary consolidation is known for a particular load increment of the test, Equation (14.8) can be used to determine the coefficient of consolidation for that load increment by re-writing it as

$$c_v = \frac{T_v}{t} h^2 \qquad (14.12)$$

If t_{50}, corresponding to 50% primary consolidation ($U = 50\%$), is used, the theoretical time factor $T_{50} = 0.197$, from Table 14.2. Therefore from Equation (14.12)

$$c_v = \frac{T_{50}}{t_{50}} h^2 = 0.197 \times \frac{h^2}{t_{50}} \qquad (14.13)$$

where h is the length of the maximum drainage path.

The coefficient c_v is usually expressed in square metres per year (m²/year), so if h is measured in mm and t in min

$$c_v = \frac{0.197 \times \left(\dfrac{h}{1000}\right)^2}{t_{50}} \times 60 \times 24 \times 365.25$$

$$= \frac{0.1036 h^2}{t_{50}} \text{ m}^2 / \text{year} \qquad (14.14)$$

In the standard oedometer consolidation test with double drainage (see Section 14.5) the height H of the specimen is equal to $2h$, so for practical purposes, Equation (14.14) becomes

$$c_v = \frac{0.026 \bar{H}^2}{t_{50}} \text{ m}^2 / \text{year} \qquad (14.15)$$

where $\bar{H}$ is the mean specimen height during the load increment, measured in mm, and t_{50} is measured in minutes.

If t_{90} from the square-root-plot is used instead of t_{50}, Equation (14.13) becomes

$$c_v = \frac{T_{90}}{t_{90}} h^2 = 0.848 \times \frac{h^2}{t_{90}}$$

Equation (14.14) is then

$$c_v = \frac{0.446h^2}{t_{90}} \ \text{m}^2 / \text{year}$$

and in terms of H

$$c_v = \frac{0.112}{t_{90}} \bar{H}^2 \ \text{m}^2 / \text{year} \tag{14.16}$$

If is preferable to calculate c_v from t_{50} rather than from t_{90} because the middle of the laboratory settlement curve is the portion which agrees most closely with the theoretical curve.

Typical values of the coefficient of consolidation as derived from oedometer tests on specimens of uniform soil, related to their approximate plasticity ranges, are indicated in Table 14.6 (see Section 14.4.5).

14.3.9 Voids ratio

The terms voids ratio and degree of saturation, and the relationship between them and moisture content and dry density, are explained in Volume 1 (third edition), Section 3.3.2. The following equations derived there are relevant to the consolidation test:

$$\text{Voids ratio } e = \left(\frac{\rho_s}{\rho_D} \right) - 1 \tag{14.17}$$

$$\text{Degree of saturation } S = \frac{w\rho_s}{e} \times 100\% \tag{14.18}$$

Where ρ_s = particle density (Mg/m³); ρ_D = dry density of soil (Mg/m³); and w = moisture content of soil (%).

The volume change which occurs during consolidation takes place only in the voids. The change in height, ΔH, from an initial height H_0 (see Figure 14.12), corresponds to a change in voids ratio Δe from an initial voids ratio e_0.

The change in voids ratio denoted by Δe, and the change in height denoted by ΔH, refer to the overall change of e and of H, with respect to the initial values, e_0 and H_0. (The actual change in height of the specimen, ΔH, is derived from the measured settlement after allowing for deformation of the apparatus, as explained in Section 14.5.6, item 5). Therefore, by proportion

$$\frac{\Delta H}{H_0} = \frac{\Delta e}{1 + e_0}$$

$$\text{i.e.} \quad \Delta e = \frac{1 + e_0}{H_0} \Delta H \tag{14.19}$$

$$\text{Or } \Delta e = \frac{\Delta H}{H_s} \tag{14.20}$$

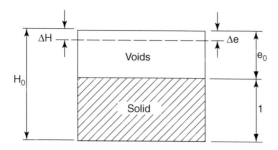

Figure 14.12 Representation of voids ratio change

where H_s is the equivalent height of solid particles, which depends only on the initial conditions of the test specimen and remains constant. It is defined by the relationship

$$H_s = \frac{H_0}{1+e_0}$$

When the initial voids ratio e_0 is known, the voids ratio e at any stage of the test can be calculated from the change in height by using Equation (14.20), and then using the equation

$$e = e_0 - \Delta e \qquad (14.21)$$

A graph of voids ratio e against applied pressure p (to a log scale), known as the e–log p curve, can then be drawn as in Figure 4.40 (see Section 14.5.7).

14.3.10 Compressibility coefficients

Three coefficients which are derived from consolidation tests to indicate the compressibility of soils, from which an estimate of the amount of settlement due to primary consolidation can be made, are

Coefficient of compressibility, a_v
Coefficient of volume compressibility, m_v
Compression index, C_c

The first is rarely used. The second is normally calculated for each load increment and values are presented as part of the results of a laboratory consolidation test. The third may be derived by the engineer from the e–log p curve, or empirically, but its determination is not usually considered to be part of the laboratory test. The coefficient m_v is generally applied to overconsolidated clays and C_c to normally consolidated clays.

Another coefficient, the swell index C_s, gives an indication of the expansion of a soil on unloading.

Coefficient of compressibility

For a particular load increment this is equal to the change in voids ratio δe for that increment, divided by the incremental pressure δp.

The change in voids ratio denoted by δe, and the change in pressure denoted by δp, refer to incremental changes, that is the change with respect to the immediately preceding values of e and p, as distinct from cumulative changes related to initial conditions denoted by Δe and Δp referred to in Section 14.3.9.

$$\therefore \quad a_v = \frac{e_2 - e_1}{\delta p} = -\frac{\delta e}{\delta p} \tag{14.22}$$

where e_1 and e_2 are the voids ratios at the beginning and end of consolidation under the load increment, respectively. The negative sign appears because e decreases as p increases.

The coefficient a_v is equal to the (negative) slope of the voids ratio–pressure curve (see Figure 4.13) assuming it is linear over the pressure increment range. The units are the reciprocal of the stress units, i.e. m²/kN in customary SI units.

Coefficient of volume compressibility

A more useful parameter than a_v is one which indicates the compressibility per unit thickness of the soil. This is known as coefficient of volume compressibility or sometimes as the modulus of volume change. It is denoted by m_v and is defined by

$$m_v = \frac{a_v}{1+e_1} = \frac{1}{1+e_1}\left(-\frac{\delta e}{\delta p}\right) \tag{14.23}$$

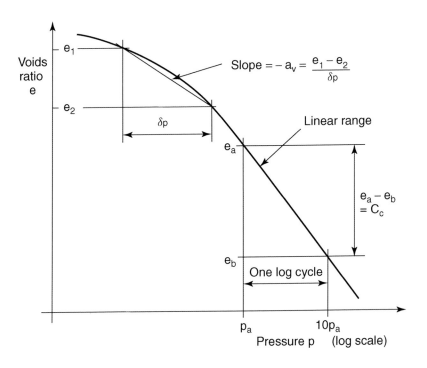

Figure 14.13 Log-pressure–voids ratio curve (e–log p curve)

where e_1 is the voids ratio at the start of the load increment δp. The units are the same as for a_v, but are usually multiplied by 1000 to express m_v in m²/MN so as to avoid inconveniently small numerical values. Therefore, if δp is measured in kPa

$$m_v = \frac{1000}{1+e_1}\left(-\frac{\delta e}{\delta p}\right) \text{ m}^2/\text{MN} \tag{14.24}$$

The values of m_v calculated for each load increment of the consolidation test are the *laboratory* values and should be reported as such (see Section 14.3.14).

Some typical values of the coefficient of volume compressibility for a number of types of clay are given in Table 14.5 (see Section 14.4.5). A similar coefficient, related to swelling, may be obtained from the unloading curve.

Compression index

The compression index, C_c, is equal to the slope of the field consolidation curve plotted to a logarithmic scale of pressure p, in the linear range. This straight line is represented by the following equation:

$$e = e_0 - C_c \, \log_{10} \frac{p_o + \delta p}{p_o} \tag{14.25}$$

Numerically, C_c is equal to the change in voids ratio for one log cycle of pressure change, as shown in Figure 14.13. It is a dimensionless number.

The compression index has been found to be related to the liquid limit (w_L) of clay to a reasonable degree of approximation by the equation (Skempton, 1944)

$$C_c = 0.009(w_L - 10\%) \tag{14.26}$$

subject to the limitations referred to in Section 14.4.5.

For a remoulded clay the corresponding compression index C_c' is given approximately by the equation

$$C_c' = 0.007(w_L - 10\%) \tag{14.27}$$

Swell index

The swell index, C_s, is equal to the slope of the swelling (unloading) curve of e plotted against log p. It is obtained in a similar manner to the compression index. Its value also increases with increasing liquid limit.

14.3.11 Coefficient of permeability

When the parameters c_v and m_v have been obtained it is possible to calculate the coefficient of permeability, k, by using Equation (14.5), which can be rewritten as

$$k = c_v m_v \rho_w g \tag{14.28}$$

Table 14.3 Units for permeability relationship

Symbol	Practical unit	Consistent unit	Multiplying factor
(1)	(2)	(3)	(4)
k	m/s	m/s	1
c_v	m²/year	m²/s	$(365.25 \times 24 \times 3600)^{-1}$
m_v	m²/MN	m²/N	10^{-6}
ρ_w	Mg/m³	kg/m³	10^3 ($\rho_w = 1$ Mg/m³)
g	m/s²	m/s²	1 ($g = 9.81$ m/s²)

The practical units normally used are shown in column 2 of Table 14.3, and the units required to make the above equation consistent in column 3. Multiplying factors are shown in column 4. Therefore in terms of practical units

$$k = \left(\frac{c_v}{365.25 \times 24 \times 3600} \right) \left(\frac{m_v}{10^6} \right) (1 \times 10^3) \times 9.81 \text{ m/s}$$

$$= c_v m_v \times 0.3109 \times 10^{-9} \text{ m/s}$$

or for practical purposes

$$k = c_v m_v \times 0.31 \times 10^{-9} \text{ m/s} \tag{14.29}$$

if c_v is expressed in m²/year and m_v is expressed in m²/MN, then

$$c_v \times m_v \times \rho_w \times g = k$$

$$\frac{\text{m}^2}{\text{s}} \times \frac{\text{m}^2}{\text{N}} \times \frac{\text{kg}}{\text{m}^3} \times \frac{\text{m}}{\text{s}^2} = \text{m/s (since N = kg m/s}^2)$$

14.3.12 Compression ratios

The relative magnitudes of the three phases of consolidation, described in Section 14.3.6, can be expressed in terms of compression ratios, designated as follows:

Initial compression ratio: r_o
Primary compression ratio: r_p
Secondary compression ratio: r_s

The derivation of these ratios was a requirement of BS 1377:1975, but is not included in the 1990 revision. The symbols used to represent compression gauge readings at various time intervals from the start of a consolidation stage are listed in Table 14.4.

The total observed compression during a loading increment, after allowing for the deformation of the apparatus (see Section 14.5.6, item 5), is $(d_c - d_f)$. The compression ratios are the proportions of this amount contributed by each of the phases of consolidation and are calculated as follows (see Figure 14.9):

$$\text{Initial compression ratio: } r_o = \frac{d_c - d_o}{d_c - d_f} \tag{14.30}$$

Table 14.4 Symbols representing gauge readings at various times

Gauge	Symbol
Initial reading for load increment (zero time)	d_I
Initial reading corrected for deformation of apparatus	d_c
Corrected zero point:	
0% primary consolidation	d_o
50% primary consolidation	d_{50}
90% primary consolidation	d_{90}
100% primary consolidation	d_{100}
Final reading (end of increment)	d_f

$$\text{Primary compression ratio: } r_p = \frac{d_o - d_{100}}{d_c - d_f} \tag{14.31}$$

$$\text{or } r_p = \frac{10}{9} \times \frac{d_o - d_{90}}{d_c - d_f} \tag{14.32}$$

(if calculated from d_{90} on square-root-time plot)

$$\text{Secondary compression ratio:} r_s = \frac{d_{100} - d_f}{d_c - d_f} \tag{14.33}$$

$$\text{or } r_s = 1 - (r_o + r_p) \tag{14.34}$$

The compression ratios are dimensionless numbers and are usually reported to the second decimal place (or the nearest whole number if expressed as percentages). The sum of the three is equal to 1 (or 100%).

For most inorganic clays the primary compression ratio r_p is by far the largest of the three.

14.3.13 Secondary compression

Secondary compression effects are often disregarded for inorganic clays, in which the primary consolidation phase is responsible for the majority of settlement. However, secondary compression is a much more significant factor in the settlement of organic soils, especially peats, and increases as the applied load increases.

The estimation of settlements due to secondary compression is less reliable than that based on primary consolidation. Secondary compression is conventionally considered to start immediately after the end of the primary consolidation phase, although in fact there is some overlap. Secondary compression is usually evident as a linear relationship on a log-time settlement plot, as shown by EF in Figure 4.9. The coefficient of secondary compression C_{sec}, on which calculations are based, is equal to the slope of this line in terms of strain per log cycle of time. It is a dimensionless number.

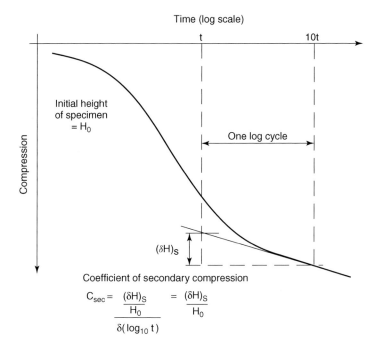

Figure 14.14 Determination of coefficient of secondary compression Csec

A graphical method for determining C_{sec} is shown in Figure 14.14. The linear portion of the secondary compression graph is extended so that it covers one complete cycle of log time, and the compression gauge readings at the beginning and end of the cycle (e.g. at 1000 and 10,000 min) are noted. If the change in height of the specimen is $(\delta H)_s$ mm over one log cycle, and H_o mm is the initial height of the test specimen, the coefficient of secondary compression is given by the equation

$$C_{sec} = \frac{(\delta H)_s}{H_o} \tag{14.34}$$

It may be multiplied by 100 to express it as a percentage.

The value of C_{sec} has been found to be independent of applied stress for inorganic clays loaded above the preconsolidated pressure. But for highly organic soils and peats the value tends to increase with increased applied stress. The ranges of typical values of C_{sec} are given in Table 14.7 (see Section 14.4.5).

14.3.14 Normally consolidated and overconsolidated clays

When considering their consolidation properties, naturally occurring clays may be divided into two main types.

        Normally consolidated clays
        Overconsolidated clays

Normally consolidated clays

These are clays which have never been subjected to an effective stress greater than the present effective overburden pressure. They are usually soft for a considerable depth. Examples are geologically recent alluvial deposits from which no subsequent overburden has been removed. Sediments which are still in the process of being formed, such as very recent marine or estuarine muds, and tailings in settling ponds, could be described as 'underconsolidated'.

Normally consolidated clays are sensitive to the effects of disturbance, which can influence the relationship between voids ratio and pressure derived from a consolidation test. Hence, extreme care is needed in the preparation of test specimens.

The effect of sample disturbance on the e–log p curve is illustrated in Figure 14.15, in which the line A represents the field loading curve (virgin compression curve) for the natural soil in-situ. Curve B indicates the form of the 'ideal' loading curve for a truly undisturbed sample. The solid curve C is typical of a laboratory test curve for an undisturbed sample of average quality. This becomes linear as it converges towards the line A. The dashed curve D represents a test on a completely remoulded clay.

Because there is a difference between the field curve and a typical laboratory curve, the values of m_v calculated from a laboratory test (see Section 14.3.10) usually differ from the field values on which settlement computations are based. They must therefore be referred to as *laboratory* values.

The field loading curve can be derived from the laboratory curve by several procedures (e.g. Schmertmann (1953)), but this forms part of the engineering analysis and is beyond the scope of this book. The analysis relies on there being sufficient load increments in the laboratory test to obtain three points in a straight line on the e–log p curve, and enough points to define

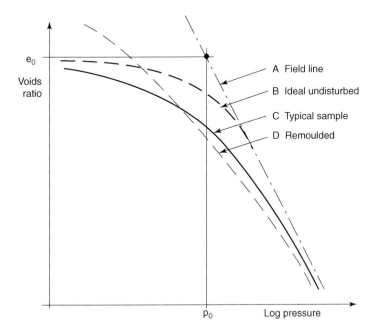

Figure 14.15 Effect of disturbance on e–log p curve for normally consolidated clay

the swelling (unloading) curve. The particle density should be measured and the Atterberg limits determined as a cross-check on the compression index C_c (see Section 14.3.10).

Overconsolidated clays (precompressed clays)

These are clays which in the past have been subjected to an effective pressure greater than the present effective overburden pressure. This may be due to the clay having been covered either by deposits of soil or rock, perhaps several kilometres thick, which were subsequently eroded away in the course of geological time; or by great thicknesses of ice during periods of glaciation. Overconsolidation is the result of a reduction in effective pressure, which can also be caused by a rise in the ground water table. Weathering and partial drying are other factors that can produce an effect of preconsolidation. Overconsolidated clays are often stiff or hard, but they can be softer if the excess pressure was small. They swell and soften readily when allowed free access to water, but if restrained from swelling they can exert considerable swelling pressures.

The maximum previous effective stress to which the soil has been subjected is known as the preconsolidation pressure, and is denoted by p_c. The ratio of the preconsolidation pressure to the existing effective pressure p_o is known as the OCR, i.e.

$$OCR = \frac{p_c}{p_o}$$

Overconsolidated clays are less sensitive to mechanical disturbance than softer, normally consolidated clays, but they are susceptible to the effects of stress relief resulting from the removal of the sample from the ground. This is especially true for fissured clays.

The process of overconsolidation can be illustrated by the e–log p curves in Figure 14.16. The line ABCH represents the virgin compression curve in-situ as the clay was

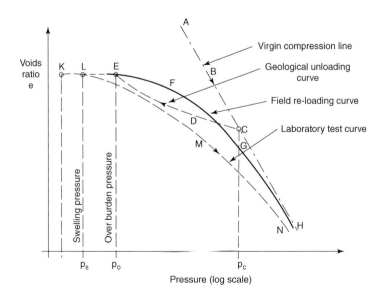

Figure 14.16 Effect of overconsolidation on e–log p curve

consolidated under the pressure of superimposed deposits. The maximum pressure reached was p_c, represented by the point C. Erosion of part of the overburden or removal of ice during the course of geological time reduced the pressure to p_o, the present overburden pressure. Swelling occurred along the unloading curve CDE to the point E, where the OCR is equal to p_c/p_o. If the clay is now re-loaded by the imposition of foundation loads it will reconsolidate along a new field curve EFG. Initially this curve is much flatter than the virgin curve (i.e. the clay now has a much lower compressibility), but it becomes much steeper (the clay becomes more compressible) if a pressure equal to p_c is reached, and it then approaches the virgin curve at H.

Removing a sample from the ground and setting it up for a consolidation test without access to water unloads the sample to a very small pressure represented by the point K, with no change in the voids ratio. The tendency to swell when water is added to the oedometer cell is resisted by applying the swelling pressure or equilibrium load, p_s, denoted by the point L. A laboratory consolidation test then gives the dashed curve LMN, although in many cases the test would terminate at a point M where the pressure is appreciably less than p_c, due to limitations of the capacity of the apparatus.

The laboratory curve LMN is quite different in form from the field curve EFG. However, the laboratory curve can be used to derive the preconsolidation pressure p_c and the OCR, using the procedure devised by Casagrande (1936). There are also recognised procedures, such as those due to Schmertmann (1954) and Leonards (1962), among others, which enable the field consolidation curve to be constructed. These procedures are part of the engineering analysis and therefore beyond the scope of this book. They require certain data from the laboratory which are listed in Section 14.6.5.

14.3.15 Swelling characteristics

Swelling is the reverse of the consolidation process; it is the increase in volume of a soil due to absorption of water within the voids when the applied stress is reduced. It is represented by the unloading curve (also known as a decompression curve) marked MNP in Figure 14.17. The compression due to consolidation is never fully recoverable on unloading and a load–unload–reload cycle produces a hysteresis loop of the form denoted by LMNPQR in Figure 14.17.

Swelling occurs when overconsolidated clays are allowed free access to water on unloading because of their great affinity for water. An overconsolidated clay when unloaded possesses very high suction tensions within the soil skeleton. These draw water into the voids causing the volume of voids to increase and the soil to swell, and eventually, often rapidly, to disintegrate. However, swelling can be prevented by constraining the clay to maintain its original volume. The pressure required to prevent swelling is known as the swelling pressure, and can be appreciable. In very heavily overconsolidated clays the author has measured swelling pressures of well over 1 MPa.

14.3.16 Temperature

The rate of consolidation of clay depends upon its compressibility and its permeability. The latter is related not only to the size of the pores but also to the viscosity of the water in the pore spaces (see Table 14.1). Viscosity depends on temperature (see Volume 1 (third edition), Table 4.14), and the viscosity of water at 35°C is about half that at 5°C. The coefficient of consolidation, from which the rate of consolidation is evaluated, is therefore dependent on temperature.

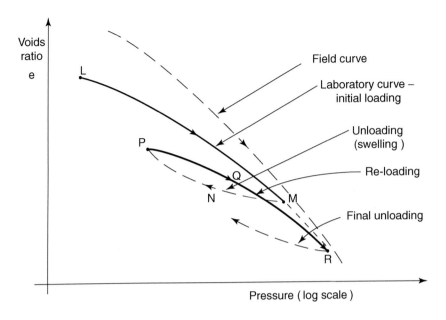

Figure 14.17 Unloading and re-loading in oedometer test

Consolidation tests are usually carried out at an ambient laboratory temperature of 20–25°C. The average temperature of soil in-situ in the UK is around 10°C, and a temperature correction to laboratory test data may be appropriate before applying them to field conditions. The correction factor graph given in Figure 14.18 provides a convenient way of doing this. This curve may also be used to convert laboratory test results to standard 20°C values, when the laboratory test has been carried out at a significantly different temperature.

Although an increase in temperature increases the rate of consolidation, it does not affect the amount of consolidation during the primary stage, but it can increase the amount of secondary compression. This is probably negligible for inorganic soils, but it can be more significant for organic soils, especially peats.

For the above reasons it is good practice to maintain continuous records of laboratory temperature adjacent to oedometer presses throughout the duration of consolidation tests. However, corrections for temperature are normally made more for the standardisation of results than for the apparent increase in accuracy of c_v values, which in any case are little more than an indication of an order of magnitude.

14.3.17 Voids ratio of peat

For the calculation of voids ratios of peats from Equation (14.17), the particle density should be determined with a reasonable degree of accuracy, because it is more variable than that of mineral soils, in the range of about 1.5–2.5 Mg/m³. Direct measurement by the density bottle method requires the use of kerosene (see Volume 1 (third edition), Section 3.6.2), and can be difficult and time-consuming. A preferable method when a large number of peat

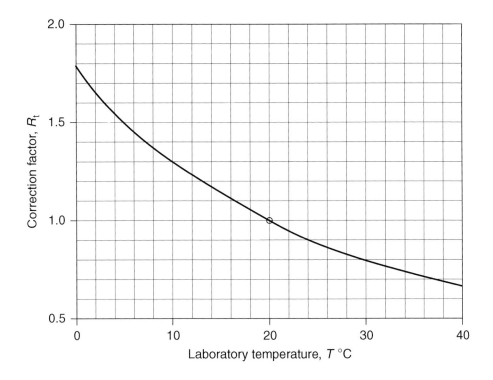

Figure 14.18 Temperature curve for coefficient of consolidation (reproduced from Figure 4 of BS 1377:Part 5:1990)

samples is being tested is to make use of the loss on ignition (Volume 1 (third edition), Section 5.10.3) as an index property which can be related to organic content and particle density, as outlined below.

The relationship between specific gravity and organic content of peaty soils was stated by Skempton and Petley (1970) as follows (using their terminology):

$$G = \frac{G_s G_p}{(G_s - G_p)P + G_p} \tag{14.37}$$

in which G = mean specific gravity of peat sample; G_s = specific gravity of mineral particles; G_p = specific gravity of organic matter; and P = proportion of organic matter by dry mass.

For most practical purposes if the factor 1.04 in Skempton and Petley's equation relating organic matter to ignition loss is replaced by unity, it can be assumed that the organic matter content is equal to the loss on ignition if a furnace temperature of 550°C is used. Thus P is approximately equal to $N/100$, where N is the ignition loss expressed as a percentage.

The specific gravity of mineral soil particles is usually about 2.7 and that of organic matter is typically 1.4. Substituting these values in Equation (14.37), using particle density in place of specific gravity, and making the above assumption, the mean particle density, ρ_{mp}, of a peat sample can be obtained from the ignition loss, $N\%$, by using the equation

$$\rho_{mp} = \frac{3.78}{\left(1.3 \times \dfrac{N}{100}\right) + 1.4} \tag{14.38}$$

Nevertheless, a few check tests by the direct method should be done to confirm this relationship.

14.4 Applications

14.4.1 Foundations for structures

Consolidation tests are carried out on specimens prepared from undisturbed samples taken from clay strata not only from immediately below the foundation level but also from considerable depth. Data obtained from these tests, together with classification data and a knowledge of the loading history of the clay, enable estimates to be made of the behaviour of foundations, as outlined below.

1. The amount of settlement which will ultimately take place for the structure as a whole can be calculated (see, for instance, Terzaghi (1939); MacDonald and Skempton (1955); Skempton and Bjerum (1957)).
2. Variations in long-term settlements between individual footings can be estimated. Differential settlements are usually more critical than the overall settlement, and must be kept within limits to avoid structural damage (Skempton and Macdonald, 1956).
3. Non-uniform ground conditions can cause differential settlements which result in tilting of the structure as a whole and distortions within the structure. Analysis based on proper investigation and testing can guard against this occurrence.
4. The most famous example of tilting is the campanile tower (the 'leaning tower') in the Italian city of Pisa, which was built during the twelfth century and was still settling with an alarmingly increasing tilt until recently. (Terzaghi, 1934; Mitchell *et al.*, 1977; Wheeler, 1993). However, the structure was made safe and further tilting halted in 2002 as a result of the foundation stabilisation measures that were carried out to the design of Professor Burland (Burland, 2001).
5. The settlement of piled foundations due to the presence of a deep-seated stratum of compressible clay can be estimated.
6. The approximate rate of consolidation can be estimated, from which it can be seen whether settlements will be substantially completed during the construction period or whether appreciable settlements will continue and, if so, for how long after completion of construction .
7. If long-term settlements are indicated, a settlement–time graph can be drawn to show the duration of the significant part of the settlements, which can be compared with the economic life of the structure (but this information must be used with caution: see Section 14.4.4).
8. From the settlement–time relationships it can be ascertained whether unacceptable differential settlements are likely to develop, either in the long term or at any time during or after the construction period.

14.4.2 Soft ground and fill

Soft soils, such as alluvial silts and clays, are too weak to carry any but the lightest of foundation loads unless the shear strength is first increased by consolidation, which can be effected by pre-loading the ground with a surcharge of temporary fill. Laboratory consolidation tests can be used to estimate the extent of the resulting settlement, but the rate of settlement is usually under-estimated (see Section 14.4.4). Field tests are more reliable for indicating whether provision of means for accelerating the consolidation, such as the installation of sand drains, is justifiable.

Fill placed on soft ground prior to construction will cause the soft strata to consolidate, and settlement of the fill may continue for a long period. If piles are driven through the soft material to transmit the building loads to a deeper firm stratum, continued downward movement of both the fill and the soft layer could throw additional loads on to the piles due to 'negative skin friction'. Knowledge of the consolidation characteristics provides a basis for safeguarding against overloading of piles due to this effect.

An embankment or earth dam consolidates under its own weight and the resulting increase in effective strength is made use of in the analysis of its long-term stability. The same applies to the soil strata beneath the embankment or dam.

Estimates of the amount and rate of consolidation of peat are possible from laboratory tests. But because of the relatively high proportions of both initial and secondary compression which occurs in peat, these estimates are more approximate than those made for inorganic soils and a completely different approach is used (see Section 14.7).

14.4.3 Effect of ground water

Consolidation can take place in clay as a result of lowering the ground water table, because the effective stress is thereby increased. A fall of 1 m in the ground water level will increase the effective stress in the whole of the clay deposit beneath the water table by about 10 kPa. The amount of consolidation depends on the change of effective stress in the soil, and settlements due to de-watering can be estimated from laboratory tests in the same way as those due to loading.

14.4.4 Limitations and advantages

Although more sophisticated consolidation tests using larger samples are now available, the laboratory oedometer test is still recognised as the standard test for determining the consolidation characteristics of homogeneous clays. In general, for inorganic clays the test provides a reasonable estimate of the amount of settlement. However, the rate of settlement is often under-estimated, that is, a given percentage of the ultimate consolidation is actually reached in a shorter time than that predicted from the test data using the theory of consolidation. This effect is largely due to limitations imposed by the small size of the specimen, which make it impracticable to represent many of the natural features such as laminations, fissures, and other discontinuities (collectively referred to as the soil fabric), and their profound effect on drainage conditions (Rowe, 1972). The most reliable means of obtaining c_v values, on which calculations for the rate of settlement are based, is to determine m_v from laboratory oedometer consolidation tests and to measure the permeability, k, in the field, and then to use Equation (14.29) to calculate c_v.

Attempts are sometimes made to take horizontal drainage into account in oedometer consolidation tests, either by fitting a pervious lining inside the sampling ring and sealing

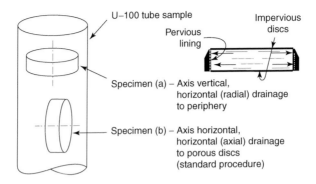

U–100 tube sample

Pervious lining

Impervious discs

Specimen (a) – Axis vertical, horizontal (radial) drainage to periphery

Specimen (b) – Axis horizontal, horizontal (axial) drainage to porous discs (standard procedure)

Figure 14.19 Test specimens to provide horizontal drainage in standard oedometer

the specimen ends (see Figure 14.19(a)), or by trimming a specimen in a vertical plane (see Figure 14.19(b)). Neither method is as satisfactory as testing larger specimens under hydraulic loading in a cell designed for the provision of horizontal drainage (Rowe, 1966). This type of test will be covered in Volume 3.

Another limitation of the test is that there is no means of measuring excess pore pressures, the dissipation of which controls the consolidation process. The extent of consolidation is based solely on measurements of the change in height of the specimen.

However, the limitations referred to above are frequently outweighed by the practical advantages of the test, which may be summarised as follows.

- The procedure and calibrations have been standardised so that they are straight forward and can be easily reproduced.
- The test provides a reasonable indication of the amount of settlement if the results are properly interpreted.
- The test is applicable to a wide range of soil types; silts and peats, as well as clays
- Tests can be carried out on undisturbed specimens trimmed from standard 100 mm diameter tube or piston samples.
- Several tests can be conveniently run simultaneously in a row of adjacent oedometer presses.
- Because of the relatively small specimen thickness, testing times are not excessively long. One day for each stage is normal, and a load–unload cycle is usually completed within two weeks. A test can easily be extended into a long-term test if the secondary compression characteristics are required.

14.4.5 Typical values of consolidation coefficients

The range of values of the coefficient of volume compressibility (m_v) for typical British soils, classified on the basis of compressibility, are given in Table 14.5. This coefficient is usually applied to overconsolidated clays.

The usual range of values of the coefficient of consolidation (c_v) obtained from laboratory oedometer tests is indicated in Table 14.6, together with values of the compression index, C_c. Skempton's empirical equation (Equation (14.26)) relating C_c to the liquid limit (LL) of normally consolidated clays does not apply to highly organic clays or where the LL exceeds 100% or where the natural moisture content exceeds the LL. The coefficient C_c is usually applied to normally consolidated clays.

Some typical values of the coefficient of secondary compression C_{sec} are given in Table 14.7.

Table 14.5 Some typical values of coefficient of volume compressibility

Description of compressibility	Coefficient of volume compressibility, m_v (m²/MN)	Clay types
Very high	Above 1.5	Very organic alluvial clays and peats
High	0.3–1.5	Normally consolidated alluvial clays (e.g. estuarine clays)
Medium	0.1–0.3	Fluvio-glacial clays Lake clays Upper 'blue' and weathered 'brown' London Clay
Low	0.05–0.1	Boulder clays Very stiff or hard 'blue' London Clay
Very low	Below 0.05	Heavily overconsolidated 'boulder clays' Stiff weathered rocks

Table 14.6 Typical range of values of coefficient of consolidation and compression index for inorganic soils (from Lambe and Whitman, 1979)

Soil type	Plasticity index range	Coefficient of consolidation c_v (m²/year)		Compression index C_C
		undisturbed	remoulded	
Clays—montmorillonite				Up to 2.6
high plasticity	Greater than 25	0.1–1		
medium			About 25–50% of undisturbed values	
plasticity	25–5	1–10		0.8–0.2
low plasticity		10–100		
Silts	15 or less	above 100		

Table 14.7 Typical values of C_{sec} (from Lambe and Whitman, 1979)

Soil type	C_{sec}
Normally consolidated clays	0.005–0.02
Very plastic clays	0.03 or higher
Organic clays	0.03 or higher
Overconsolidated clays (Overconsolidation ratio greater than 2)	Less than 0.001

14.5 Consolidation test (BS 1377:Part 5:1990:3, and ASTM D 2435)

14.5.1 General

The test which is described in Sections 14.5.3–14.5.7, generally follows the procedure given in BS 1377:Part 5:1990. Notes on the ASTM test (D 2435) are given separately in Section 14.5.8. The apparatus referred to, incorporating a fixed-ring type of oedometer cell with a test specimen of approximately 75 mm diameter and 20 mm high, is typical of that which is commercially available in the UK for routine testing in accordance with the BS.

The soil is assumed to be a virtually saturated, normally consolidated inorganic clay, of a consistency ranging from soft to stiff. The procedure also applies to inorganic silty clays and silts. The specimen is allowed to drain freely from the top and bottom surfaces. Special procedures for other types of soil and for different test conditions are dealt with separately in Section 14.6.

The size of the largest particle in the test specimen permitted by the BS is one-fifth of the height of the specimen. For a specimen 20 mm high this means that particles up to 4 mm are permissible, but such particles should occur only occasionally. A maximum size of one-tenth of the specimen height seems more satisfactory.

In order to carry out the necessary calculations from the test data, the particle density of the soil is required. This should be measured by using the procedure given in Volume 1 (third edition), Section 3.6.2, unless a reliable value can be assumed. For peats this may be supplemented by the indirect method given in Section 14.3.17.

Analysis of the graphical data obtained from the test is described in Section 14.5.6, using the conventional procedure applicable to clays. The time–settlement curves for silty clays and silts may depart from the theoretical relationships, and the analysis of these soils is presented separately. Peats are dealt with in Section 14.7.

14.5.2 Features of apparatus

Present design

Several designs of consolidation cell and loading frame have been introduced in the past, some of which were referred to in Section 14.1.4. Of these the fixed-ring cell is the type now generally preferred, and a loading yoke arrangement on the loading frame virtually eliminates any tendency towards eccentric loading or side thrust on the specimen as the beam deflects. Use of standardised equipment is a major factor in obtaining consistent and reproducible results.

Deformation characteristics

When weights are added to the load hanger of a consolidation press, some deformation of the apparatus is unavoidable because of the elasticity of the frame itself and bedding effects on contact surfaces. The compression dial gauge registers this deformation in addition to the settlement of the specimen itself, but the deformation occurs as soon as the load is applied. It can be allowed for by applying a calibration correction, which is obtained as described in Section 14.8.1. A calibration curve should be obtained for each load frame before it is put into use. If several load frames of a similar type are found to have very similar calibration curves, a mean curve may be appropriate for all of them.

Calibration corrections are of relatively greater significance for stiff clays, for which the amount of total settlement is small, than for firm or soft clays.

Size of specimen

The size of the test specimen can influence the results obtained from a consolidation test. Ideally the specimen should be as large as possible, so that it can be representative of the soil fabric as well as the material itself. Large specimens are less susceptible to disturbance than small ones, and the greater the specimen height the better is the relative accuracy of settlement readings. However, the effect of side friction between the specimen and containing ring increases with the height:diameter ratio (*H:D* ratio), so this value should not be too large. For practical reasons the specimen should be of a size suitable for easy preparation from standard undisturbed samples.

The BS specifies that the specimen diameter (*D*) should be between 50 mm and 105 mm, and the height (*H*) from 18 mm to 0.4 times the diameter; i.e.

$$50 \text{ mm} \le D \le 105 \text{ mm}$$
$$18 \text{ mm} \le H \le 0.4\,D$$

These dimensions provide a reasonable compromise between the above requirements. The *H:D* ratio for a specimen 75 mm diameter and 20 mm high is 1:3.75 and this size allows a reasonable trimming margin from the outer edge of a 100 mm diameter tube or piston sample. The ASTM standard requires a minimum *H:D* ratio of 1:2.5, and specimens not less than 50 mm diameter and 12.5 mm thick. For specimens of about these dimensions, Cooling and Skempton (1941) showed that observed behaviour agreed closely with that predicted from the Terzaghi theory, and that the assumptions made were reasonably valid.

Some types of loading frame can accommodate specimens of several different sizes, for which a range of cells and fittings is available. This enables the largest practicable size of specimen to be used as standard procedure, while providing for the use of smaller specimens when the size of sample is limited, or when stresses higher than usual are to be applied. It also allows for the testing of specimens conforming to standards prevailing in many parts of the world. Some of the recognised specimen sizes are listed in Table 14.8.

14.5.3 Apparatus for BS consolidation test

Preparation of test specimens

Most of the items listed below are described elsewhere, as indicated.[T/S, list 1–6 (continues later as 7–15), each item is punctuated at the end]

1. Flat glass plate, such as is used for the liquid limit test (see Volume 1 (third edition), Section 2.6.4).
2. Cutting tools and straight-edge for specimen trimming (see Section 9.1.2).
3. Jig for holding the consolidation ring in place while jacking out the sample from a U-100 tube (see Section 9.2.2 and Figure 9.16).
4. Watch glass 100 mm diameter, or metal tray, to hold the consolidation ring.
5. Balance, and measuring instruments, for weighing and measuring the specimen and ring (see Volume 1 (third edition), Sections 1.2.2 and 1.2.3).
6. Apparatus for determining moisture content (see Volume 1 (third edition), Section 2.5.2).

Table 14.8 Oedometer specimen sizes and beam ratios

Specimen		Beam ratio	Load (kg)	Stress	Typical maximum stress	Stress for 1 kg	Application
diameter (mm)	area (mm²)						
50.0	1963	10:1	1	50 kPa	8 MPa	50 kPa	High pressure (SI)
50.5	2003	10:1	10	5 kgf/cm²	80 kgf/cm²	0.5 kgf/cm²	Metric 'technical'
63.5 (2.5 in)	3167 (4.909 in²)	10:1	1.55	1000 1bf/ft²	103200 1bf/ft²	645 1bf/ft²	ASTM
71.4	4004	10:1	2	0.5 kgf/cm²	40 kgf/cm²	0.25 kgf/cm²	Metric 'technical'
75.0	4418	9:1	1	20 kPa	3.2 Mpa	20 kPa	BS 1377:Part 5:1990
112.0	9852	10:1	1	10 kPa	1.6 Mpa	10 kPa	Large diameter (SI)
3 in (76.2)	7.069 in² (4560)	11:1	10 1b	1 tonf/ft² (107.3 kPa)	16 tonf/ft²	0.22 tonf/ft² (23.7 kPa)	BS imperial units

Apparatus for the test

7. Consolidation cell, consisting of

(a) Consolidation ring (cutting ring), typically 75 mm internal diameter and 20 mm high, of stainless steel or plated brass or gunmetal, rigid, with smooth internal surface and a cutting edge.

(b) Cell body and base of corrosion-resistant material, and watertight, with a retaining ring in which the specimen ring can be fitted with a push fit so that it is laterally restrained. The materials of the components of the whole assembly must not be corrodible by electro-chemical reaction with each other.

(c) Loading cap (pressure pad) which transmits the vertically applied force to the specimen via a central seating (usually a ball or hemispherical seating).

(d) Two porous discs of ceramic ware, sintered bronze or sintered fused aluminium oxide, and free-draining, with plane upper and lower surfaces. They must be capable of withstanding the maximum vertical pressure applied to the specimen. The upper disc should be 0.5 mm smaller in diameter than the inside diameter of the ring, and chamfered to prevent jamming. The lower disc should be large enough to support the ring.

The cell assembly is shown diagrammatically in Figure 14.3. The separate components can be seen in Figure 14.20(a) and the assembled cell in Figure 14.20(b).

Filter papers are not placed between the specimen and the porous discs, because fine soil particles can become enmeshed in the fibres of filter paper, causing clogging of the pores and impairing the drainage. However, proper attention must be given to the porous discs. They should be inspected before each test, to check that they allow free drainage of water. After each use the discs should be scarified with a bristle or nylon brush, and boiled. Do not use a metal brush or steel wool. Grinding may be necessary periodically to remove surface clogging. Clogged discs were examined by Baracos (1976).

Saturate the pores of the discs either by boiling in distilled water, or by immersion in distilled water under vacuum (pressure about 20 mm of mercury, i.e. about 2.5 kPa), for at least 20 minutes. Keep the discs immersed in distilled water until required for use, unless

Figure 14.20 Typical oedometer consolidation cell: (a) component parts; (b) assembled cell

the soil to be tested readily absorbs water, in which case allow them to air-dry at room temperature.

8. Micrometer dial gauge (referred to as the compression gauge), 10 mm travel, readable to 0.002 mm. A 'backward reading' gauge is convenient, i.e. one on which the reading increases as the stem is extended, as seen in Figure 14.21(a). Alternatively a displacement transducer may be used (see Section 8.2.1, subsection on 'Conventional and electronic measuring instruments') (see Figure 14.21(b)).

9. Consolidation press or load frame, the main features of which are as follows:

    (e) Rigid beam supported in suitable bearings to provide a convenient magnification ratio which enables the intensity of pressure applied to the test specimen to be determined to an accuracy within 1%, or 1 kPa, whichever is greater. When working in SI units of pressure, a 9:1 beam ratio is suitable for a 75 mm specimen using kilogram weights. Other possibilities are given in Table 14.8.

    (f) Adjustable counterbalance weight on beam.

    (g) Loading yoke assembly, to apply a vertical force to the specimen loading cap through a spherical seating on the loading stem.

    (h) Screw jack support for beam.

    (i) Rigid bed to support consolidation cell.

    (j) Vertical specimen compression movement capability of at least 15 mm (or at least 75% of the specimen thickness if greater than 20 mm).

    (k) Rigid support and mounting for compression gauge.

    (l) Weight hanger for slotted weights.

    (m) Pan for smaller loose weights.

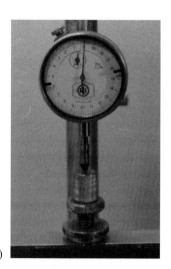

(a)   (b)

Figure 14.21 Measurement of compression in oedometer test using: (a) dial gauge; (b) displacement transducer

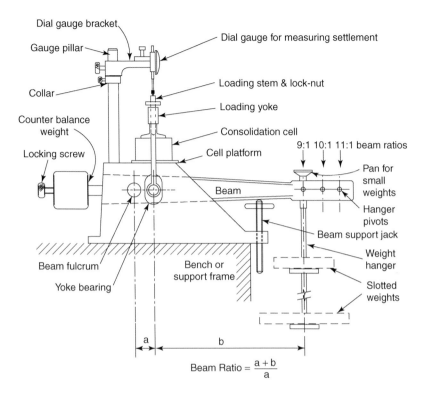

Figure 14.22 General arrangement of a typical oedometer press

The principle of an oedometer press is shown in Figure 14.22, and a group of presses in operation is shown in Figure 14.23. The deformation characteristics of the load frame and cell assembly should be determined as described in Section 14.8.1.

10. Rigid stand or bench, to which the load frame (or several frames) can be bolted. It is essential to bolt down the back of the frame, to prevent it overturning when the hanger is fully loaded. The stand or bench must be bolted to the floor, or provided with a shelf on which counterbalance weights can be placed (see Figure 14.24).

If a number of frames are mounted on one bench, it must be strong enough to carry the frames and the full complement of weights without distortion, and must be made secure against overturning when all frames are fully loaded. The stand or bench should preferably be placed on a solid floor at ground level, but if installation on a suspended floor is unavoidable the feet should be placed on spreader pads to avoid highly concentrated point loads on the floor.

11. Calibrated masses, the exact value of which should be known to an accuracy of 1%:

Slotted weights		Separate weights	
15 no. × 10 kg		*1 no.* × 500 g	
1 no. × 5 kg		2 no. × 100 g	
2 no. × 2 kg		1 no. × 50 g	
1 no. × 1 kg		1 no. × 10 g	

Figure 14.23 Oedometer presses with cells under test (photograph courtesy of Soil Mechanics Ltd)

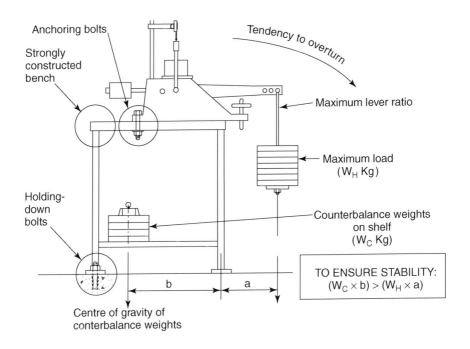

Figure 14.24 Stability requirements for oedometer press and bench

When not in use, slotted weights for the load hangers should be stored carefully (see Section 8.3.6). Each load frame should be allocated its own set of weights. Small loose weights should be kept in a box on the bench beside the load frame
Stop-clock or timer, readable to 1 s
Maximum and minimum thermometer
Wash-bottle or beaker containing water at room temperature
Silicone grease

14.5.4 Procedural stages

1. Prepare and check apparatus
2. Weigh and measure consolidation ring
3. Cut and trim specimen into ring
4. Determine moisture content and particle density from trimmings
5. Weigh specimen in ring
6. Assemble specimen in consolidation cell
7. Fit cell in load frame
8. Set up loading yoke
9. Adjust beam
10. Set dial gauge
11. Add first load increment to hanger
12. Apply load to specimen
13. Saturate specimen
14. Record settlement readings
15. Plot readings
16. Decide whether to proceed to next load increment
17. Apply next load increment
18. Further loading stages, as required
19. Unload
20. Plot readings
21. Unloading stages, as required
22. Drain cell
23. Remove specimen
24. Weigh specimen
25. Dry and weigh
26. Calculate moisture content of specimen
27. Analyse settlement graphs (graphical analysis is described in detail in Section 14.5.6)
28. Calculate values of e, m_v, c_v (see Section 14.5.7)
29. Plot voids ratio curve (see Section 14.5.7)
30. Report results

14.5.5 Test procedure

1. *Preparation of apparatus*
Check that the consolidation ring is clean and not distorted, the inside face is smooth, and the cutting edge is sharp with no burrs. If the cell body and base are separate components, ensure that the 'O' ring seal is in good condition and fits properly, and apply a little silicone grease to it before screwing the base tight home. Ensure that the cell components are clean

and dry, and assemble them to see that they fit together properly in accordance with the manufacturer's instructions. Take care not to damage the cutting edge of the consolidation ring. Dis-assemble the components and moisten the inside surfaces of the cell body. The porous discs should be prepared as described in Section 14.5.3, item 7 (d). Fit the lower disc into place at the centre of the cell base.

Check that the beam moves freely and that the weight hanger is fitted to give the required lever ratio (see Table 14.8).

With the loading yoke in the vertical position, adjust the counterbalance weight as necessary so that the beam and hanger assembly is just in balance. Lock the counterbalance weight in position, and re-check. Ensure that screwed connections on the loading yoke, weight hanger and gauge supports are all tightly secured.

Detailed adjustments and checks will vary with the type of equipment used, and the manufacturer's instructions should be carefully followed.

2. Measurement of consolidation ring

Measure the internal diameter of the ring, using internal vernier callipers, to an accuracy of 0.1 mm in two directions at right angles. The mean diameter is denoted by D (mm).

Measure the height of the ring at several points using vernier callipers or a micrometer, to an accuracy of 0.05 mm. Alternatively, the measurements may be made on a flat glass plate using a dial gauge mounted on a comparator stand. The average measurement, rounded to the nearest 0.1 mm, is the initial specimen height H_o. Weigh the ring to the nearest 0.01 g (m_R). Lubricate the inner face lightly with silicone grease, or apply a light coating of polytetrafluorethylene (PTFE) compound from an aerosol spray can, to minimise side friction. Weigh the watch glass or metal tray to the nearest 0.01 g (m_T).

3. Cutting specimen and trimming into ring

The procedure depends upon the type of sample, and three possibilities are described in Chapter 9.

Preparing an undisturbed specimen from a tube sample (see Section 9.2.2)
Cutting an undisturbed specimen from a block sample (see Section 9.3.1)
Making up a recompacted specimen (see Sections 9.5.3 or 9.5.5)

4. Determination of initial moisture content and particle density

The trimmings taken from immediately adjacent to the test specimen are placed into one or more moisture content containers. Fit the lid or lids without delay and determine the moisture content ($w_o\%$) of the trimmings by the usual method (see Volume 1 (third edition), Section 2.5.2). Some of the trimmed material can be used for the determination of the particle density of the soil (see Volume 1 (third edition), Section 3.6.2).

5. Weighing specimen

Place the specimen in the consolidation ring on the watch glass or tray, and weigh to the nearest 0.01 g (m_1). The mass of the specimen, m_o, is determined from

$$m_o = m_1 - (m_R + m_T) \text{ grams}$$

(m_R and m_T are as measured in stage 2).

6. *Assembly of consolidation cell*

With the lower porous disc located centrally on the base of the cell, lower the consolidation ring and specimen (cutting edge uppermost) centrally onto the disc.

Fit the ring retainer and cell body around the ring so that it is securely held and tighten the fixing nuts progressively. In some types of cell the body itself acts as the ring retainer.

Place the upper porous disc centrally on top of the specimen, checking that the clearance is equal all round.

Locate the spigot on the loading cap into the recess in the upper disc, so that the cap fits centrally.

7. *Fitting cell in load frame*

With the loading yoke swung forward and resting on the beam, place the cell centrally in position on the platform of the machine base (see Figure 14.22), locating with the spigot if there is one. An intermediate adaptor may be necessary if a non-standard cell is used. Raise the beam to just above the horizontal position and hold it there with the screw jack support. A small weight (10 g) placed on the top weight pan should be enough to prevent the beam from 'floating'.

8. *Setting up loading yoke*

Lift the end of the beam to allow the loading yoke to be raised to the vertical position (see Figure 14.25) and adjust the loading stem by screwing it downwards until the end engages closely in the recess on the top of the loading cap. Tighten the locking nut on the loading stem (see Figure 14.26). In some presses a separate ball-bearing is provided to transmit the load from the stem to the cap.

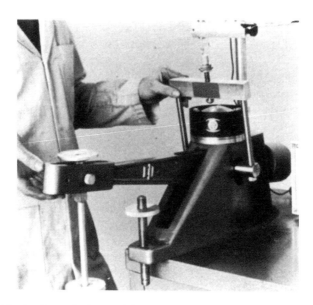

Figure 14.25 Raising loading yoke into position

Figure 14.26 Tightening locking nut after adjusting loading stem to bear on top cap

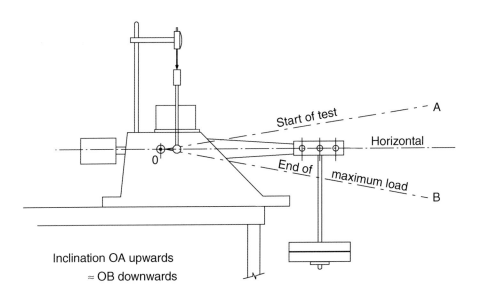

Figure 14.27 Ideal limits of beam inclination during a test

9. *Adjustments to beam*

The ideal position for the beam initially is to be inclined above the horizontal at about the same angle as it will be below the horizontal at the end of the maximum load increment (see Figure 14.27). The exact position is not critical, especially for stiff soils for which the amount of compression is small, and can be estimated from experience. Inclination adjustment requires only a small rotation of the loading stem.

If a large depression of the beam occurs during a test on a soft soil, the beam can be raised at the end of a loading increment by adjusting the loading stem. Any movement of the compression dial gauge as a result of this operation should be recorded and accounted for in the subsequent calculations, but it may be possible to re-set the gauge to give the same reading as before adjustment. After adjustment tighten the locking nut on the stem (see Figure 14.26) and check that contact with the cap is still maintained.

10. *Setting dial gauge*

Attach the compression dial gauge or transducer to the arm on the support post. If the top surface of the load stem is flat, the dial gauge should be fitted with a ball anvil; if spherical, a flat anvil should be used (see Section 8.3.2). Adjust first the height of the dial gauge, and then adjust and lock the gauge stem seating stud (see Figure 14.28) to give a convenient initial reading. If a dial gauge is used a small final adjustment may be made by rotating the bezel slightly (see Section 8.3.2). The gauge stem should be near the top end of its travel, but not fully compressed. This reading is the initial zero reading of the compression gauge. If the screw jack support is wound down slightly for an instant, a fractional downward indication of the dial gauge confirms that all members of the assembly are properly seated. Re-adjust the support so that the dial gauge is set at the initial zero reading (denoted by d_i) again.

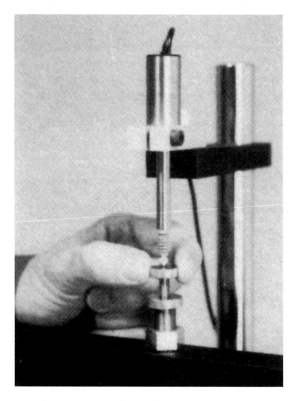

Figure 14.28 Adjustment of gauge stem seating stud

11. Addition of weights to hanger

Add weights carefully to the load hanger to give the required initial pressure (see Figure 14.23), and remove the small weight from the top pan. There should be little or no deflection indicated by the compression gauge because the additional load is carried by the screw jack support.

The initial pressure should be appropriate for the type of soil. Values suggested as a general working guide are given in Table 14.9, and are based on those given in the BS. The symbol σ'_{vo} indicates the estimated effective overburden pressure at the horizon from which the specimen was taken.

Weights should be placed on the hanger systematically. The main (lower) weight pan is intended for the largest (10 kg) slotted weights and the upper one for smaller slotted weights (5, 2 and 1 kg) (as in Figure 14.23). The small pan at the top, if fitted, is intended for small loose weights. The suggested disposition of weights for recommended pressures is summarised in Table 14.10.

Table 14.9 Suggested initial pressures for consolidation test

Consistency of soil	Initial pressure
Stiff	Equal to σ'_{vo}, or swelling pressure. Nearest pressure on recommended loading sequence (see Table 14.12) may be used for convenience
Firm	Somewhat less than σ'_{vo} (or next lower recommended pressure)
Soft	Appreciably less than σ'_{vo}; usually 20 kPa or less
Very soft, including peat	Very low; 6 or 10 kPa. Consolidation under initial very small load will give sufficient added strength to prevent soil squeezing out through clearance between ring and upper porous disc when next increment is added

Table 14.10 Suggested disposition of hanger weights (beam ratio: 9:1; specimen 75 mm diameter)

Pressure (kPa)	Total weights (kg)	Disposition of weights		
		Pan (g)	Upper hanger (kg)	Lower hanger (kg)
3	0.15	100 + 50		
6	0.3	200 + 100		
10*	0.5	500		
12	0.6	500 + 100		
20*	1		1	
25	1.25	200 + 50	1	
50	2.5	500	2	
100	5		2 + 2 + 1	
200	10		2 + 2 + 1	5
400	20		2 + 2 + 1	5 + 10
800	40		2 + 2 + 1	5 + (3 × 10)
1600	80		2 + 2 + 1	5 + (7 × 10)
3200	160		2 + 2 + 1	5 + (15 × 10)

*Not consistent with doubling sequence
Note: See Section 8.5.2 regarding the stacking of slotted weights on a hanger

12. *Application of load to specimen*

Wind down the beam support and at the same time start the clock. The compression gauge should indicate an immediate downward movement.

13. *Soaking of specimen*

Add water at room temperature to the cell without delay, so that the specimen and upper porous disc are completely submerged. If swelling is indicated, increase the load to the next higher pressures, or to give sufficient pressure to overcome the swelling tendency and cause consolidation. If it is required to determine the swelling pressure, the procedure given in Section 14.6.1 should be followed from the outset.

14. *Recording settlement readings*

Observe the compression gauge readings and the clock, and record the readings on a consolidation test form (see Figure 14.29(b)) at the selected time intervals.

Convenient intervals are those given in the left-hand half of Table 14.11, which are based on those given in the BS. These give approximately equal intervals when plotted on a log-time scale. Corresponding values of square-root-time (minutes) are included and if these are printed on the test form as in Figure 14.29 it facilitates plotting the square-root-time graph.

An alternative series of time intervals, which provides equal intervals on a square-root-time scale, is given in the right-hand half of Table 14.11. But in practice these time intervals are less easy to remember, and plotting to a log-time base is more difficult than with the other series.

At the end of the day record the 8 h reading or as close to it as possible. Record the 24 h reading next day. Record the daily maximum and minimum temperatures, to the nearest 1°C, in the vicinity of the test.

The exact time of reading is not critical, especially after the first hour, provided that the actual time is recorded as well so that the true time interval can be plotted.

15. *Plotting readings*

Plot the readings of the compression gauge against time to a logarithmic scale using five-cycle semi-logarithmic paper (see Figure 14.30). A graph of gauge readings against square-root-time may also be plotted on a separate sheet (normal graph paper), as in Figure 14.31, in which case it is preferable to make the ordinate scale of gauge readings the same on both graphs. Plotting should be started as soon as convenient after the application of the load and then kept up to date as the test proceeds.

16. *Decision regarding next load increment*

After plotting the 24 h reading, the decision must be taken whether or not to apply the next load increment. If the log-time graph shows a flattening out from the steep part of the curve to a straight line which is less steeply inclined, as in Figure 14.30, it indicates that the primary consolidation phase is complete and that the next load increment may be applied. However, if the inclined straight line representing secondary compression has not yet been established, the load should be left unchanged for another 24 h. The only additional readings necessary are at about 28 h, 32 h and 48 h from the start of the increment, i.e. two during the day and

Consolidation Test — Settlement Readings

Location: Dulston Operator: M. B. J Loc. No. 3824 Sample No. C2 25 Cell no. 3 Ring no. 3 Dia. 74.9 mm Height 20.1 mm Date started 19.5.80

LOAD & UNLOAD — Incr. no, Date:

Stage	Load	Pressure	Date
(1)	2.5 kg	50 kN/m²	20.5.80
(2)	5 kg	100 kN/m²	21.5.80
(3)	10 kg	200 kN/m² (Swelling pressure)	21.5.80
(4)	20 kg	400 kN/m²	23.5.80
(5)	10 kg	200 kN/m²	26.5.80
(6)	2.5 kg	50 kN/m²	27.5.80

Gauge readings in units of ×10⁻³ mm; ΔH in ×10³ mm.

hr	min	sec	mins	√t	(1) Clock	(1) Gauge 000	(2) Clock	(2) Gauge	(3) Clock	(3) Gauge	(4) Clock	(4) Gauge	(5) Clock	(5) Gauge	(6) Clock	(6) Gauge
0	0	0	0	0	0920	0	0927	124	09/2	384	0917	793	0925	1309	0918	1149
		6	.10	.32		21		157		452		855		1281		1094
		10	.17	.41		23		163		458		862		1280		1091
		15	.25	.50		25		167		463		870		1278		1090
		30	.50	.71		29		174		468		889		1273		1083
	1		1	1.0	0921	35		188		482		906		1267		1076
	2		2	1.41		41		209		499		927		1261		1070
	4		4	2.0		49		232		518		962		1251		1058
	8		8	2.83		58		260		546		1003		1237		1045
	15		15	3.9		66		284		573		1044		1225		1024
	30		30	5.5		75	0957	311		620		1098		1205		999
1			60	7.75	1020	86	1027	332	0942	661	1017	1153	1025	1183	1018	957
2			120	11.0	1120	95		349	1012	707		1211	(1147)	1162		895
			(142)	(11.9)												
4			240	15.5	1320	107	1327	364	1355	749		1260		1157		826
			283	16.8												
8			480	21.9	1720	115	1727	375	1712	768	1717	1289	1725	1155		783
			770	27.7					2202	778						
24			1440	38.0	0915 (21/5)	124	0902 (22/5)	384	0910 (23/5)	793	0930 (24/5)	1309	0915 (27/5)	1149	0910 (29/5)	763
48			2880	53.7							0905 (26/5)	1320			0915 (29/5)	759
3 days			4320	65.7												

	(1)	(2)	(3)	(4)	(5)	(6)
ΔH (×10³ mm)	.124	.384	.793	1.309	1.149	.759
Cumulative correction	18 / .018	24 / .024	31 / .031	40 / .040	31 / .031	18 / .018
Net total settlement	.106	.360	.762	1.269	1.118	.741

LOAD — stages (1)–(4); UNLOAD — stages (5)–(6).

Figure 14.29 Typical settlement readings from an oedometer consolidation test of six stages (4 loadings, 2 unloadings) recorded on printed form

Table 14.11 Time intervals for recording compression readings

(h)	Logarithmic intervals (BS)			Square-root intervals		
	t (min)	(s)			t (min)	(s)
	0.167	10	0.409	0.09	5.4	0.3
	0.25	15	0.5	0.25	15	0.5
	0.5	30	0.707	0.49	29	0.7
	1		1	1	60	1
	2		1.41	2.25	135	1.5
	4		2	4		2
	8		2.83	9		3
	15		3.87	16		4
$\frac{1}{2}$	30		5.48	25		5
				36		6
1	60		7.75	64		8
				90.5		9.5
2	120		10.95	121		11
4	240		15.49	240		15.5
8	480		21.91	484		22
24	1440		37.95	1444		38
28	1680		41.0			
32	1920		43.8			
2 days	2880		53.7			
3 days	4320		65.7	Extended tests only		
4 days	5760		75.9			
5 days	7200		84.8			
6 days	8640		92.9			
7 days	10080		100.4			

one the next day. It is rarely necessary to extend a loading stage any further, unless data on secondary compression are required, in which case one week (about 10,000 min) or longer may be required.

The duration of every load increment throughout the test should be the same. Normally, for convenience, this will be 24 h. Prolonged secondary compression can affect the primary consolidation characteristics of succeeding stages. If a longer period than normal under one load cannot be avoided (e.g. over a weekend), and appreciable secondary compression occurs during this additional time, this should be allowed for when calculating the data for the voids ratio change graph (see Section 14.5.6, item 2, and Figure 14.34).

If it is evident that the primary consolidation phase is completed within the normal working day, the next load increment may be applied immediately.

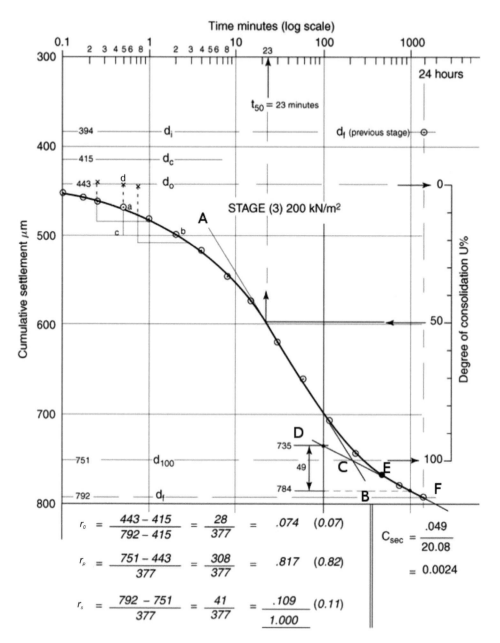

Figure 14.30 Log-time–settlement curve from readings in stage 3 of Figure 14.29, with calculations of compression ratios and secondary compression coefficient (compare with Figure 14.9)

17. *Application of next load increment*

When it has been established that the loading stage may be terminated and the next load increment applied, wind up the screw jack unit until it just touches the beam. A fractional

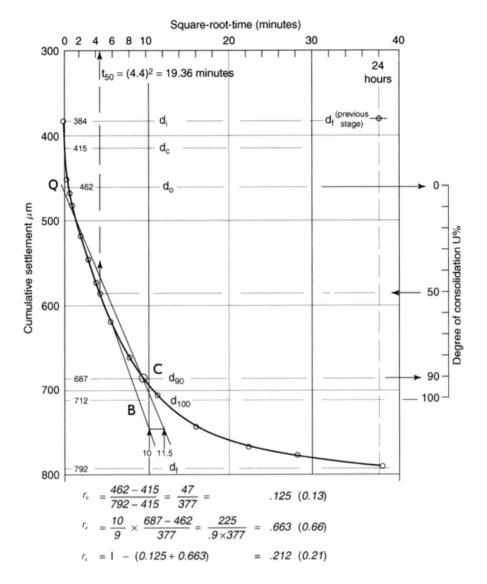

Figure 14.31 Square-root-time–settlement curve from stage 3 of Figure 14.29 (compare with Figure 14.10)

movement of the compression gauge may be indicated. Re-set the timer clock to zero. Do not re-set the compression gauge.

Place additional weights on the hanger to give the required new pressure. Normal procedure is to double the pressure at each new stage. The recommended sequence of loading in BS 1377 is given in Table 14.12. Loading increments are discussed further in Section 14.8.2.

If small weights need to be removed in exchange for larger weights, add the larger weights first so that the pressure on the sample is not momentarily diminished. All the additional load will be carried by the beam support jack. See stage 11 for the disposition of weights.

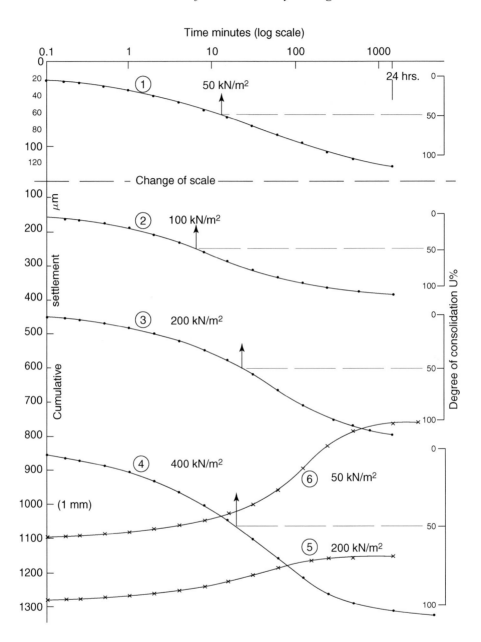

Figure 14.32 Cumulative log-time–settlement curves (loading and unloading) for all stages of Figure 14.29

To start the next loading stage, wind down the beam support and simultaneously start the clock, as in stage 12. Take readings, and plot the settlement–time graphs, as in stages 14 and 15.

Table 14.12 Recommended loading stages

Pressure on specimen for load ratio = 2, $\left(\text{i.e } \dfrac{\partial \rho}{\rho} \right)$ (kPa)	
6	Extended range for very soft soils
12	
25	
50	
100	'Normal' range
200	
400	
800	
1600	Extended range for stiff or overconsolidated clays
3200	

18. *Further loading stages*

Repeat stages 17, 14 and 15 for each successive load increment, after first complying with the recommendations of stage 16. The number of stages to apply depends on the purpose of the test and is discussed in Section 14.8.2. Settlement readings are plotted cumulatively against time, as shown in Figure 14.32.

When the end of the primary consolidation phase under the maximum required pressure has been reached, as indicated by the beginning of the secondary compression line, the specimen is considered to be fully loaded and consolidated.

19. *Unloading*

Unloading should take place not all at once but in a series of decrements. Usual practice is to unload and allow swelling in about half the number of stages as were applied during consolidation, with not less than two unloading stages. For instance if the loading sequence was 25, 50, 100, 200, 400, 800 kPa, a suitable unloading sequence would be 400, 100, 25 kPa.

Before taking off any weights from the hanger, set the clock to zero, and wind up the beam support so that it just touches the beam. Hold the beam down firmly against the support while the weights are removed; this requires a second person, except when only small weights are being removed. Check that the dial gauge shows little or no movement.

Release the beam, and at the same instant start the clock. There is no need to wind down the beam support because the beam will rise as the specimen swells. Ensure that the upper loading cap remains covered with water. Take readings of the compression gauge and temperature exactly as during the consolidation stages, and record them as shown in Figure 14.29 under increments 5 and 6.

20. *Plotting readings*

Plot the compression gauge readings against log-time on the same graph sheets as for the consolidation stages. The swelling curves will cut across the consolidation curves but will be easily distinguishable because they slope the other way, as shown in Figure 14.32 (curves 5 and 6). Square-root-time graphs are not usually plotted for the swelling stages.

Completion of swelling under a particular load is indicated by a flattening of the graph. If this is not evident after 24 h, allow to swell for a further 24 h before removing the next stage of load.

21. Further unloading stages

Each load decrement is removed as described in stage 19 and readings are plotted as in stage 20, making sure that swelling is virtually complete at each stage. Sometimes it might be possible to remove two stages of loading in one day.

When the pressure is back to the initial pressure (or the swelling pressure if applicable), a period longer than 24 h may be needed to allow completion of swelling. It is essential that equilibrium be established before finally unloading and removing the specimen.

22. Draining cell

When equilibrium is achieved, remove the water from the cell, either by draining off if there is a suitable connection or by siphoning it out. Allow to stand for 15–30 min so that the porous discs can drain. Observe any further movement of the compression gauge.

23. Removal of specimen

Take off the remaining weights from the load hanger and move the compression gauge to one side, and the loading yoke forwards, so that the cell can be removed.

Dismantle the cell, and take out the consolidation ring and specimen. Remove the porous discs carefully; any soil adhering to them should be scraped off and returned to the specimen. Wipe the outside of the ring dry.

24. Weighing specimen

Place the specimen and ring on the weighed watch glass or tray, and weigh to the nearest 0.01 g (m_2). Calculate the final mass of the specimen (m_f) from

$$m_f = m_2 - (m_R + m_T)$$

where m_R and m_T are as measured in stage 2.

25. Drying and weighing

Place the tray with specimen and ring in the oven overnight or long enough to ensure that the specimen has dried to constant mass. Allow to cool in a desiccator and weigh (m_3).

Calculate the dry mass (m_d) of the specimen from

$$m_d = m_3 - (m_R + m_T)$$

The dry mass is the same at the end of the test as at the beginning if no material has been lost.

26. Calculation of moisture contents

The calculated initial moisture content w_i is given by

$$w_i = \frac{m - m_d}{m_d} \times 100\%$$

which provides a check against the moisture content w_o obtained from the trimmings (stage 4).
The final moisture content (w_f) is calculated from the equation

$$w_f = \frac{m_f - m_d}{m_d} \times 100\%$$

27. Analysis of settlement graphs

The analysis of the log-time–settlement and square-root-time–settlement graphs is detailed in Section 14.5.6.

28. Calculations

Calculations for each stage of the test are set out in Section 14.5.7.

29. Plotting voids ratio curve

See Section 14.5.7.

30. Reporting results

See Section 14.5.7.

14.5.6 Graphical analysis

The plotting of settlement–time graphs, both to a log-time base and to a square-root-time base, for each loading stage, was described in Section 14.5.5, stage 15. These graphs are used to determine the values of the coefficient of consolidation, c_v, by performing the curve-fitting analyses referred to in Section 14.3.7, and described below. The form of the curve obtained depends upon the type of soil. For saturated clays, on which the theory of consolidation is based, conventional curves similar to those derived from the theory are usually obtained. These are regarded as 'standard' and their analysis follows routine procedures.

Other types of inorganic soil such as silts give curves which appear to depart from the conventional curves, but fundamentally they are similar and the differences are only in degree resulting from the much higher permeability of these soils. Soils which are not fully saturated may give non-standard curves for other reasons. The analysis of non-standard curves sometimes causes difficulties, and suggestions for dealing with them are outlined below.

Four main types of soil are considered, namely:
1. Clay soils ('standard')
2. Clayey silts
3. Silts
4. Unsaturated soils

Other aspects of graphical analysis which are included are common to all types, and are as follows:

5. Allowance for deformation of equipment
6. Calculation of compression ratios
7. Determination of coefficient of secondary compression

The analysis of results from tests on peats, for which values of c_v are not relevant, is described in Section 14.7.

1. *Clays ('Standard' curves)*

The conventional methods of analysing settlement-time curves are given below for

  log-time curves

  square-root-time curves

These methods are similar to those given in the BS and in most textbooks.

The present authors suggest that in most cases the log-time curves should be used for deriving the theoretical 100% point, and the square-root-time curves for the theoretical 0% point.

(a) *Log-time method*

Theoretical 0% This construction applies to the initial convex-upwards portion of the curve, which is usually clearly defined for clay soils. The procedure is illustrated in Figure 14.9, and a typical example is shown in Figure 14.30.

Select two points on the curve whose time values are in the ratio 1: 4, e.g. 0.5 min and 2 min (points a and b in Figure 14.30). Measure a distance ad equal to ac (the vertical distance between a and b) upwards from point a to obtain the point d. Repeat the process once or twice more, say at 0.25 min and 1 min and at 0.75 min and 3 min, provided that all points lie on the convex-upwards part of the curve. Draw a horizontal line at the mean level of the points d so determined. This represents the theoretical $U = 0\%$ line and its intersection with the compression scale gives the conceptual gauge reading corresponding to the corrected point denoted by d_o (see Figures 14.9 and 14.30).

Theoretical 100% The point of inflexion of the log-time–settlement curve, i.e. the point at which the curvature changes direction, occurs at about 75% consolidation. Draw the tangent at this point, which is the tangent common to both the upper and the lower branches of the S-shaped curve (line AB in Figure 14.9.) Draw the tangent to the straight line portion at the end of the curve, and produce it backwards (line DEF) to meet AB at the point C. Draw a horizontal line through C; this represents the theoretical $U = 100\%$ line and its intersection with the compression scale gives the gauge reading of the corrected 100% primary consolidation point, denoted by d_{100} (see Figures 14.30 and 14.9).

A percentage consolidation scale, from $U = 0\%$ to $U = 100\%$ can now be constructed from d_o to d_{100} on the right-hand side of the settlement curve. Hence the abscissa d_{50} corresponding to 50% primary consolidation, which lies at a compression gauge reading equal to $0.5(d_o - d_{100})$, can be drawn. At the intersection of this horizontal line with the settlement curve the time for 50% primary consolidation, denoted by t_{50} (min), can be read off as shown in Figure 14.30.

(b) *Square-root-time method*

Theoretical 0% The procedure is indicated in Figure 14.10, and a typical example is shown in Figure 14.31. Extend the straight-line portion of the settlement curve downwards, and also upwards to intersect the zero time ordinate at Q (see Figure 14.10). This point represents theoretical $U = 0\%$, denoted by d_o, which lies below the initial reading d_i.

Theoretical 100% From the point Q draw the line which at any level has an abscissa 1.15 times that of line QB. An easy way of doing this is to find the point q on the line QB at which $\sqrt{t} = 10$, and draw a horizontal line through q to intersect the zero time ordinate at p. Make pr = $1.15 \times$ pq (i.e. the point r corresponds to $\sqrt{t} = 11.5$ in Figure 14.31). Join Qr, which intersects the settlement curve at C. The level of this point gives the theoretical 90% consolidation point, d_{90}. Read off the corresponding value of $\sqrt{t_{90}}$ and multiply it by itself to give t_{90} (min).

The 100% primary consolidation point d_{100} can be found by dividing the vertical distance between d_o and d_{90} into nine equal spaces, and extrapolating below d_{90} by a distance equal to one space. The percentage consolidation scale from $U = 0\%$ to $U = 100\%$ can be marked on the right-hand side of the graph (see Figure 14.31). The d_{50} point can then be found and the value of $\sqrt{t_{50}}$ read off, which multiplied by itself gives the 50% consolidation time t_{50} (min).

2. Clayey silts

This example relates to a soil for which the log-time–settlement curve is of the form shown in Figure 14.33. The latter part of this curve is similar to that shown in Figure 14.30, and two tangents can be drawn for the establishment of the d_{100} point (100% primary consolidation) as explained above.

The shape of the early part of the curve differs from the curve shown in Figure 14.30, which invalidates the construction described above for the determination of the d_o point (theoretical 0% primary consolidation). It is reasonable to assume that settlement occurred so rapidly immediately after loading that the initial convex-upwards portion of the primary curve was passed before any readings could be taken, as indicated by the broken curve to the left of the 0.1 min ordinate in Figure 14.33. In some cases it is possible to make a reasonable estimate of the d_o point from the square-root-time–settlement curve, using the construction described above. The d_o point is then transferred to the log-time curve and the conventional analysis can be carried out as before.

If the position of d_o is not evident from the square-root-time curve, a reasonable estimate is to assume that it lies somewhere within the middle third of the range between the initial reading at the start of the increment, d_i, and the earliest observed reading. These limits are marked on the graph, as shown in Figure 14.33, and the probable limits of d_{50} can then be drawn, from which the range of possible values of t_{50} can be read off.

The mid-point of this range can be used to obtain an approximate value of c_v, which should be reported as such to one significant figure, i.e. to the nearest 10 m^2/year if in the range 10–100 m^2/year. This degree of accuracy is adequate because a c_v of this magnitude indicates that settlements would occur quite rapidly and would not be expected to cause long-term problems.

If the primary stage has clearly been completed by the time 100 min has been reached, as in Figure 14.33, it would be feasible to apply a second increment almost immediately instead of having to wait for 24 h. It may even be possible to complete several increments within one day. But if a subsequent load increment is sustained for a longer period, such as overnight, the consequent additional secondary compression occurring during this stage should not be included, in order to avoid a discontinuity in the e–log p curve. This is illustrated in Figure 14.34, in which the point P, instead of point Q, is used to compute the e value at the end of increment No. 3.

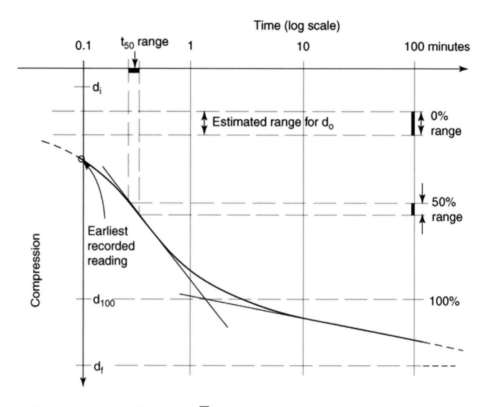

Mean height during load stage $\bar{H} = 19.8$ mm

$t_{50} = 1$ min gives $c_v = \dfrac{0.256 \times 19.8^2}{1} = 10$ m²/year

$t_{50} = 0.1$ min gives $c_v = \dfrac{0.256 \times 19.8^2}{0.1} = 100$ m²/year

∴ If t_{50} is between 1 and 0.1 min,
   c_v lies between 10 and 100 m²/year
   and is reported to the nearest 10 m²/year

Figure 14.33 Derivation of likely range of c_v from log-time–settlement curve for clayey silt

3. *Silts*

A typical log-time–settlement curve for a relatively rapid draining soil, such as silt, is of the form shown in Figure 14.35. In this instance the recorded data give a curve which is concave-upwards from the start. The point of inflexion has been passed earlier than 0.1 min, as indicated by the hypothetical dashed curve. The d_{100} (100% primary) point cannot be determined by the conventional method, and the square-root-time curve (see Figure 14.36) is of little use for determining the d_o (0% primary) point because there is no linear portion evident. However, it is reasonable to assume that the d_{50} (50% consolidation) point will lie in a zone around the mid-way point for the test, i.e. about halfway between the d_c and d_f points. If this zone

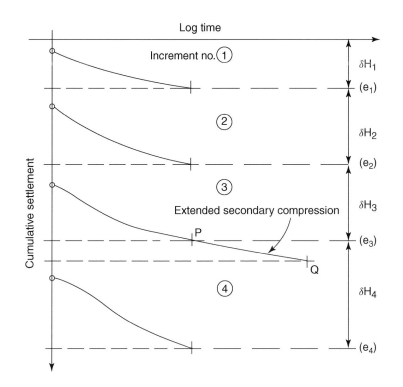

Figure 14.34
Evaluation of
voids ratios for
a load increment
which includes
extended
secondary
compression

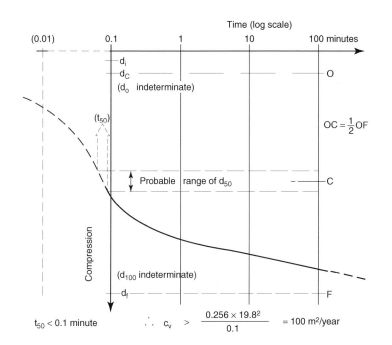

Figure 14.35
Derivation of
probable c_v value
from log-time–
settlement curve
for silt

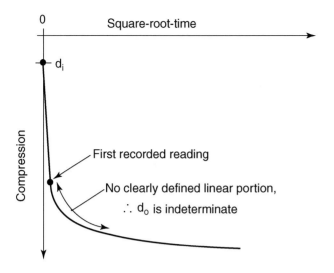

0   Square-root-time

d_i

Compression

First recorded reading

No clearly defined linear portion,
∴ d_o is indeterminate

Figure 14.36 Square-root-
time–settlement curve for silt
with a high c_v

clearly lies above the beginning of the laboratory curve, the value of t_{50} must be less than 0.1 min, as shown in Figure 14.35. For a specimen of the standard 20 mm thickness, the value of c_v will therefore be greater than about $0.0256 \times 20^2/0.1 = 102.4$ m²/year. The coefficient of consolidation would then be reported as being greater than 100 m²/year, i.e. $c_v > 100$ m²/year. This indicates very rapid consolidation, and a more explicit result may not be needed.

If the probable zone of the d_{50} point lies within the range of settlement readings, an estimated value of t_{50} can be obtained from which an approximate value of c_v can be calculated and reported to one significant figure as in item 2 above.

To obtain a more definite value of c_v a test using a larger sample, such as the use of a Rowe consolidation cell (to be covered in Volume 3) would be needed. Alternatively, a single-drainage test can be carried out in the standard oedometer cell by placing an impervious membrane (such as a disc cut from a triaxial test rubber membrane) between the specimen and the lower porous disc. Drainage then takes place upwards only, and the length of drainage path, h, is equal to the specimen height, H. The time to achieve a given percentage of consolidation is increased by a factor of 4. Using Equation (14.14) the value of c_v is calculated from the equation

$$c_v = \frac{0.104 \times (\bar{H})^2}{t_{50}} \text{ m}^2/\text{year}$$

for the single-drainage case. Other calculations are the same as those for double drainage.

4. *Unsaturated clays*

Clays which are not fully saturated contain pockets or bubbles of gas (usually air) in the voids between the solid particles. This results in two significant departures from the assumptions given in Section 14.3.4.

The pore fluid is compressible

The permeability changes under the influence of applied stress

These effects are more likely to be significant in compacted clays, even when compacted at or slightly wet of the optimum moisture content, than in naturally occurring unsaturated clays.

The characteristic features of the time–settlement curves obtained from oedometer consolidation tests on unsaturated clay are generally as follows:

1. A large initial compression
2. A log-time–settlement curve which is somewhat flatter than the theoretical curve in the primary consolidation phase
3. A square-root-time–settlement relationship which is continuously curved, instead of showing an initial linear portion
4. A steeper secondary compression line

These features are illustrated in Figure 14.37. Allowances may be made for these effects on an empirical basis when analysing the curves, on the lines indicated in point 3 above. A detailed theoretical study was made by Barden (1965), but as yet no standard curve-fitting procedure has been devised.

5. *Calibration corrections*

Allowance has to be made for the deformation of the apparatus when calculating voids ratio changes and the compression ratios (see Section 14.8.1). The corrections are usually insignificant for very compressible soils.

The method of calculating voids ratio changes given in Section 14.3.9 uses the cumulative displacement for each stage, with reference to the start of the test. The cumulative deformation of the apparatus, denoted by Δa, is subtracted from the observed cumulative displacement at the end of each stage to obtain the net change in specimen height, denoted

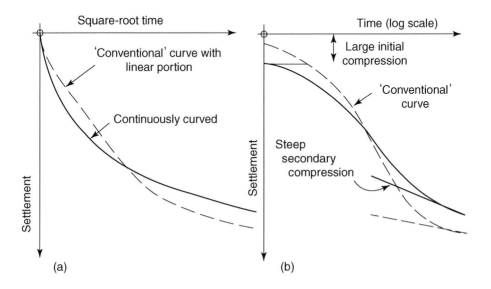

Figure 14.37 Features of time–settlement curves for partially saturated clay: (a) square-root time; (b) log-time

by ΔH. Values of Δa are read directly from a calibration curve of the type shown in Figure 14.54, obtained as described in Section 14.8.1.

Compression ratios

The compression ratios (Section 14.3.12) are calculated for each loading stage individually. The incremental deformation of the apparatus, denoted by δ, due to the additional load placed on the hanger, is added to the initial displacement gauge reading d_i for the stage (which is the same as the final reading d_f at the end of the previous stage) to give the corrected initial reading d_c from which the actual compression of the specimen is measured. The procedure is illustrated in Figure 14.9. The value of δ is obtained from calibration data for the load frame tabulated in the manner shown in Figure 14.54 (see Section 14.8.1).

6. *Compression ratios*

The compression ratios r_o, r_p, r_s are calculated by using Equations (14.30)–(14.34) after applying the correction referred to above. An example is included in Figures 14.30 and 14.31. (Reporting of compression ratios is not a requirement of BS 1377: 1990.)

7. *Coefficient of secondary compression*

The coefficient of secondary compression, C_{sec}, is derived from the straight-line portion of the log-time–settlement curve, as follows. The procedure is illustrated in Figure 14.14. The initial height of specimen at start of test = H_o mm. The compression in the linear range over one log cycle of time = $(\Delta H)_s$ mm. This is the same as the slope of the line, related to one log cycle. The coefficient of secondary compression is

$$C_{sec} = \frac{(\Delta H)_s}{H_o}$$

An example is shown in Figure 14.30. C_{sec} is a dimensionless number and the result is expressed to two significant figures.

14.5.7 Calculations and results

Calculations are summarised below, generally in the form of equations, grouped under the main aspects of the test, i.e. initial conditions; final conditions; parameters from each load increment stage. They are illustrated by the examples shown in the typical worksheets of Figures 14.38 and 14.39, using the settlement gauge readings tabulated in Figure 14.29 which are plotted graphically in Figure 14.32.

The symbols used in the equations, some of which have appeared earlier, are summarised in Table 14.13. Plotting of the voids ratio–log pressure curve, and reporting of results are given at the end of this section.

Initial conditions

Area of specimen $A = \dfrac{\pi D^2}{4}$ mm^2

Initial volume $V_o = \dfrac{A \times H_o}{1000}$ cm^3

CONSOLIDATION TEST – CALCULATION SHEET

Date Started.......... 19.5.80   Loc. No. 3824
Job .... Dulston   Sample No........ C2 – 25
Soil Type .... Soft to firm grey & brown   Cell No. 3
.... mottled silty CLAY   Ring No. 3

BEFORE TEST

Moisture content from trimmings 22.9

Particle density (Measured) 2.66 Mg/m³

Weight of sample + ring + tray.	439.35 g	Diameter D .... 74.9 mm
Weight of ring + tray	260.43 g	Area A .... 4406 mm²
Weight of sample	178.92 g	Thickness H_0 .... 20.1 mm
Weight of dry sample	145.35 g	Volume .... 88.56 cm³
Weight of initial moisture	33.57 g	Density ρ .... 2.02 Mg/m³
Initial Moisture Content m_0	23.1 %	Dry Density ρ_D .... 1.64 Mg/m³

Initial Void Ratio $e_0 = \dfrac{G}{\rho_D} - 1 = \dfrac{2.66}{1.64} - 1 = 0.622$

Initial Saturation $S_0 = \dfrac{m_0 \times G}{e_0} = \dfrac{23.1 \times 2.66}{0.622} = 98.8$ %

Equivalent height of solids $H_S = \dfrac{H_0}{1 + e_0} = \dfrac{20.1}{1.622} = 12.39$ mm

AFTER TEST

Weight of sample + ring + tray	436.94 g	Overall Settlement .... 0.741 mm
Weight of dry sample + ring + tray	405.36 g	Volume Change .... 3.26 cm³
Weight of ring + tray	260.43 g	Final Volume .... 85.30 cm³
Weight of wet sample	176.51 g	Final Density .... 2.07 Mg/m³
Weight of dry sample	144.93 g	Final Dry Density .... 1.70 Mg/m³
Weight of moisture	31.58 g	Final Void Ratio e_f 0.562
Final Moisture Content m_f	21.8 %	

Final Saturation $S_f = \dfrac{m_f \times G}{e_f} = \dfrac{21.8 \times 2.66}{0.562} = 103.2$ %

Figure 14.38 Typical data sheet for oedometer test specimen details

Initial mass $m_o = m_1 - (m_R + m_T)$ g

Dry mass (remains constant) $m_d = m_3 - (m_R + m_T)$ g

Moisture content $w_o = \dfrac{m_o - m_d}{m_d} \times 100\%$

Density $\rho = \dfrac{m_o}{V_o} \, \text{Mg}/\text{m}^3$

Dry density $\rho_D = \rho \times \dfrac{100}{100 + w_o} \, \text{Mg}/\text{m}^3$

449

CONSOLIDATION TEST – DATA FOR e/log p CURVE

Location __Dulston__ 3824   Sample no. __C2 - 25__
Date __28.5.80__

Increment no.	Voids ratio				Volume compressibility				Coefficient of consolidation				
	Pressure p kN/m²	Settlement ΔH mm	$\Delta e = \dfrac{\Delta H}{H_s}$ $H_s =$ 12.39mm	$e = e_o - \Delta e$ $e_o =$	Incremental changes δe	δp kN/m²	$(1+e_1)$	$m_v = \dfrac{\delta e}{\delta p} \times \dfrac{1000}{1+e}$ m²/MN	t_{50} min.	$H = H_o - \Delta H$ $H_o =$ 20.10 mm	$\bar{H} = \dfrac{H_1 + H_2}{2}$ mm	$(\bar{H})^2$ mm²	$c_v = \dfrac{.026 \times (\bar{H})^2}{t_{50}}$ m²/year
–	O	O	O	0.622	O	O	–	–	–	20.10	–	–	–
1	50	.106	.0086	.613	.0086	50	1.622	0.106	13	19.99	20.05	402	0.804
2	100	.360	.0291	.593	.020	50	1.613	0.248	6.3	19.74	19.87	395	1.63
3	200	.762	.0615	.561	.032	100	1.593	0.201	23	19.34	19.54	382	0.432
4	400	1.269	.1024	.520	.041	200	1.561	0.131	19	18.83	19.08	364	0.498
5	200	1.118	.0902	.532	-.012	-200							
6	50	.741	.0598	.562	-.032	-150							

Figure 14.39 Oedometer consolidation test calculation sheet, using data from Figures 14.29 and 14.38

Voids ratio $e_o = \dfrac{\rho_s}{\rho_D} - 1$

Degree of saturation $S_o = \dfrac{w_o \times \rho_s}{e_o}\%$

Equivalent height of solid particles

$$H_s = \dfrac{H_o}{1 + e_o} \text{ mm}$$

Final conditions

Mass $m_f = m_2 - (m_R + m_T)$ g

Moisture content $w_f = \dfrac{m_f - m_d}{m_d} \times 100\%$

Height of specimen $H_f = H_o - (\Delta H)_f$ mm

Density $\rho_f = \dfrac{m_f}{AH_f} \times 1000 \text{ Mg}/\text{m}^3$

Table 14.13 Symbols used for oedometer test calculations

Measurement or reading	Units	Before test	During test				After test
			Start of stage	End of stage	Cumulative change	Incremental change	
Specimen mass*	g	m_o					m_f
Dry mass*	g	m_d					m_d
Diameter	mm	D					D
Height	mm	H_o	H_1	H_2	ΔH	δH	H_f
Mean height	mm		$H = 1/2$ $(H_1 + H_2)$				
Compression gauge	μm	G_o	G_1	G_2			
Area*	mm²	A					A
Volume	cm³	V_o					
Pressure	kPa		p		Δp	δp	
Swelling pressure	kPa	p_s					p_s
Voids ratio	—	e_o	e_1	e_2	Δe	δe	e_f
Saturation	%	S_o					S_f
Moisture content	%	w_o					w_f
Coefficient of volume compressibility	m²/MN		m_v				
Coefficient of consolidation	m²/year		c_v				
Permeability	m/s		k				
(Mass of specimen + ring + container)	g	m_1					m_2
Mass of ring*	g	m_R					m_R
Density	Mg/m³	ρ					ρ_f
Dry density	Mg/m³	ρ_D					ρ_{Df}
Particle density*	—	ρ_s					ρ_s
Correction for deformation of apparatus	μm				Δa	δ	

Values which remain constant

Dry density $\rho_{Df} = \rho_f \times \dfrac{100}{100 + w_f}$ Mg/m^3

Voids ratio $e_f = \dfrac{\rho_s}{\rho_{Df}} - 1$

or $\quad e_f = e_o - (\Delta e)_f$

$\left.\begin{array}{c}\ \\ \ \end{array}\right\}$ two methods provide cross-check

Degree of saturation $S_f = \dfrac{w_f \times \rho_s}{e_f}\%$

(often gives rather more than 100%)

End of each loading stage

The total observed settlement from the beginning of the test (settlement dial gauge reading G_o) to the end of a given load increment stage (reading G_2) is (G_2-G_o) µm. From this must be subtracted the cumulative correction for the deformation of the apparatus, denoted by Δa (see Section 14.5.6, item 5), to obtain the net compression of the specimen ΔH, as shown by the last two lines in Figure 14.29. The difference on the bottom line is divided by 1000 to convert to mm and is transferred to the appropriate line in the ΔH column of the calculation sheet, Figure 14.39, for the following calculations:

Voids ratio change (cumulative):

$$\Delta e = \frac{\Delta H}{H_s}$$

Voids ratio after an increment:

$$e = e_o - \Delta e$$

Voids ratio change during an increment:

$$\delta e = e_1 - e_2$$

Coefficient of volume compressibility for an increment:

$$m_v = \frac{\delta e}{\delta p} \times \frac{1000}{1 + e_1} \text{ m}^2/\text{MN}$$

Coefficient of consolidation during an increment:

$$c_v = \frac{0.026 \times (\bar{H})^2}{t_{50}} \text{ m}^2/\text{year}$$

Coefficient of permeability during an increment:

$$k = c_v m_v \times 0.31 \times 10^{-9} \text{ m/s}$$

If this is calculated it must be reported as the *calculated* permeability under the stated pressure.

If the average laboratory temperature during an increment stage differs from 20°C by more than ± 2°C, multiply the calculated values of c_v and k by the appropriate correction factor obtained from Figure 14.18 (see Section 14.3.16) to obtain the equivalent values at 20°C.

Coefficient of secondary compression during the secondary compression stage of an increment (if required):

$$C_{sec} = \frac{(\Delta H)_s}{H_o}$$

Voids ratios are calculated for the unloading as well as the loading stages. Calculations of m_v, c_v and k are made for the loading stages only.

The calculation of the compression ratios, r_o, r_p, r_s is described in Section 14.3.12, and an example is referred to in Section 14.5.6, item 6 (see Figure 14.30).

Plotting voids ratio curve

From the calculated data derived as in Figure 14.39, plot a graph of voids ratio against log pressure for the loading and unloading stages, as shown in Figure 14.40. This is known as the *e*–log *p* curve.

Values of c_v can also be plotted against log pressure, in the manner shown by the graph below the *e*–log *p* curve in Figure 14.40. The c_v values can alternatively be plotted midway between each pressure value, because c_v and m_v relate to the increase from one load to the next. In Figure 14.40, the c_v and m_v values are tabulated in this way. These coefficients are laboratory values and should be so described. They may not be directly applicable to settlement calculations (see Sections 14.3.8 and 14.4.4).

Reporting results

A complete set of results from an oedometer consolidation test comprises the following:

Specimen identification, location, depth
Soil description
Specimen dimensions
Bulk density, moisture content, dry density
Initial voids ratio, degree of saturation (if calculated degree of saturation exceeds 100%, report it as 100%)
Particle density, if required, and whether measured or assumed
Swelling pressure (to two significant figures) if applicable
Plot of voids ratio, or vertical compression (%), against log pressure, for loading and unloading
Load increments and decrements applied (kPa)
Coefficient of volume compressibility (to two significant figures)
Coefficient of consolidation (to two significant figures). These coefficients are reported for each load increment, with the remark that they are laboratory calculated values
Coefficient of permeability (to two significant figures) (not normally reported unless requested, and then as a laboratory calculated value)
Coefficient of secondary compression (to two significant figures) for the secondary phase of each increment (if appropriate)

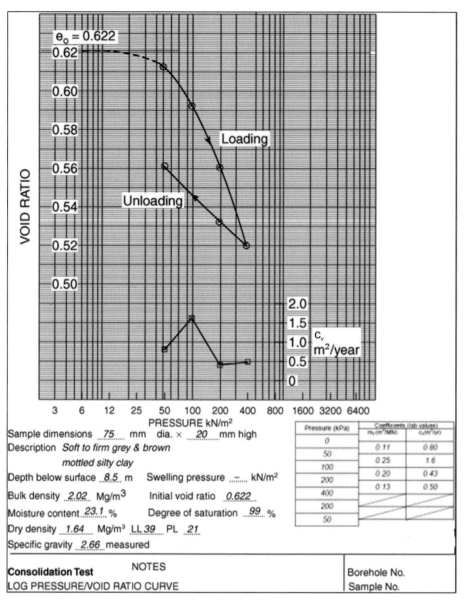

Figure 14.40 Voids ratio–log pressure curve and results summary from Figure 14.39

Plots of compression against log time, and/or compression against square-root-time, as
appropriate, for each load increment

Method of curve fitting used

Mean laboratory temperature during the test

Method of test, i.e. the oedometer test for determination of one-dimensional consolidation
properties, carried out in accordance with Clause 3 of BS 1377:Part 5:1990

14.5.8 ASTM consolidation test

The one-dimensional consolidation test specified by ASTM D 2435 is very similar in principle to the BS test, but there are differences in some details, the most significant of which are summarised as follows:

1. The consolidation cell may be either the fixed-ring type or the floating-ring type.
2. The minimum specimen dimensions are 2 in (50 mm) diameter and 0.5 in (12.5 mm) high. The minimum diameter: height ratio is 2.5:1, with preference for a ratio exceeding 4:1.
3. A disc of copper or hard steel is used for the deformation calibration of the apparatus, with the porous stones moistened.
4. An initial seating load of 5 kPa is applied to the specimen unless it is a very soft soil, in which case 2 or 3 kPa is applied.
5. Standard loading pressures (in kPa) are as follows:

$$5, \quad 12, \quad 25, \quad 50, \quad 100, \quad 200, \text{ etc.}$$

Smaller increments may be used on very soft soils. Alternatively a loading, unloading and re-loading sequence which reproduces the in-situ stress changes may be used.

Calculations, plotting, graphical analysis and reporting of results are similar to those described for clay soils in Section 14.5.6 or 14.5.7. Graphical analysis includes a method for evaluating the preconsolidation pressure from the voids ratio–log pressure curve.

14.6 Special purpose tests

This section describes test procedures other than the standard consolidation test given in Section 14.5. Most of them are extensions of that test or relate specifically to particular soil types.

14.6.1 Measurement of swelling pressure (BS 1377:Part 5:1990:4.3 and ASTM D 4546)

This test is applicable to overconsolidated clays or to other soils which are susceptible to swelling when they are allowed free access to water (see Section 14.3.15). These could include recompacted soils that have been very heavily compacted. The following procedure for the determination of swelling pressure continues from stage 13 of Section 14.5.5 if it is known beforehand that a swelling pressure measurement is required; or from stage 12 if swelling is indicated after a load increment is applied for a standard consolidation test. The procedure was developed by the original author and is included in BS 1377:1990. A similar procedure is given as part of Method C of ASTM D 4546, except that the specimen is loaded to the in-situ vertical pressure before inundation.

As soon as swelling is indicated, add a small weight to the pan at the top of the weight hanger to bring the compression gauge back to the zero reading, or to within two or three divisions of it. If this is not enough, add more weights; if too much, leave for a little while in case swelling resumes. Continue to observe the compression gauge and add more weights as necessary to maintain the reading as close as possible to zero. Record the amount of each load increment and the time from the start when it was added.

As the weight on the load hanger increases, the deformation of the apparatus has to be taken into account. The calibration is described in Section 14.8.1 and reference to the calibration curve (such as Figure 14.55) will indicate the appropriate deformation correction. This correction is added to the original zero reading to give the gauge reading at which to aim. The zero 'target' therefore moves with every adjustment of the load.

Several hours may be needed to reach equilibrium with an overconsolidated clay. If the specimen has to be left overnight, load the hanger with excess weights with the beam resting on its support jack, and with the compression gauge at the existing corrected zero reading. Swelling pressure can then continue to develop but upward movement will be prevented as long as the hanger weights provide an excess pressure. The excess will be carried by the support jack. Next morning remove excess weights very carefully, wind down the support jack, and adjust the hanger load as before until the compression gauge indicates the correct zero reading appropriate to that load, which will be the swelling pressure at that time. Make further adjustments as necessary until equilibrium is indicated. This procedure could perhaps subject the specimen to unknown internal stresses overnight, but it is preferable to leaving the specimen to swell without enough restraint, which would result in the specimen being unusable for the consolidation test.

Plot a graph of the pressure on the specimen (kPa) against square-root-time, as in Figure 14.41. Flattening of the curve indicates that equilibrium is virtually reached. The pressure required to maintain the specimen at its original height is known as the swelling pressure (p_s) and is reported as such to two significant figures. When equilibrium is established the dial gauge reading should be the same as that shown on the calibration curve for that hanger load, to within 0.01 mm.

The consolidation test (see Section 14.5.5) is then started at stage 13 by adding weights to the hanger to bring the total pressure on the specimen to the next pressure above the swelling pressure on the 'standard' loading sequence (see Table 14.12). The compression gauge should not be re-set. Thereafter, a normal loading sequence is followed.

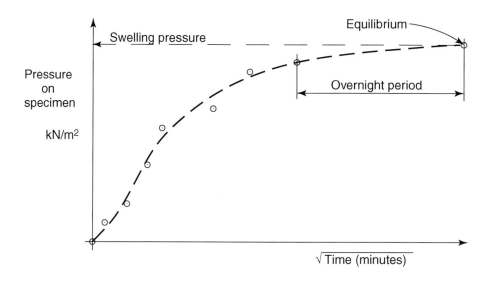

Figure 14.41 Swelling pressure test curve

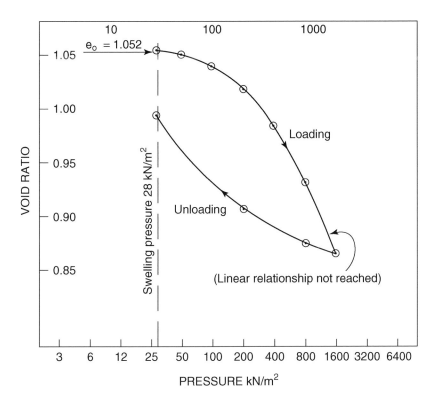

Figure 14.42 Typical *e*–log *p* curve for overconsolidated clay

Unloading is carried out as for a standard test, but the final unloading pressure should be equal to the swelling pressure. Unless provision is made for a swelling test (see Section 14.6.2) the pressure should not be reduced below that value until the specimen is removed. After reaching equilibrium at the swelling pressure, the cell is drained and the specimen removed as described in Section 14.5.5, stage 22 onwards.

On the *e*–log *p* curve (Figure 14.42) the swelling pressure is indicated by a vertical line. The first point on the curve is the intersection of this line with the horizontal line representing the initial voids ratio, because during the swelling pressure test the volume, and therefore the voids ratio, remains constant. The whole of the *e*–log *p* curve lies to the right of the vertical line representing the swelling pressure.

14.6.2 Swelling test (BS 1377:Part 5:1990:4.4)

It is important to appreciate the difference between the *swelling pressure test*, described above, in which the pressure to *prevent* swelling is measured, and the *swelling test*, described below, in which swelling is allowed to take place and is measured.

When the swelling characteristics are to be measured, special provision must be made at the specimen preparation stage. The specimen height must be less than the height of the consolidation ring so as to ensure that the specimen remains laterally confined as it swells. A difference in height of 3–5 mm is usual.

Apparatus

The only item required in addition to that listed in Section 14.5.3 is a flanged disc of corrosion-resistant metal, with flat and parallel faces, of the type shown in Figure 14.43(e). The diameter D_1 should be about 1 mm less than the specimen diameter D, and the height of the upstand, t, is equal to the difference between the required specimen height and the height of the oedometer ring: typically 3–5 mm. The flange diameter D_2 should be a few millimetres larger than the outside diameter of the ring. Alternatively a glass or metal disc of diameter D_1 and thickness t could be used (see Figure 14.43(a)), but the absence of a flange might make it difficult to remove from the inside of the oedometer ring.

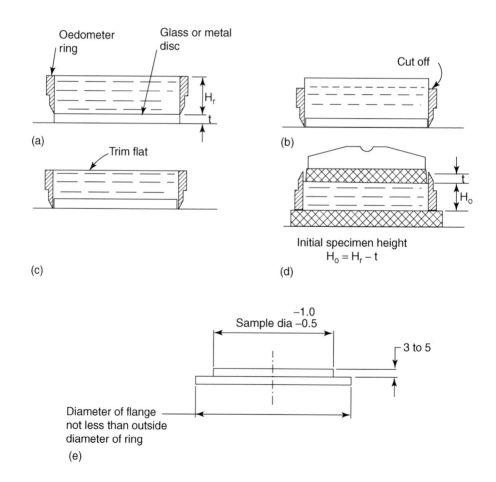

Figure 14.43 (a)–(d) Stages in the preparation of specimen for oedometer swelling test; (e) flanged disc (reproduced from Figure 1(b) of BS 1377:Part 5:1990)

Procedure

Measure the thicknesses of the disc and flange to the nearest 0.01 mm, using a micrometer. The difference is the height of the upstand.

Prepare the apparatus and the specimen in the consolidation ring as described in Section 14.5.5, stages 1–3, and measure the moisture content of trimmings, stage 4.

Place the disc on the flat glass plate, and place the specimen and ring, cutting edge downwards, centrally over the disc (see Figure 14.43(a)).

Push the ring steadily downwards, without tilting, until the cutting edge is firmly in contact with the glass plate or flange (see Figure 14.43(b)). Without moving the specimen, cut off the extruded portion and trim the specimen flat and flush with the end of the ring (see Figure 14.43(c)).

Weigh the specimen in the ring, assemble in the oedometer cell and set up in the load frame, as in Section 14.5.5, stages 5–10. The difference now is that the top of the specimen is a few millimetres below the top edge of the ring, as shown in Figure 14.43(d). Adjustment of the loading stem will allow for this. The initial height of the specimen is equal to the height of the ring less the upstand height or disc thickness.

The swelling pressure is then determined as in Section 14.6.1.

When equilibrium has been established, the swelling test is performed by unloading the specimen in stages from the swelling pressure as the starting point. The procedure is as described in Section 14.5.5, stage 19, and log-time graphs are plotted as in stage 20. Loads are usually halved at each decrement; i.e. if the swelling pressure is denoted by p_s, the unloading sequence could be $0.5p_s$, $0.25\,p_s$, $0.125\,p_s$ etc. and so on, down to the smallest load required.

Alternatively the load may be reduced in equal decrements by removing for instance 20 kPa at each stage. Each stage should be continued until the graphical plot indicates that equilibrium has been reached.

If the amount of swell approaches the thickness of the disc, no further swelling should be allowed, otherwise the top of the specimen will no longer be confined by the ring. If this happens a repeat test should be carried out using a thinner specimen.

Re-loading back to the equilibrium load is carried out in the same stages as for unloading. If the consolidation characteristics are also required, the test may then be continued as a standard consolidation test. Otherwise the specimen can be removed after equilibrium is reached, as described in stages 22 onwards, in Section 14.5.5.

In either case the complete cycle of unloading and re-loading is plotted in the form of an e–log p curve. Values of m_v and c_v are derived for the loading stages only.

Results are presented in the same way as for an ordinary consolidation test (see Section 14.5.5, stage 30). The test is described as a 'swelling test' or a 'swelling and consolidation test', as appropriate.

14.6.3 Settlement on saturation (BS 1377:Part 5:1990:4.5, and ASTM D 4546)

For soils that are not completely saturated, the effect of sudden inundation is sometimes significant. For instance, silt or sand at a low relative density, whether naturally deposited or recompacted, may show a sudden decrease in volume on saturation due to collapse of the grain structure (Capps and Hejj, 1968). Fly ash (PFA) is a material which is susceptible to this effect. (This phenomenon is in contrast to the behaviour of overconsolidated soils

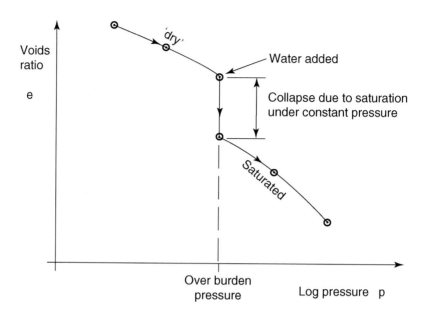

Figure 14.44 Effect of settlement due to saturation on *e*–log *p* curve

which tend to swell on being allowed access to water, described above). A procedure for investigating collapse potential is included in BS 1377:1990, and the procedure given under Method B of ASTM D 4546 is similar in principle.

The specimen is set up in the oedometer cell, using porous discs which have been air-dried after de-airing. The specimen is loaded in suitable pressure increments, up to a pressure equal to the in-situ overburden pressure, without the addition of water to the cell. During this operation the cell should be covered with a damp cloth under a piece of polythene sheet to prevent the specimen drying out. When equilibrium has been reached, water is added to the cell so that the sample becomes totally submerged while the pressure remains unaltered. If collapse occurs, this will be indicated by movement of the compression dial gauge, which should be observed and recorded in the same way as for a normal consolidation stage until equilibrium is re-established. On the *e*–log *p* curve this stage will be plotted as a vertical line, because the voids ratio will change under a constant pressure (see Figure 14.44). Further loading and unloading stages may be carried out, with the specimen remaining saturated, as appropriate. An example is quoted by Kezdi (1980).

14.6.4 Measurement of expansion

Expansion index of soils (ASTM D 4829)

This test is for the determination of the amount of swell, expressed as the 'expansion index', of a compacted soil when inundated with water.

A mould with extension collar securely holds a specimen ring 102 mm diameter and 25.4 mm high. The soil at the appropriate moisture content is compacted into the mould to give a specimen about 50 mm high, using a 2.5 kg compaction rammer applying 15 blows on each of two layers. The specimen is then trimmed to the height of the ring and placed in an oedometer loading frame.

A vertical pressure of 6.9 kPa is applied to the specimen, and after 10 min the compression gauge is set to read zero. The specimen is then flooded with water and readings of the gauge are taken until equilibrium is established, the minimum period being 3 h.

The change in height, ΔH (mm), from the initial height H_1 (mm) is calculated from the final reading. The expansion index (EI), is calculated from

$$EI = \frac{\Delta H}{H_1} \times 1000$$

and is reported to the nearest whole number. The degree of saturation is also reported, from which the value of EI corresponding to 50% saturation can be estimated.

Expansion of ferrous slags

Most slags generated as by-products from the ferrous industries provide satisfactory and stable fill material. Exceptions are steel slags, which have potential expansive properties which it is important to detect if they are to be used as confined fill. A laboratory procedure for assessing the degree of expansion was described by Emery (1979) and is now published as ASTM standard D 4792.

The material is compacted into a compaction mould (a CBR mould could also be used), and at the same time a non-expansive control material is compacted in a similar manner into a separate mould. Perforated bases are fitted, and the samples are immersed in a water bath maintained at $82 \pm 1°C$. Appropriate surcharge weights are applied, and the swell is observed by means of a dial gauge mounted in a manner similar to that used for measuring the swell of a soaked CBR sample (see Figure 14.45). Readings are continued over a period of 7 days, and the amount of expansion (relative to the control sample) is expressed as a percentage of the initial height. The temperature of 82°C accelerates the swelling, which at 7 days has been found to be about twice that observed after more than a year at 20°C.

Fig 14.45 Accelerated slag expansion tests

14.6.5 Overconsolidated clays

Some of the methods of analysis of consolidation test data for overconsolidated clays were referred to in Section 14.3.14. Those procedures are beyond the scope of this book, but the data required for the analysis includes the following which can be provided from laboratory tests.

1. The e–log p curve, extended to as many load increments as the capacity of the apparatus will allow. Three points on a straight line do not necessarily indicate that the virgin compression line has been reached.
2. The unloading curve back to the p_o value or to the swelling pressure p_s. This is just as important as the loading curve and should be well defined by means of several decrements of load.
3. The Atterberg limits of the clay, for the derivation of C_c empirically.
4. Measured particle density.
5. Some analysis procedures make use of a laboratory load–unload–reload cycle. The type of curve obtained is indicated in Figure 14.17, which can be used to obtain a field curve similar to BCDEFG in Figure 14.16. The re-loading curve eventually merges with the continuation of the initial loading curve.

14.6.6 Permeability measurement

Some oedometer cells are provided with a means of carrying out a direct permeability measurement while the specimen is under load. The essential features of such a cell are a bottom inlet, which can be connected to a standpipe such as a burette; sealing rings to prevent water finding its way around the specimen and containing ring; and an upper overflow outlet. A cell of this type, arranged for a falling head permeability test, is shown diagrammatically in Figure 14.46.

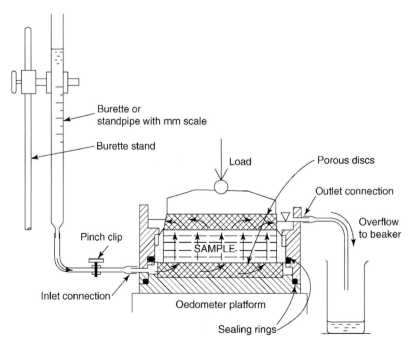

Figure 14.46 Arrangement for falling head permeability test in oedometer consolidation cell

Burette or standpipe with mm scale

Burette stand

Load

Porous discs

Outlet connection

Overflow to beaker

Pinch clip

SAMPLE

Inlet connection

Oedometer platform

Sealing rings

When equilibrium has been reached under the pressure at which the permeability is to be measured, connect the bottom inlet to a burette or to a suitably calibrated glass tube, by means of a length of rubber tubing fitted with a pinch clip. Connect a length of rubber tubing to the upper outlet leading to a beaker. Fill to near the top of the burette with de-aired water without trapping any air, and fill the cell up to the overflow level. Open the pinch clip, and start the clock when the level in the burette reaches a given mark. Record the time taken for the level in the standpipe tube to fall to a second mark. Repeat two or three times.

From each set of readings calculate the permeability of the soil specimen as for the falling head permeability procedure (see Section 10.7.2). The height of specimen used in the calculations is the height of specimen at the end of the load increment.

Report the average value, to two significant figures, as the *measured* permeability under that particular effective stress.

14.7 Consolidation tests on peat

14.7.1 Properties and behaviour of peat

General characteristics

The general characteristics and classification of peat were outlined in Volume 1 (third edition), Section 7.6.2. Unconsolidated peats usually have very high water contents (typically 75% to 95% by volume, or several hundreds percent, even over 1000%, by dry mass); high voids ratios (typically 5–20), and organic contents ranging from about 30% in fen peats to 98% in bog peats. The voids also include gas which is generated during humification. Peats are also very variable within very short distances, as is evident from the variability of moisture content distribution and general character of the peat. A very detailed account of the properties and engineering behaviour of peat was given by Hobbs (1986), which is essential reading for anyone involved in the testing of peat and in the interpretation of test data.

As a result of these characteristics, peats have a very high compressibility, low shear strength, high initial permeability, very rapid initial rate of consolidation and a high degree of secondary compression. These properties can change dramatically under increasing load.

The conventional Terzaghi theory of consolidation (see Section 14.3.5) is not applicable to peats. Departures from the basic assumptions on which the Terzaghi theory is based, listed in Section 14.3.4, may be summarised as follows:

1. The solid material is itself compressible
2. Permeability changes considerably during a load increment stage
3. Vertical displacements are large compared with the material thickness, resulting in a moving boundary which invalidates the conventional boundary assumptions
4. As a result of these large deformations, appreciable structural re-arrangements take place within the material during consolidation

Interpretation of tests on peat requires a complete reversal of the traditional view of the consolidation process. The conventional approach based on the determination of m_v and c_v is not valid.

Terminology

The following terminology (some of which is relevant only to this section) is used.

'Primary' consolidation (c_p): The total compression accompanied by the dissipation of excess pore water pressure which takes place during a loading stage from the end of the

'primary' consolidation of the previous stage (or, in the first stage of loading, from the instant of application of the load) to the end of the 'primary' phase of the stage considered. The end of 'primary' is ascertained as described below.

This definition of 'primary' encompasses the 'initial' compression (see below), and any secondary compression from the previous loading stage.

Time t_p: The time elapsed from the start of the load increment to the end of the 'primary' phase.

Initial compression (c_i): The amount of compression which occurs from the instant of loading ($t = 0$) to the arbitrarily selected time $t = 15$ s (0.25 min), being the time at which the first sensible settlement reading can usually be observed.

Secondary compression: Compression which becomes apparent only after the 'primary' phase, and showing a linear relationship with log time.

Compression ΔH_p: The cumulative compression of the specimen up to time t_p.

Coefficient of secondary compression (C_{sec}): The ratio of the change in height of the specimen over one cycle of log time during the secondary phase, to the original height of the specimen.

Primary compression index (strain-related) (C_c^*): The slope of the primary consolidation line when strain is plotted against log pressure. The relationship to the conventional compression index, C_c, is given by the equation

$$C_c^* = \frac{C_c}{1 + e_o}$$

Isochronous compression index (C_{ci}^*): The slope of compression isochrones on a plot of strain against log pressure.

Most of the above definitions are illustrated in Figure 14.48 (see section below) and some differ from those given earlier for conventional tests. That is why 'primary' is here written in quotation marks.

Consolidation behaviour

In conventional analysis of the compressibility of clays, primary consolidation is the dominant process and secondary compression is an appendage which is often dismissed as being of little significance. In peats, however, the reverse applies, and secondary compression is the dominant factor. Immediately after loading the compression process is temporarily distorted by the time lag associated with primary consolidation. Hobbs (1986) elegantly describes the primary effect as 'a mere aberration at the outset of the stately march through logarithmic time'. This concept is illustrated in Figure 14.47, in which the secondary compression line has been extended backwards virtually to the start of loading. Primary consolidation in itself is not important except that it is used to locate a point of reference on the secondary compression line.

Because of the dominant role of secondary compression, which is time-dependent, time has to be introduced into the relationship between deformation and log pressure. One of the main objectives in testing peat is to derive an array of lines representing secondary compression strain at various times, referred to as isochrones, which are fundamental for the interpretation of consolidation test data. Interpretation is beyond the scope of this book but the laboratory procedures for obtaining the necessary data and graphical plots are

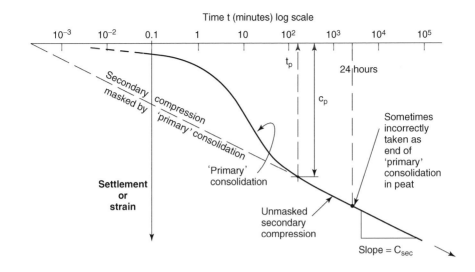

Figure 14.47 Typical relationship between strain and log time for peat, showing initial distortion from linearity due to 'primary' time lag

outlined below. A thorough study of the paper by Hobbs (1986), which includes methods of interpretation, should be made before embarking on the testing of peat.

The division of the consolidation process into initial, 'primary' and secondary phases is arbitrary, and follows the practice normally used in the UK. Some Canadian engineers prefer to consider only an 'immediate' or 'initial' phase, followed by a 'long-term' (secondary) phase.

The initial compression, c_i, forms a very large proportion of the 'primary' consolidation, c_p, at first, but the proportion decreases with increasing load. It also takes place very rapidly. The parameter c_i is useful because it provides the engineer with an indication of the proportion of settlement which will occur in the field quite rapidly relative to the rate of construction loading.

The time t_p for 'primary' consolidation is initially very small but increases with increasing load as the permeability decreases. It can be determined more accurately if pore water pressure is measured, either by using a Rowe cell or by the simple method outlined below for a test in a sample tube.

The coefficient of secondary compression, C_{sec}, depends on several factors and the relationships are very complex. As far as test procedures are concerned, the following factors are relevant to the assessment of C_{sec}:

1. Load increment ratios should be equal to or greater than unity
2. Each load increment should be sustained for a time that is long enough to eradicate the influence of previous loads in multiple-increment tests
3. The extent to which C_{sec} depends on the applied load should be determined, but this is likely to be small for pressures exceeding the critical pressure p_c, (indicated in Figure 14.49)

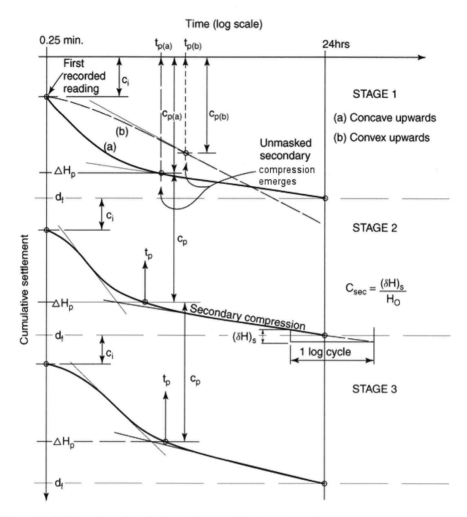

Figure 14.48 Illustration of symbols used in consolidation analysis for peat

The amount of both primary and secondary compression is approximately proportional to the thickness of the specimen or the layer in-situ, as for clays. Calculations of the amount of settlement are based on strains, not on values of m_v, and therefore a plot of cumulative strain $(H_p/H_o = \varepsilon)$ against log pressure is required as well as an e–log p plot.

The proportionality between the time for primary consolidation, t_p, and the square of the length of drainage path, h (see Section 14.3.5) needs modification for peats because the vertical permeability of peat in the field is greater than that measured on a laboratory specimen. The relationship stated by Hobbs (1986, Equation (11)) is

$$\frac{t_f}{t_s} = \left(\frac{h_f}{h_s}\right)^2 \cdot \frac{k_s}{k_f}$$

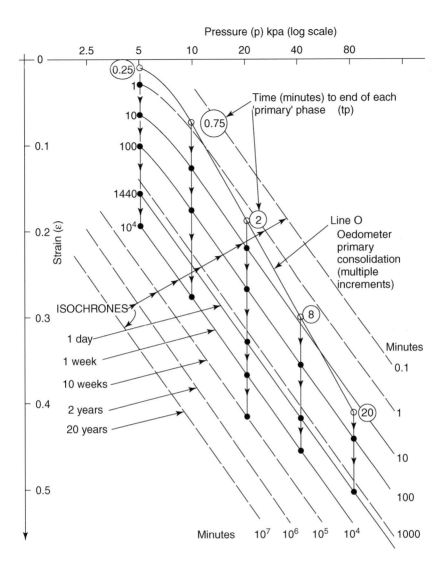

Figure 14.49 Derivation of isochrones of 'primary' and secondary compression for a lightly humified peat

where t_f, t_s are the times for field and laboratory consolidation, respectively; h_f, h_s are the field and laboratory lengths of drainage path; and k_f, k_s are the field and laboratory vertical permeabilities.

14.7.2 Derivation of isochrones

The way in which points defining isochrones of primary and secondary compression are obtained is illustrated in Figure 14.49 for a lightly humified peat (H3), in which deformation is plotted against log pressure in terms of strain. This diagram relates to Figures 27 and 29 of the paper by Hobbs (1986), and the examples of elapsed times are taken from the latter. An idealised set of five single-increment oedometer loading tests (in this example 5, 10, 20, 40, 80 kPa) on identical specimens is represented.

The strain at the end of each primary consolidation phase (at time t_p) is plotted against the pressure as an open circle. Typically, t_p might be less than one minute at small pressures, increasing with pressure up to more than 1 hour depending upon the degree of humification. The actual time examples are written against each point. Each pressure is sustained so that secondary compression continues, and strains are recorded and plotted at selected times from the start of loading. Intervals forming log cycles of time are convenient, viz. 100 min, 1000 min, 10,000 min (1 week), in addition to 1440 min (24 h). Continuation to about 10 weeks (if practicable) would give 10^5 min. Temperature control of the laboratory to within ± 2°C would be necessary for long-duration tests.

The strain corresponding to each time duration is shown as a solid black circle in Figure 14.48. Lines joining sets of points of equal time form the secondary compression isochrones (full lines). Interpolated or projected isochrones are shown as dashed lines. Within the range of pressures normally applicable to peat these lines are sensibly straight, but they are not necessarily parallel or uniformly spaced. Their slope (change in strain per log cycle of pressure) is equal to C_{ci}^*, and the vertical spacing between isochrones separated by a factor of 10 is equal to C_{sec}. The primary consolidation line labelled O is not parallel to isochrones because of the variability of the time t_p with pressure.

Isochrones of this type are essential for sensible interpretation of data from tests on peat. The derivation of isochrones is described in more detail in Hobbs (1986), which also describes their application.

14.7.3 Samples and test specimens

Sample tubes containing undisturbed peat samples should be stored on end, not lying down. Tubes which are not full should be topped up with water taken from the site—not with tap water, because the permeability of peat can be affected by the water chemistry.

The length of each undisturbed sample should be measured and compared with the distance the tube was driven when the sample was taken. The 'recovery ratio' can then be calculated, i.e. the ratio of the length of sample recovered to the corresponding thickness in-situ.

Specimens should be prepared as described in Chapter 9 (Sections 9.2.2 or 9.3.1), with attention to the following additional details.

Examination of the sample, together with von Post test (see Volume 1 (third edition), Section 7.6.2), should indicate whether the peat contains lumps of woody matter or other hard material. 'Woody' peats can cause difficulties such as tilting and possible jamming of the loading cap or the indication of a marked reduction of compressibility at some stage. A thick specimen should therefore be tested wherever possible. In any case the specimen should be closely examined and probed with a pin to locate any pieces of woody or hard material, which should be removed and replaced by peaty material before weighing. If excessive disturbance is caused the specimen should be discarded and a new one prepared. However it is better to accept some disturbance than to test a thin specimen containing a hard lump.

A difficulty which can arise with a peat or soil containing organic matter is that gas in the pore water may cause the specimen to expand, causing a trimmed flat surface to swell up to a rounded profile. If this is observed it should be reported and, if practicable, the amount of swell should be measured over a period of time while preventing the specimen from losing moisture. A second specimen should be prepared for the consolidation test and placed in the cell and loaded as quickly as possible.

In addition to measuring the bulk density of the specimen, the following tests should be carried out on the whole test specimen, wherever practicable, even though preliminary classification tests may have already been done, in order to provide data for correlation of the relevant properties of peat:

> von Post classification (see above)
> Moisture content (oven temperature not exceeding 105°C)
> Loss on ignition (550°C)
> Liquid limit (after thorough mashing, using a liquidiser, and mixing)
> Plastic limit (if practicable)
> Particle density (for determination of degree of saturation, and relating to ignition loss (see Section 14.3.17))

14.7.4 Apparatus for consolidation tests

Conventional oedometer consolidation tests using the normal size of specimen may not be suitable for peats unless modified procedures are used. Ideally, test specimens should be as large as possible, and use of Rowe consolidation cells (250 mm diameter for block samples, 150 mm diameter for piston samples) is preferable to a standard oedometer. Uniform strain loading should be applied, allowing vertical drainage. This apparatus will be described in Volume 3. Consolidation of a block sample may also be carried out in a large shearbox, using a specimen 300 mm^2 and 150–200 mm thick. The contact surfaces between the two halves of the box should be well greased and tightly clamped together, to make the box watertight. A settlement dial gauge with 50 mm travel may be necessary.

If a standard oedometer is the only suitable apparatus available, provision should be made for testing a specimen thicker than the normal 20 mm if one can be accommodated. The duration of the 'primary' consolidation phase can be extended by a factor of 4 if a single-drainage test is carried out, as referred to in Section 14.5.6, item 3. A settlement dial gauge with 25 mm travel, instead of the usual 12 mm, should be used.

Porous discs of high permeability should be used. They should be clean, and separated from the peat by a disc of Whatman No. 54 filter paper to prevent clogging of the pores.

$$\text{Pore water pressure (p.w.p.)} = \frac{y}{100} \text{ kN/m}^2 \text{ approximately}$$

Another arrangement is to consolidate a portion of an undisturbed tube sample while it is still in the sampling tube. A U-100 tube can be cut so that a convenient length, up to 100 mm, is available for testing. The sample can be loaded either in a load frame or by means of dead weights suspended from a counterbalanced hanger, as indicated in Figure 14.50. Adequate clearance should be provided between the loading piston and tube wall to avoid the risk of jamming. A porous disc should be placed between piston and sample to permit drainage of water. Consolidation settlement is observed by means of a long travel dial gauge and the overall settlement can be checked with a steel ruler. The volume of water squeezed

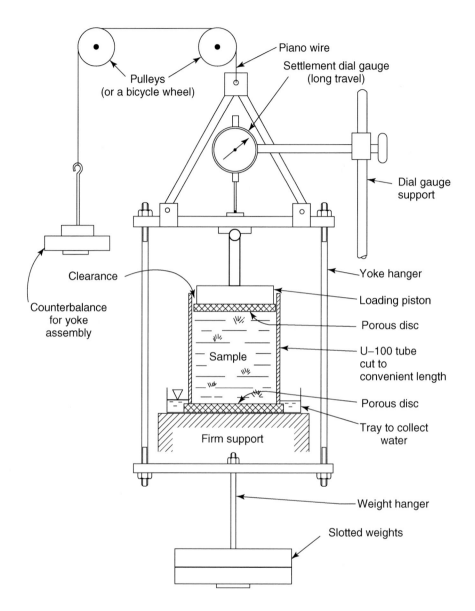

Figure 14.50 Consolidation of peat sample in U-100 tube

out can also be collected and measured. A pressure of 100 kPa on a 100 mm diameter sample requires a load of about 80 kg, so this method is practicable only if the maximum desired loading is not too high.

A disadvantage of this procedure is the relatively large amount of side friction, which could amount to about 10% of the applied load for a specimen with an *H:D* ratio of 1:1. An appropriate correction can be made to the applied load.

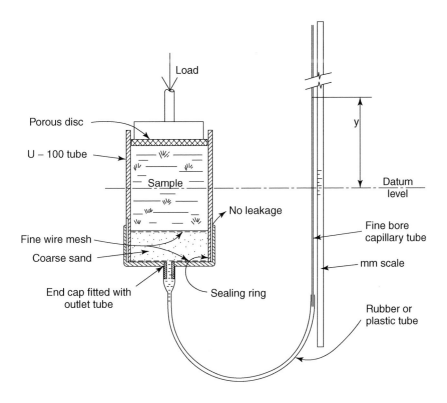

Figure 14.51 Simple method for measuring pore water pressure using capillary tube Elevation of water level y mm indicates excess pore pressure at base of sample

On the other hand it is possible to measure the change in pore water pressure by connecting a long capillary tube to the base of the specimen, as indicated in Figure 14.51. The lower end of the sample tube includes a layer of coarse sand, which must be saturated and completely free of air, contained between two layers of fine wire mesh.

Initially the water in the capillary tube should be about level with the mid-height of the specimen and this is taken as the datum level. The height of the water level *y* in the capillary tube above datum is observed at the same time as each settlement reading and is plotted in the same way. About 1 m of capillary tube is required above datum level for every 10 kPa of pressure to be applied. A fine bore tube is necessary to limit the volumetric displacement to a negligible amount.

The test procedures to be described may be carried out by using any of the above apparatus.

14.7.5 Test procedures

Types of test

Prediction of the field behaviour of peat from laboratory tests would be difficult enough if peat were a reasonably uniform material. But prediction is further complicated by the large variability of peat over small distances, both horizontally and particularly vertically between different strata. Every test method has its disadvantages and the engineer has to do the best he can with the available capabilities and the data obtained. Averaging of results from multiple tests is usually necessary.

Three types of test, which are a function of the rate of application of loading irrespective of the type of apparatus used, are described below.

1. Conventional loading in multiple increments, as described in Section 14.5.5
2. Rapid loading in multiple increments
3. Single-increment loading

Multiple-increment tests in which the likely field loading is simulated will not model the field behaviour if the in-situ permeability is significantly greater than that measured in the laboratory, and if the field value of C_{sec} differs from the laboratory value. Accurate predictions of settlement are not possible without large scale field trials, however elaborate the test procedures.

Procedure details that are common to all three types of test are as follows.

The highest pressure to be applied need not be much in excess of the maximum estimated field loading. The initial loading stage should be at a small pressure, to guard against material being squeezed out past the loading piston.

Because of the very large initial settlement immediately after loading, the correction for the deformation of the apparatus (see Section 14.8.1) may be neglected.

Throughout the test the clearance between the loading plate (including the porous stone) and the oedometer ring, or cell or tube wall, should be checked with a feeler gauge or piece of razor blade. If rubbing against the side is detected the test should be discontinued and the reason recorded on the test form.

At the end of the test any free water in contact with the specimen should be removed as in the standard test (see Section 14.5.5, stage 22). The specimen can then be finally unloaded and removed, and weighed and dried for the determination of moisture contents as in stages 23–26. The oven drying temperature should not exceed 105°C; even so some oxidation of organic matter may occur. The whole test specimen should be dried for the moisture content measurement and then used for the loss on ignition test.

Initial voids ratio, degree of saturation, moisture content, bulk density and dry density are calculated from the final dry mass.

The graphical analyses described below makes use of plots of settlement against log time. However, settlement readings should also be plotted against square-root-time, and against time to an arithmetical scale. These curves are sometimes helpful in locating the end of the 'primary' phase, and they can provide a useful check as to whether the selected point (which in any case is arbitrary) looks sensible.

Conventional multiple-increment loading test

The conventional test described in Section 14.5.5, using a load increment ratio of unity and extending each increment to 24 h, can be used if the data are plotted as described below and interpreted as described by Hobbs (1986).

On the log–time settlement curve for each load increment the t_p point at which the 'primary' consolidation ends is identified as follows. If the point of inflection of the curve is evident (i.e. double curvature), as in Figure 14.48, stages 2 and 3, two tangents are drawn in the same way as described in Section 14.5.6, stage 1(a). A horizontal line drawn through the intersection of the tangents intersects the settlement curve at the t_p point, as shown in Figure 14.48.

If a reversal of slope is not evident (e.g. curves (a) or (b) of stage 1 in Figure 14.48), it might be reasonable to assume that the t_p point is located where the linear part of the graph (the secondary compression line) can first be identified. A curve similar to (a) is likely when the load increment ratio exceeds unity.

For the first load increment the 'primary' compression c_p is that which occurs from the instant of application of the load, time $t = 0$, to time $t = t_p$. For subsequent loading stages, c_p is measured from one t_p point to the next and includes the secondary compression of the previous increment, as shown in Figure 14.48. This is a simplifying assumption which gives a more realistic result than would be obtained from the conventional analysis.

The initial compression, c_i, is the displacement which takes place during the first 15 s after application of the load increment, as shown in Figure 14.48.

The time to the end of each 'primary' phase, t_p, and the corresponding cumulative settlement, ΔH_p, are recorded. Values of ΔH_p are used for the calculation of voids ratio at the end of each 'primary' stage in the equations given in Section 14.5.7.

The coefficient of secondary compression, C_{sec}, is obtained from the slope of the secondary compression line for each loading stage, and is calculated on the basis of the initial specimen thickness as shown in Figure 14.48.

The e–log p curve is drawn and the compression index, C_c, is determined (see Equation (14.25) or Figure 14.13). The results should also be plotted in terms of strain (ε), as this is the recommended procedure (Hobbs, 1986) for estimating strains in the field.

The following data are tabulated:

Incremental time (t_p) to the end of each 'primary' phase

Cumulative time from the start of the first increment to each t_p point

Cumulative time to the end of each loading stage (normally multiples of 1440 min)

Cumulative strains at each t_p point and at the end of each 24 h increment

The value of C_{sec} for each increment, calculated as described in Section 14.5.7

Cumulative strain for each t_p point is plotted against log pressure, as shown in Figure 14.52. This does not give the true primary consolidation (as was obtained in Figure 14.49) except for the first increment because each subsequent plotted strain includes strain due to secondary compression in previous increments. Cumulative 24 h strains may also be plotted, as shown by the heavy broken line in Figure 14.52. The cumulative time is written against each plotted point. Isochromes can then be constructed, in the manner described by Hobbs (1986).

Rapid multiple-increment test

This procedure was described by Hobbs (1986) and also by Macfarlane (1969), who recommended a specimen *H:D* ratio of about 1:3, and not greater than 1:2.5.

The specimen is set up and loaded as for a conventional test. As the specimen consolidates, a curve of settlement against log time is plotted. As soon as the end of the 'primary' phase is detected by the curve running into the linear secondary phase (see Figure 14.53) or by virtual dissipation of excess pore water pressure, the next load increment is applied. This process is repeated for each loading stage except the last, which may be extended to obtain a reliable value of the secondary compression coefficient at that pressure. If a normal oedometer specimen is used it may be possible to complete all the loading stages within one day, and to unload and remove the specimen on the second day.

If values of C_{sec} are required at lower pressures, these should be determined from separate tests on adjacent specimens. Isochrones can be derived as indicated above.

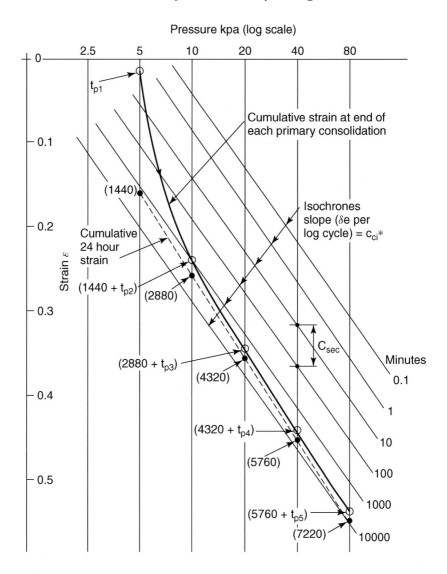

Figure 14.52 Plot of strain–log pressure for single increment and conventional multiple increment tests

At the end of each loading stage the cumulative strain is plotted against log pressure, as described above. The test need not be extended beyond the maximum field loading pressure unless a straight-line relationship between strain and log p has not been reached.

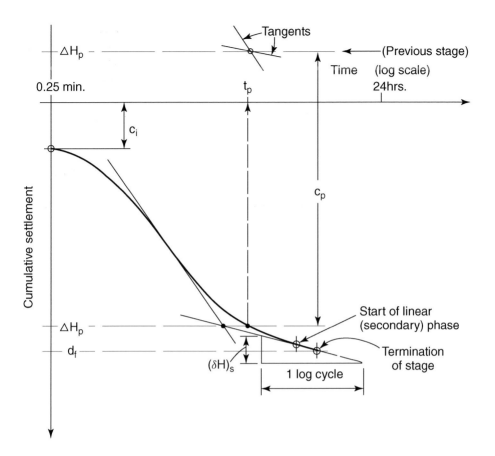

Figure 14.53 Rapid loading test on peat: log-time–settlement curve for one stage

Single-increment test

This procedure was referred to by way of illustration of the derivation of secondary compression isochrones (see Figure 14.49). One increment of loading is applied to each of a set of similar specimens so that the applied pressures cover the required range. Each load is sustained for long enough to obtain the value of C_{sec} and to provide readings from which isochrones can be constructed. Extension of the test would enable any variation of C_{sec} with time to be observed.

Since all individual specimens are likely to have different initial void ratios, individual results should be combined by plotting curves of log pressure–log time against strain, not against void ratio. A re-plot against void ratio can be made if the initial void ratio is taken as the average of the individual values.

A high pressure should not be applied all at once, otherwise squeezing out of material might occur. A small load is applied at first (say 6 kPa), followed by further small increments at perhaps 1 min intervals, depending on the time required for the initial very rapid settlement to take place. Timing of settlement readings is measured from the instant of placing the

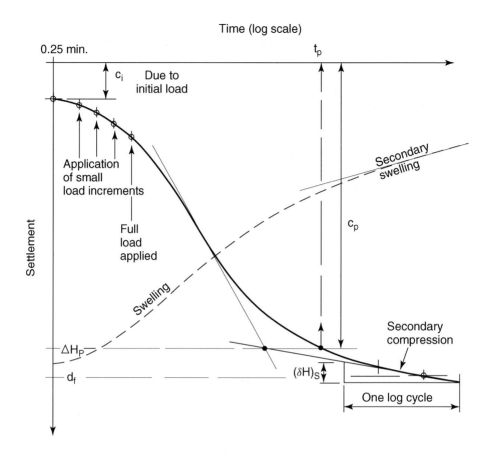

Figure 14.54 Single load increment test on peat: log-time–settlement and swelling curves

initial load (see Figure 14.54). The full load is applied when it is judged that the sample has become stiff enough not to be squeezed out past the loading plate.

Unloading

Swelling due to removal of part of the applied load can be assessed by modelling the field loading, surcharging and unloading sequence, as described by Hobbs. When the pressure on a specimen is reduced, there is first an immediate heave (the rapid primary swell phase), followed by a long-term secondary swelling which becomes linear on a log-time plot. This then flattens out, and if the test is continued long enough the resumption of secondary compression under the existing load might be detected.

Before removing a specimen from the consolidation cell it should be unloaded to a small pressure, left long enough to reach equilibrium, and any free water drained off. It should be dried at a temperature not exceeding 105°C and weighed, to provide data for calculating initial void ratio and degree of saturation.

Test results

The following data should be presented for each loading increment, and unloading decrement where appropriate:

Time–settlement curve

Log-time–settlement curve

Square-root-time–settlement curve.

Time to end of 'primary' consolidation, t_p

Cumulative settlement to end of 'primary' phase, ΔH_p. Specimen thickness at end of 'primary' phase, $(H_o - \Delta H_p)$

Cumulative strain, $\Delta H_p / H_o$

Voids ratio at end of 'primary' phase, e

Coefficient of secondary compression, $\overline{C}_{sec}$

Initial compression, c_i

'Primary' compression, c_p

Total change in height during load increment, δH

Initial compression ratio,

$$\frac{c_i}{\delta H} \times 100\%$$

'Primary' compression ratio, $\dfrac{c_p}{\delta H} \times 100\%$

If pore water pressure is measured, it should be plotted against time, square-root-time, and log time.

The following data should be reported for the whole test:

Plot of $\Delta H_p / H_o$ against log pressure, with derived isochrones

Plot of voids ratio, e, against log pressure

Plot of C_{sec} against log pressure

Initial and final moisture contents, $w_o\%$ and $w_f\%$

Initial and final voids ratios, e_o and e_f

Initial bulk density and dry density

Compression index, C_c

Specimen diameter, thickness and depth

Liquid limit

Plastic limit (if possible)

Loss on ignition

Particle density

Initial degree of saturation

Degree of humification and decomposition on the von Post scale

Visual description of specimen

14.8 Calibration and use of equipment

14.8.1 Calibration of load frame

The necessity for obtaining the deformation characteristics of the apparatus was explained in Section 14.5.2.

For the calibration test a disc of metal (preferably brass or copper) of the same dimensions as the test specimen is required. The end faces must be machined flat, smooth and parallel.

The load frame and cell are calibrated as one unit. The cell is assembled and placed in the load frame exactly as for a standard test (see Section 14.5.5, stages 6–10), except that the metal disc is placed between the two porous discs in place of the soil specimen and ring. No water is added to the cell.

After setting the dial gauge to its initial reading, add weights to the hanger to provide a pressure of 12 kPa, as described in Section 14.5.5, stages 11 and 13. Read the compression gauge; the deflection will take place almost instantaneously. Continue adding load increments in the usual stages, doubling at each stage, up to the maximum for the apparatus, and record the compression gauge reading Δa under each load. Reduce the load by the same stages to 12 kPa, again reading the gauge. Repeat the load–unloaded cycle twice more.

Plot the compression gauge readings against load on hanger (kg), the latter to a log scale, as shown in Figure 14.55. Draw a smooth mean curve through the points obtained for the loading stages and use this as the calibration curve for that apparatus. Read off the gauge readings at each standard load from the calibration curve and tabulate them. Incremental deflections from one standard load to the next, denoted by δ, are obtained by difference and can also be tabulated (see Figure 14.55).

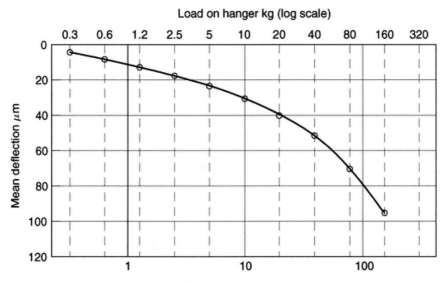

Mean calibration of oedometer presses nos 5, 6, 7 & 8

Load on hanger		kg	0.3	0.6	1.2	2.5	5	10	20	40	80	160
Cumulative correction	Δa	μm	5	8	13	18	24	31	40	52	70	96
Incremental correction	δ	μm	5	3	5	5	6	7	9	12	18	26

Calibrated by ..D.R.E.   Date 17.7.11

Figure 14.55 Typical calibration data for oedometer presses

Load frames and cells of the same type usually give very similar calibration curves, so an average curve can be drawn up for all such frames. However, if a non-standard cell is fitted or different porous discs are used, a fresh calibration will be necessary.

The cumulative correction Δa is subtracted from the cumulative settlement of the specimen at the end of each load increment (see Section 14.5.6, item 5, and Figure 14.29). The incremental correction δ is applied to the settlement/time graphs for each loading stage in order to obtain the d_c points (see Figure 14.9).

Reference to the calibration curve is necessary during a swelling pressure test (see Section 14.6.1).

14.8.2 Load increments

The number of loading increments and/or decrements is dependent on the geotechnical properties to be derived over the particular loading or unloading circumstances under consideration. The number of load decrements should not be less than half the number of increments, because the unloading curve may be needed as part of the analysis for establishing the field compression curve (see Section 14.6.5). As a minimum, the number of increments can be reduced to four, followed by two decrements.

The usual procedure is to apply increments of load such that the pressure is doubled at each increment, i.e. a load ratio of 2. (Another way of expressing this is that the load increment, δp, is equal to the load, p, already applied, i.e. $\delta p/p = 1$, a *load increment ratio* of unity). This is the loading sequence recommended in Table 14.12 (see Section 14.5.4, item 17).

Throughout at test the load ratio should be kept constant, preferably at a value of 2, because a change of ratio from one increment to the next can affect the c_v values. If a closer than normal spacing of points is required to define the e–log p curve, a load ratio of $\sqrt{2}$ is convenient because alternate loads will be the same as the recommended loads, giving a sequence 6, 8.4, 12, 17, 25, 30, 50, 71, 100…kPa. If a load ratio of 1.5 was used, most loads would be non-'standard' and would require inconvenient combinations of weights.

14.8.3 Multiple tests

In a commercial soil laboratory, consolidation tests are rarely carried out singly, but are more usually run several at a time. Consolidation frames can be mounted side by side on a bench, provided that the bench is strong enough and that provision has been made to prevent overturning when all frames are fully loaded (see item (10) of Section 14.5.3).

One person can look after several consolidation tests, but the intervals between starting one test and the next should be chosen with care to avoid two or more readings at standard time intervals occurring simultaneously. Convenient intervals for starting eight tests are shown in Table 14.14. This is only a consideration if tests are being read manually. If electronic measuring and recording equipment systems are being used, then a larger number of samples can be loaded and automatically read with a greater degree of accuracy than that obtained by dial gauges and manual readings.

The pattern can be repeated for subsequent tests. After the first half hour from the start, the difference of a few seconds in taking a reading is hardly noticeable. When large weights are to be moved on or off the load hanger the assistance of a second person may be needed.

Table 14.14 Starting times for multiple tests

Test No.	Starting time (min)	Interval (min)
1	0	5
2	5	5
3	10	11
4	21	5
5	26	5
6	31	11
7	42	5
8	47	

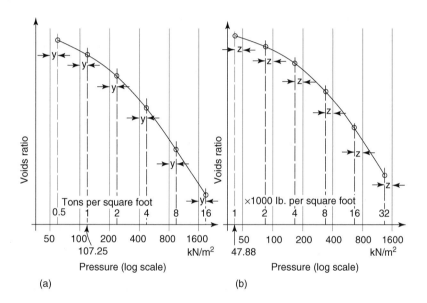

(a)

(b)

Figure 14.56 Plotting e–log p curves from test data in imperial units of pressure: (a) loadings in tons per square foot; (b) loadings in pounds per square foot (y and z are exaggerated for clarity)

14.8.4 Use of obsolescent load frame

If a load frame designed for imperial units is used for tests in SI units, it is doubtful whether it is worth attempting to carry out any conversions to the apparatus. It is easier to use the same specimen size and hanger weights as before, and to make an adjustment when plotting the e–log p curve. For instance the standard apparatus previously used in the UK had a specimen 3 in diameter and a beam ratio of 11:1, so that 10 1b on the hanger gave 1 ton/ft^2 on the specimen. The apparatus can be used exactly as before, but when plotting the e–log p curve and when calculating pressures the conversion 1 ton/ft^2 = 107.25 kPa should be used.

Since the pressure scale is logarithmic, every point corresponding to the standard sequence of loading will be displaced to the right of an exact multiple of 100 kPa by the same small distance y (see Figure 14.56(a)). The e–log p curve will be identical in form to that which would be obtained if pressures in exact multiples of 100 kPa were used. The displacement y, on a typical A4 graph sheet, is about 1.5 mm.

Similarly, a load frame designed to ASTM Standards, for which the loadings are in multiples of 1000 lb/ft^2, can be used without modification by using the conversion 1000 lb/ft^2 = 47.88 kPa. Every multiple of 1000 lb/ft^2 will be displaced to the left of an exact multiple of 50 kPa by the same small distance z (see Figure 14.56(b)). The displacement z on a typical sheet is about 1 mm.

The specimen dimensions should be expressed in millimetres (3 in = 76.2 mm), and a compression dial gauge measuring millimetres should be used. Then, all other calculations are exactly like those described in Section 14.5.7.

References

ASTM D 2435-04. Standard test method for one-dimensional consolidation properties of soils. American Society for Testing and Materials, Philadelphia, PA, USA

ASTM D 4546-08. Standard test method for one-dimensional swell or settlement potential of cohesive soil.American Society for Testing and Materials, Philadelphia, PA, USA

ASTM D 4829–08A. Standard test method for expansion index of soils. American Society for Testing and Materials, Philadelphia, PA, USA

ASTM D 4792-00 (reapproved 2006). Standard test method for potential expansion of aggregates from hydration reactions. American Society for Testing and Materials, Philadelphia, PA, USA

Baracos, A. (1976). Clogged filter discs. Technical note, *Géotechnique*, Vol. 26, No. 4

Barden, L. (1965). 'Consolidation of compacted and unsaturated clays'. *Géotechnique*, Vol. 15, No. 3

Burland, J. B. (2001). The stabilisation of the Leaning Tower of Pisa, *Ingenia*, 2001, pp. 10–18

Capps, J. F. and Hejj, H. (1968). Laboratory and field tests on a collapsing sand in northern Nigeria. Technical Note, *Géotechnique*, Vol. 18, No. 4

Casagrande, A. (1932). The structure of clay and its importance in foundation engineering. *J. Boston Soc. Civ. Eng.*, Vol. 19

Casagrande, A. (1936). The determination of the pre-consolidation load and its practical significance. *Proc. 1st Int. Conf. Soil Mech.*, Cambridge, Mass., Vol. 3.

Cooling, L. F. and Skempton, A. W. (1941). Some experiments on the consolidation of clay. *J. Int. Civ. Eng.*, Vol. 16

Davis, E. H. and Poulos, H. (1965). The analysis of settlement under three-dimensional conditions. *Symp. on Soft Ground Eng.*, Inst. Eng. Australia, Brisbane

Emery, J. J. (1979). Assessment of Ferrous Slags for Fill Applications. Paper F1, Conference on Reclamation of Contaminated Land. *Proc. Soc. of Chemical Industry Conference*, Eastbourne, UK, October 1979.

Gilboy, G. (1936). Improved soil testing methods. *Eng. News Rec.*, 21 May 1936

Hobbs, N. B. (1986). Mire morphology and the properties and behaviour of some British and foreign peats. *Quarterly Journal of Eng. Geology*, London, Vol. 19, No. 1

Kezdi, A. (1980). *Handbook of Soil Mechanics, Vol. 2, Soil Testing*. Elsevier Scientific Co. (English translation from the Hungarian, *Talajmechanika I*, Budapest, 1960)

Lambe, T. W. (1951). *Soil Testing for Engineers*, Wiley, New York

Lambe, T. W. and Whitman, R. V. (1979). *Soil Mechanics, SI Version*. Wiley, New York

Leonards, G. A. (ed.) (1962). *Foundation Engineering*, Chapter 2. McGraw-Hill, New York

MacDonald, D. H. and Skempton, A. W. (1955) A survey of comparisons between calculated and observed settlements of structures on clay. Paper No. 19. *Conf. on Correlation Between Calculated and Observed Stresses and Displacements in Building. Inst. Civ. Eng.*, London

Mitchell, J. K., Vivatrat, V. and Lambe, T. W. (1977). Foundation performance of the tower of Pisa. *Proc. ASCE. Geotech. Eng. Div.*, Vol. 103, No. GT3

MacFarlane, I. C. (1969). *Muskeg Engineering Handbook*, Chapter 4, University of Toronto Press, Toronto, Canada

Rowe, P. W. (1966). A new consolidation cell. *Géotechnique*, Vol. 16, No. 2

Rowe, P. W. (1972). The relevance of soil fabric to site investigation practice. 12th Rankine lecture, *Géotechnique*, Vol. 22, No. 2

Rutledge, P. C. (1935). Recent developments in soil testing apparatus. *J. Boston Soc. Civ. Eng.*, Vol. 22, No. 4

Schmertmann, J. H. (1953). Estimating the true consolidation behaviour of clay from laboratory test results. *Proc. ASCE*, Vol. 79, Separate No. 3111

Schmertmann, J. H. (1954). The undisturbed consolidation behaviour of clay. *Trans. ASCE*, Vol. 120, Paper 2775

Scott, C. R. (1974). *An Introduction to Soil Mechanics*. Applied Science Publishers, Barking, UK

Simons, N. E. and Menzies, B. K. (1977). *A Short Course in Foundation Engineering*. Newnes-Butterworth, London

Skempton, A. W. (1944). Notes on the compressibility of clays. *Q. J. Geol. Soc.*, Vol. C

Skempton, A. W. and MacDonald, D. H. (1956). The allowable settlements of buildings. *Proc. Inst. Civ. Eng.*, Vol. 5, No. 3, Part 3

Skempton, A. W. and Bjerum, L. (1957). A contribution to the settlement analysis of foundations on clay. *Géotechnique*, Vol. 7, p. 168

Skempton, A. W. and Petley, J. (1970). Ignition loss and other properties of peats and clays from Avonmouth, King's Lynn and Cranberry Moss. *Géotechnique*, Vol. 20, No. 4

Taylor, D. W. (1948). *Fundamentals of Soil Mechanics*, Wiley, New York

Taylor, D. W. (1942). Research on consolidation clays. Report No. 82. Department of Civil and Sanitary Engineering, Massachusetts Institute of Technology, Cambridge, MA, USA

Terzaghi, K. (1925). *Erdbaumechanik auf bodenphysikalischer Grundlage*. Deuticke, Wien

Terzaghi, K. (1934). Die Ursachen der Schiefstellung des Turmes von Pisa. *Der Bauingenieur*, Berlin. Reprinted (1960). in *From Theory to Practice in Soil Mechanics*, Wiley, New York

Terzaghi, K. (1939). Soil mechanics—a new chapter in engineering science. James Forrest Lecture, *J. Inst. Civ. Eng.*, London, Vol. 12, No. 7

Terzaghi, K. (1943). *Theoretical Soil Mechanics*. Wiley, New York

Terzaghi, K. and Fröhlich, O. K. (1936). *Theorie der Setzung von Tonschichten; eine Einführung in die analytische Tonmechanik*. Deuticke, Leipzig

Terzaghi, K. and Peck, R. B. (1967). *Soil Mechanics in Engineering Practice*. Wiley, New York

Tschebotarioff, G. P. (1951). *Soil Mechanics, Foundations and Earth Structures*, Chapter 6. McGraw-Hill, New York

Wheeler, P. (1993). Academic leanings—Field trials at Pisa. *Ground Engineering*, Vol. 26, No. 6

Further reading

Hobbs, N. B. (1987). A note on the classification of peat. Technical Note, *Géotechnique*, Vol. 37, No. 3

Landva, A. O. and Pheeney, P. E. (1980). Peat, fabric and structure. *Can. Geotech. J.*, Vol. 17, No. 3, pp. 416–435

Padfield, C. J. and Sharrock, M. J. (1983). *Settlement of Structures on Clay Soils*. CIRIA Special Publication 27. Property Services Agency, Department of the Environment, London

von Post, L. (1924). Das genetische System der organogenen Bildungen Schwedens. Int. Comm. Soil Sci., IV Commission

Appendix: Units, symbols, reference data

B1 Metric (SI) units

Customary SI units used throughout Volumes 1 and 2, which are also generally accepted for use in soil mechanics and foundation engineering, are summarised in Table B1. Standard multiplying prefixes are given in Table B2. A selection of factors for converting British, US and CGS units to SI, and vice versa, is given in Table B3, generally to four significant figures.

Definitions of some SI units, and explanatory notes, were given in the Appendix to Volume 1 (third edition), Section A.1.2, and are not repeated here. The tables given here incorporate data tabulated in the Appendix to Volume 1 (third edition).

Table B1 SI units for soil mechanics

Quantity	Unit	Unit symbol	Application	Metric equivalents
Length	millimetre	mm	Sample measure-ments, particle size	$1\ \mu m = 10^{-6}$ m
	micrometre	μm	Sieve aperture and particle size	$= 10^{-3}$ mm
Area	square millimetre	mm^2	Area of section	
Volume	cubic metre	m^3	Earthworks	
	cubic centimetre	cm^3	Sample volume	1 m$^3 = 10^6$ cm^3
	millilitre	ml	Fluid measure	
	cubic millimetre	mm^3	Sample volume as calculated	
Mass	gram	g	Accurate weighings	
	kilogram	kg	Bulk sample and approximate weights	1 kg $= 1000$ g
	megagram	Mg	Alternatively known as tonne	1 Mg $= 1000$ kg $= 10^6$ g
Density (mass)	megagram per cubic metre	Mg/m^3	Sample density and dry density	Density of water $= 1$ Mg/m^3 $= 1$ g/cm^3

(continued)

Table B1 SI units for soil mechanics (*continued*)

Quantity	Unit	Unit symbol	Application	Metric equivalents
(weight)	kilonewton per cubic metre	kN/m³	Overburden pressure	1 Mg/m³ = 9.807 kN/m³
Temperature	degree Celsius	°C	Laboratory and bath temperatures	Celsius is preferred name for Centigrade
Time	second	s	Timing of laboratory tests	1 minute = 60 s
Force	newton	N	Load ring calibrations	1 kgf = 9.807 N
			Small-magnitude forces	1 N = 101.97 gram f
	kilonewton	kN	Forces of intermediate magnitude	1 kN = 1000 N = approx. 0.1 tonne f
Pressure and stress	newton per square metre = pascal kilonewton per square metre = kilopascal	N/m² Pa kN/m² kPa	Very low pressures and stresses Pressure gauges Compressive strength and shear strength of soils	1 g/cm² = 98.07 N/m² = 98.07 Pa 1 kgf/cm² = 98.07 kN/m² 1 bar = 100 kN/m²
Pressure (vacuum)	torr	torr	Very low pressure under vacuum	1 torr = 133.3 Pa = 133.3 N/m² = 1 mmHg
Dynamic viscosity	millipascal second = millinewton second per square metre	mPas mNs/m²	Viscosity of water	1 mPas = 1cP (centipoise)
Coefficient of volume compressibility (*mv*)	square metre per meganewton	m²/MN	Settlement calculations	1 cm²/kgf = 10.20 m²/MN
Coefficient of consolidation (*cv*)	square metre per year	m²/year	Estimation of rate of settlement	1 cm²/s = 3156 m²/year
Coefficient of permeability (*k*)	metre per second	m/s	Flow of water in soils	1 cm/s = 0.01 m/s
Frequency	hertz	Hz	Rate of vibration, load repetition etc.	1 Hz = 1 s⁻¹ = 1 cycle per second (cps)

Metric sieve aperture sizes used for soil testing are listed in Table 4.5 (Volume 1 (third edition))

Table B2 Multiplying prefixes

Prefix symbol	Name	Multiplying factor
G	giga	$1\ 000\ 000\ 000 = 10^9$
M	mega	$1\ 000\ 000 = 10^6$
k	kilo	$1\ 000 = 10^3$
h	[1]*hecto	$100 = 10^2$
Da	*deca	10
d	*deci	$10^{-1} = 0.1$
c	*centi	$10^{-2} = 0.01$
m	milli	$10^{-3} = 0.001$
μ	micro	$10^{-6} = 0.000\ 001$
n	nano	$10^{-9} = 0.000\ 000\ 001$

* Not recommended in SI

Table B3 Conversion factors: British, US, CGS units to and from SI

	Others to SI				SI to others
Length	0.3048	m	:	foot (ft)	3.281
	25.4*	mm	:	inch (in)	0.03937
Area	0.09290	m²	:	square foot	10.76
	645.2	mm²	:	square inch	0.001550
Volume	0.02832	m³	:	cubic foot	35.31
	4.546	litre	:	gallon (UK)	0.2200
	3.785	litre	:	gallon (USA)	0.2642
	28.32	litre	:	cubic foot	0.03531
	16.39	ml	:	cubic inch	0.06102
	16387	mm³	:	cubic inch	
Mass	1.016	Mg (tonne)	:	ton	0.9842
	0.4536	kg	:	pound (lb)	2.205
	453.6	g	:	pound	
	28.35	g	:	ounce (oz)	0.03527
	0.9072	Mg	:	short ton (USA)	1.1023
Density (mass)	0.01602	Mg/m³(g/cm³)	:	pound per cubic foot	62.43
Density (weight)	0.1571	kN/m³	:	lb/ft³	6.366
Force	9.964	kN	:	ton force	0.1004
	4.448	N	:	pound force	0.2248
	9.807	N	:	kgf (kilopond)	0.10197
	10^{-5*}	N	:	dyne	10^{5*}
	0.1383	N	:	poundal	7.233

(*continued*)

Table B3 Conversion factors: British, US, CGS units to and from SI (*continued*)

	Others to SI				*SI to others*
Pressure	0.04788	kPa, kN/m²	:	lb f/sq ft	20.89
and stress	6.895	kN/m²	:	lb f/sq in	0.1450
	47.88	Pa, N/m²	:	lb f/sq ft	0.02089
	107.25	kPa, kN/m²	:	tonf/sq ft	0.009324
	0.10725	Mpa, MN/m²	:	tonf/sq ft	9.324
	98.07	kN/m²	:	kgf/sq cm (kp/cm2)	0.0102
	101.32	kN/m²	:	atm	0.009869
	100*	kN/m²	:	bar	0.01*
	0.1*	Pa, N/m²	:	dyne/sq cm	10*
Fluid	2.989	kN/m²	:	feet of water	0.3346
pressure	0.2491	kN/m²	:	inches of water	4.015
	9.807	N/m²	:	mm of water	0.10197
	0.009807	kN/m²	:	mm of water	101.97
Spring	0.1751	N/mm	:	lbf/in	5.710
modulus	0.009807	N/mm	:	gf/mm	101.97
Torque	112.98	N.mm	:	lbf in	0.008851
	1.356	N.m	:	lbf ft	0.7376
Coefficient of	9.324	m²/MN	:	ft²/ton	0.1072
volume compressibility	10.197	m²/MN	:	cm²/kgf	0.9807
Coefficient of	0.09290	m²/year	:	ft²/year	10.76
consolidation	339.3	m²/year	:	in²/minute	2.947×10^{-3}
	3156	m²/year	:	cm²/s	3.169×10^{-4}
Coefficient of	0.9659×10^{-8}	m/s	:	ft/year	1.035×10^{8}
permeability	0.01†	m/s	:	cm/s	100*
Rate of	0.01667	ml/s	:	ml/min	60*
flow	4546	ml/min	:	gallon/min (UK)	2.20×10^{-4}
	0.07577	litre/min	:	gallon/hour (UK)	13.20
	3785	ml/min	:	gallon/min (US)	2.642×10^{-4}
	0.02832	(cumec) m³/s	:	cu.ft/s (cusec)	35.31

* Exact value

Examples: Other units to SI: to convert feet to metres, multiply by 0.3048
SI to other units: to convert metres to feet, multiply by 3.281

B2 Symbols

Symbols generally used in Volumes 1 and 2 (third edition) are summarised in Table B4 (soil properties) and Table B5 (miscellaneous). The Greek alphabet is given in Table B6.

Table B4 Symbols for soil and water properties

Chapter reference	Measured quantity	Symbol	Unit of measurement
2–6	Moisture content	w	%
(Vol. 1)	Liquid limit	w_L	%
	Plastic limit	w_p	%
	Plasticity index	I_p	%
	Non-plastic	NP	–
	Relative consistency	C_r	–
	Liquidity index	I_L	–
	Shrinkage limit	w_s	%
	Linear shrinkage	L_s	%
	Shrinkage ratio	R	–
	Unit weight	γ	kN/m³
	Bulk (mass) density	ρ	Mg/m³
	Dry density	ρ_D	Mg/m³
	Saturated density	ρ_s	Mg/m³
	Submerged density	ρ'	Mg/m³
	Minimum dry density	$\rho_{D\,min}$	Mg/m³
	Maximum dry density	$\rho_{D\,max}$	Mg/m³
	Density of water	ρ_w	Mg/m³
	Optimum moisture content	OMC	%
	Particle density	ρ_s	Mg/m³
	Density liquid	ρ_L	Mg/m³
	Degree of saturation	S	%
	Voids ratio	e	–
	Porosity	n	–
	Percentage air voids	V_a	%
	Particle size	D	μm or mm
	Percentage smaller than D	P	%
	Effective size	D_{10}	mm
	`60% finer than' size	D_{60}	mm
	Uniformity coefficient	U	–
	Dynamic viscosity of water	η	mPas
10	Coefficient of permeability	k	m/s
	(at T°C)	k_T	m/s
	Absolute (specific) permeability	K	mm²

(continued)

Table B4 Symbols for soil and water properties (*continued*)

Chapter reference	Measured quantity	Symbol	Unit of measurement
	Specific surface	S	mm^{-1}
11	California Bearing Ratio	CBR	%
12, 13	Drained strength:		
	Cohesion	c'	kN/m^2
	Angle of shear resistance	ϕ'	degrees
	Residual strength:		
	Cohesion	c'_r	kN/m^2
	Angle of shear resistance	ϕ'_r	degrees
	Vane shear strength	τ_v	kN/m^2
13	Unconfined compressive strength	q_u	kN/m^2
	Remoulded compressive strength	q_{ur}	kN/m^2
	Sensitivity	S_t	–
14	Coefficient of compressibility	a_v	m^2/kN
	Coefficient of volume compressibility	m_v	m^2/MN
	Coefficient of consolidation	c_v	$m^2/year$
	Secondary compression index	C_{sec}	–
	Compression index	C_c	–
	Swell index	C_s	–

Table B5 Miscellaneous symbols

Measured quantity	Symbol	Unit of measurement
Length	L, l	mm
Diameter	D	mm
Height, thickness	H	mm
Area of cross-section	A, a	mm^2
Volume (solid)	V	cm^3
Volume (fluid)	Q	ml
Mass	m	g
Time	t	minutes
Temperature	T	°C
Rate of flow of fluid	q	ml/minute
Head of water	h	mm
Head difference	Δh	mm
Height above datum	y	mm
Hydraulic gradient	i	–

(*continued*)

Table B5 Miscellaneous symbols (*continued*)

Measured quantity	Symbol	Unit of measurement
Critical hydraulic gradient	i_c	–
Velocity of flow	v	mm/s
Pressure	p	kN/m² (kPa)
Pressure difference or change	Δp	kN/m² (kPa)
Load ring factor	C_R	N/division
Force	P, F	N
Shear force	S	N
Weight force	W	N
Normal stress	σ, σ_n	kN/m² (kPa)
Major and minor principal total stresses	σ_1, σ_3	kN/m² (kPa)
Effective stress	σ'	kN/m² (kPa)
Shear stress	τ, S	kN/m² (kPa)
Shear stress at failure	τ_f	kN/m² (kPa)
Strain	ε	%
Strain at failure	ε_f	%
Shear strain	γ	radians
Torque	T, T_r	Nmm
Pore water pressure	u	kN/m²
Degree of primary consolidation	U	%
Initial compression ratio	r_o	–
Primary consolidation ratio	r_p	–
Secondary compression ratio	r_s	–
Time factor	T_v	–
Length of drainage path	h	mm
Infinity	∞	–

Table B6 Greek alphabet

Upper case	Lower case	Name	Upper case	Lower case	Name
A	α	alpha	N	ν	nu
B	β	beta	Ξ	ζ	xi
Γ	γ	gamma	O	o	omicron
Δ	δ	delta	Π	π	pi
E	ϵ	epsilon	P	ρ	rho
Z	ζ	zeta	Σ	σ	sigma
H	η	eta	T	τ	tau

(*continued*)

Table B6 Greek alphabet (*continued*)

Upper case	Lower case	Name	Upper case	Lower case	Name
Θ	ϑ	theta	Y	u	upsilon
I	ι	iota	Φ	φ	phi
K	κ	kappa	X	χ	chi
Λ	λ	lambda	Ψ	ψ	psi
M	μ	mu	Ω	ω	omega

B3 Reference data

Data relating to standard test specimens referred to in Volumes 1 and 2 (areas, volumes, approximate mass) are given in Table B7. Some useful data for quick reference are given in Table B8. The approximate mass quoted in the last column of the table is based on a bulk density of 2.1 Mg/m^3.

Table B7 Specimen dimensions, area, volume, mass

Specimen type	Diameter		Height		Area	Volume	Approximate
	(inch)	(mm)	(mm)	(inch)	(mm²)	(cm³)	mass
Constant head		75	180		4418	795.2	1.67 kg
permeability	3			7	4560	810.8	1.70 kg
		114	350		10 207	3572	7.50 kg
Falling head		100	130		7854	1021	2.14 kg
permeability	4			5	8107	1030	2.16 kg
Compaction mould		105	115.5		8659	1000	2.10 kg
(BS) with collar		105	165.5		8659	1433	3.01 kg
Compaction mould (ASTM)	4			4.584	8107	944	1.98 kg
with collar	4			6.584	8107	1356	2.85 kg
CBR mould (BS)		152	127		18 146	2305	4.84 kg
with collar		152	177		18 146	3212	6.74 kg
CBR mould	6			7	18 243	3244	6.81 kg
(ASTM) with collar	6			9	18 243	4170	8.76 kg
mould less spacer	6			4.584	18 243	2124	4.46 kg
Uniaxial and		35	70		962.1	67.35	141 g
triaxial	1.4			2.8	993.2	70.64	148 g
compression		38	76		1134	86.19	181 g
	1.5			3	1140	86.88	182 g
		50	100		1963	196.3	412 g
	2			4	2027	205.9	432 g

(*continued*)

Table B7 Specimen dimensions, area, volume, mass (*continued*)

Specimen type	Diameter (inch)	Diameter (mm)	Height (mm)	Height (inch)	Area (mm²)	Volume (cm³)	Approximate mass
		70	140		3848	538.8	1.13 kg
	2.8			5.6	3973	565.1	1.19 kg
		100	200		7854	1571	3.30 kg
	4			8	8107	1647	3.46 kg
		105	210		8659	1818	3.82 kg
		150	300		17 671	5301	11.1 kg
	6			12	18 241	5560	11.7 kg
Oedometer		50	20		1963	39.27	83 g
consolidation		50.5	20		2003	40.06	84 g
	2.5			0.75	3167	60.33	127 g
		70	20		3848	76.97	162 g
		71.4	20		4004	80.08	168 g
		75	20		4418	88.36	186 g
	3			0.75	4560	86.87	182 g
		100	20		7854	157.1	330 g
		112	20		9852	197.0	414 g
		square					
Shearbox		60	20		3600	72.00	151 g
		60	(19.05)	0.75	3600	68.58	144 g
		60	(25.4)	1	3600	91.44	192 g
	2.5			0.75	4032	76.81	161 g
		100	20		10 000	200.0	420 g
	4			1	10 322	262.2	551 g
		150	75		22 500	1688	3.54 kg
	6			3	23 226	1770	3.72 kg
		300	150		90 000	13 500	28.4 kg
	12			6	92 909	14 159	29.7 kg

Table B8 Useful data

Time	1 day	=	1440 min
	1 week	=	10 080 min
	1 month (average)	=	43 920 min
	1 year	=	525 960 min
		=	31.56×10^6 s

(*continued*)

Table B8 Useful data (*continued*)

Density (20°C)	Pure water		0.99820 g/cm³
	Sea water		1.04 g/cm³
	Mercury		13.546 g/cm³
Fluid pressure	1 kN/m² = 1 kPa	=	102 mm of water
		=	7.53 mm of mercury
	1 m of water	=	9.807 kN/m²
	Mercury-to-water manometer with limb open to atmosphere:		
	1 mm difference	=	0.128 kN/m²
	1 kN/m²	=	7.81 mm difference
	Mercury-to-water manometer with limb connected to water reservoir or to constant pressure system:		
	1 mm difference	=	0.123 kN/m²
	1 kN/m²	=	8.13 mm difference
General	Circumference/diameter of a circle		
	π =		3.142
	Base of natural logarithms		
	e =		2.718
	Standard acceleration due to terrestrial gravity		
	g =		9.807 m/s²

Further reading

Anderton, P. and Bigg, P. H. (1972) *Changing to the Metric System.* National Physical Laboratory, HMSO, London

British Geotechnical Society Sub-committee on the Use of SI units in Geotechnics (1973) Report of the sub-committee. News Item, *Géotechnique*, Vol. 23, No. 4, pp. 607–610

Bureau International des Poids et Mesures (BIPM) (2006) *The International System of Units (SI).* BIPM, Sèvres, France

International Society of Soil Mechanics and Foundations Engineering (1968) *Technical Terms, Symbols and Definitions (eight languages)*, (fourth edition). Société suisse de mécanique des sols et des travaux de fondations, Zürich

Metrication Board (1976) *Going Metric: The International Metric System.* Leaflet UM1, 'An outline for technology and engineering'. Metrication Board, London

Metrication Board (1977) *How to Write Metric: A Style Guide for Teaching and Using SI Units.* HMSO, London

Page, C. H. and Vigoureux, P. (1977) *The International System of Units* (approved translation of *Le Systeme International des Unités*, Paris, 1977). National Physical Laboratory, HMSO, London

Walley, F. (1968) 'Metrication' (Technical Note). *Proc. Inst. Civ. Eng.*, Vol. 40, May 1968. Discussion includes contribution by Head, K. H., Vol. 41, December 1968

Index